Media Models to Foster Collective Human Coherence in the PSYCHecology

Stephen Brock Schafer
Pacific Rim Enterprises, USA

A volume in the Advances in Psychology, Mental Health, and Behavioral Studies (APMHBS) Book Series

IGI Global
DISSEMINATOR OF KNOWLEDGE

Published in the United States of America by
IGI Global
Information Science Reference (an imprint of IGI Global)
701 E. Chocolate Avenue
Hershey PA, USA 17033
Tel: 717-533-8845
Fax: 717-533-8661
E-mail: cust@igi-global.com
Web site: http://www.igi-global.com

Library of Congress Cataloging-in-Publication Data

Names: Schafer, Stephen Brock, 1942- editor.
Title: Media models to foster collective human coherence in the psychecology / Stephen Brock Schafer, editor.
Description: Hershey, PA : Information Science Reference, [2020]
Identifiers: LCCN 2018056283| ISBN 9781522590651 (hardcover) | ISBN 9781522590675 (ebook) | ISBN 9781522590668 (softcover)
Subjects: LCSH: Consciousness. | Conscious automata. | Digital media--Psychological aspects.
Classification: LCC BF311 .M435 2020 | DDC 153--dc23 LC record available at https://lccn.loc.gov/2018056283

This book is published in the IGI Global book series Advances in Psychology, Mental Health, and Behavioral Studies (APMHBS) (ISSN: 2475-6660; eISSN: 2475-6679)

British Cataloguing in Publication Data
A Cataloguing in Publication record for this book is available from the British Library.

Advances in Psychology, Mental Health, and Behavioral Studies (APMHBS) Book Series

Bryan Christiansen
Global Research Society, LLC, USA
Harish C. Chandan
Argosy University, USA

ISSN:2475-6660
EISSN:2475-6679

Mission

The complexity of the human mind has puzzled researchers and physicians for centuries. While widely studied, the brain still remains largely misunderstood.

The **Advances in Psychology, Mental Health, and Behavioral Studies (APMHBS)** book series presents comprehensive research publications focusing on topics relating to the human psyche, cognition, psychiatric care, mental and developmental disorders, as well as human behavior and interaction. Featuring diverse and innovative research, publications within APMHBS are ideally designed for use by mental health professionals, academicians, researchers, and upper-level students.

Coverage

- Trauma & Stress
- Human Behavior
- Personality Disorders
- Counseling & Therapy
- Substance Abuse
- Cognition
- Addiction
- Human Interaction
- Eating Disorders
- Mental Health & Disorders

IGI Global is currently accepting manuscripts for publication within this series. To submit a proposal for a volume in this series, please contact our Acquisition Editors at Acquisitions@igi-global.com or visit: http://www.igi-global.com/publish/.

The Advances in Psychology, Mental Health, and Behavioral Studies (APMHBS) Book Series (ISSN 2475-6660) is published by IGI Global, 701 E. Chocolate Avenue, Hershey, PA 17033-1240, USA, www.igi-global.com. This series is composed of titles available for purchase individually; each title is edited to be contextually exclusive from any other title within the series. For pricing and ordering information please visit http://www.igi-global.com/book-series/advances-psychology-mental-health-behavioral/110200. Postmaster: Send all address changes to above address.

Titles in this Series

For a list of additional titles in this series, please visit: www.igi-global.com/book-series

Ethical Problem-Solving and Decision-Making for Positive and Conclusive utcomes
Penelope D. Keough (National University, USA)
Information Science Reference • ©2019 • 326pp • H/C (ISBN: 9781522575825) • US $195.00

Chronic Stress and Its Effect on Brain Structure and Connectivity
Ana Starcevic (University of Belgrade, Serbia)
Medical Information Science Reference • ©2019 • 318pp • H/C (ISBN: 9781522575139) • US $225.00

Analyzing the Role of Cognitive Biases in the Decision-Making Process
Verónica Juárez Ramos (International University of Rioja, Spain)
Information Science Reference • ©2019 • 262pp • H/C (ISBN: 9781522529781) • US $190.00

The Mental Health Effects of Informal Caregiving Emerging Research and Opportunities
Debra Gavin Ware (Argosy University, USA) and Tarnue Johnson (Argosy University, USA & Kirkwood Community College, USA)
Medical Information Science Reference • ©2019 • 131pp • H/C (ISBN: 9781522574521) • US $195.00

Virtual and Augmented Reality in Mental Health Treatment
Giuliana Guazzaroni (Università Politecnica delle Marche, Italy)
Medical Information Science Reference • ©2019 • 329pp • H/C (ISBN: 9781522571681) • US $265.00

Multicultural Counseling Applications for Improved Mental Healthcare Services
Anasuya Jegathevi Jegathesan (HELP University, Malaysia) and Siti Salina Abdullah (University Malaysia Terengganu, Malaysia)
Medical Information Science Reference • ©2019 • 302pp • H/C (ISBN: 9781522560739) • US $265.00

Assessing and Averting the Prevalence of Mass Violence
Sarah E. Daly (St. Vincent College, USA)
Information Science Reference • ©2019 • 271pp • H/C (ISBN: 9781522556701) • US $185.00

Scientific Concepts Behind Happiness, Kindness, and Empathy in Contemporary Society
Nava R. Silton (Marymount Manhattan College, USA)
Information Science Reference • ©2019 • 321pp • H/C (ISBN: 9781522559184) • US $185.00

IGI Global
DISSEMINATOR OF KNOWLEDGE

701 East Chocolate Avenue, Hershey, PA 17033, USA
Tel: 717-533-8845 x100 • Fax: 717-533-8661
E-Mail: cust@igi-global.com • www.igi-global.com

Table of Contents

Detailed Table of Contents

With the enormous chaotic changes taking place today, contemporary artists are showing us a vast and mystifying range of artworks that show glimpses of nascent worlds coming into being and just as quickly disappearing into oblivion. The author explores a world that seems to be gaining some traction—the world of the posthuman. Contemporary art is showing glimpses of this still-forming world in artworks produced from a collision between or interpenetration of virtual reality and empirical reality, giving rise to weird, horrific, and sometimes strangely beautiful forms. Then the author seeks to penetrate "behind" this posthuman art to the activity of the artist, in order to find the original "bringing forth" (Heidegger) of posthuman artworks. This move reveals the fundamental place of revelation and prophecy as the origin of any artwork and thus indicates the essential nature of the posthuman world.

The media model called "belief" serves as the language by which humans converse with God in the scripture of redemption. Hypothetically, redemption is based on the same cognitive dynamic as healing with dream analysis established by Carl Jung. These dynamics promote coherence within a sentient-cognitive psyche (a combination of conscious/unconscious scales of sentience), universal reality based on unity/energism, and narrative/dramatic structure. Based upon these dynamics, this chapter establishes a correspondence between the cognitive language of the mediated dreamscape and the cognitive language of belief that allows for conversation between God and humanity. The key to understanding this cognitive conversation is the belief model of sentience. Conflicting belief systems are endemic to human experience. Differences in the languages of belief contribute to the most psychotic-hysterical misunderstandings and their resulting mayhem. Due to illiteracy relative to the emergent language of cognitive mediation, the portents for planetary survival are dire.

Section 3
Mediated Sentience

We live in a world not of science but of science fiction. Like pixel patterns from unconscious software is projected onto a monitor, unconscious archetypal quantum patterns are projected as what Carl Jung called archetypal representation (AR). Projected images are then subject to the vagaries of personal perception, so it may be stipulated that no absolute reality exists for humans. Rather, each person lives in a perceptual fiction. According to Carl G. Jung, dreams are projections from quantum-level unconscious dimensions into the cognitive dimension of "consciousness." In the language of dream analysis, Jung would have described the science fiction genre as a prospective (future-oriented) dreamscape of archetypal representations. In the media-dream model, quantum patterns are derived from research in cognitive neuroscience and physics. Contemplated as AR, the sci-fi genre is predictive of cultural futures and formats psychological motives and morality. Sci-fi has the potential to detect the psychological dynamics at work during the paradigm shift into a dreamscape of illusion.

As pervasive and vital as they are in human experience, emotions have long remained an enigma to science. This chapter explores recent scientific advances that clarify central controversies in the study of emotion, including the relationship between intellect and emotion and the historical debate on the source of emotional experience. Particular attention is given to the intriguing body of research illuminating the critical role of ascending input from the body to the brain in the generation and perception of emotions. This discussion culminates in the presentation of a new, systems-oriented model of emotion in which the brain functions as a complex pattern-matching system, continually processing input from both the external and internal environments. From this perspective, it is shown that the heart is a key component of the emotional system, thus providing a physiological basis for the long-acknowledged link between the heart and our emotional life.

Evolution is fluid, and it adapts to the shapes of its containers. This chapter contemplates the idea of liquid humanitarianism in the process of adapting to its cultural containers. True humanitarianism is evolving from the earlier stage defined by the term "humanism," which has focused primarily on human self-interest rather than geocentric interest. Collective cognitive transcendence from a humanist state to a humanitarian state constitutes initiation into an infinite cosmic reality that is based on the emerging quantum perspective of science.

Preface

It does not require a high degree of imagination to see that humanity is out of touch with the emergent reality being defined by the paradigm shift. The paradigm shift, a concept identified by the American physicist and philosopher Thomas Kuhn, is a fundamental change in the basic concepts and experimental practices of a scientific discipline. The subject of this anthology represents just such a fundamental change in scientific concepts. The idea of ***cosmic sentience*** is ancient—instantiated in every ontology and cultural mythology—but perpetually misconstrued. Cosmic sentience is a holistic ontology represented as the universal trinity of philosophy ranging from Gnosticism to Christianity.

But this is old news. New intelligence (the good news) is that humanity is capable of knowing and enacting the ethical purpose of "God". Like the grin of the Cheshire cat—divine purpose can now be seen among the quantum archetypal frequencies and vectors of quantum field theory (QFT). Within this emergent worldview, everything we thought we knew must be *re-evaluated.* (Radin, 2018, 2006, 1997; Saffell, 2015; Sheldon, 2011; Talbot, 1991) Within the context of re-evaluation, this book sets the scientific stage for re-visiting such critical matters as Divine Mind, creative tension (understood as crucifixion or stress), and human agency (heroism vs. treason) from the modern scientific perspective that includes the psychological-electromagnetic influence of the media-sphere, computer hacking as "obsession", suicidal (leading to entropy) quantum mechanics, and Capitalism as the manifest death-wish of ancient Sumerian culture.

APOLLO 13

The world was watching the sky. "Perhaps never in history has the entire world been united in such a global drama," said Walter Cronkite when, on April 17, 1970, the Apollo 13 astronauts miraculously returned. Cronkite had announced that, "The [US] house and senate passed resolutions last night calling upon the American people to ***pray*** for the astronauts." In NYC thousands have gathered in Time Square to hear updates on the mission. In Rome, Pope Paul asked fifty thousand people to pray for the safe return of the astronauts, and in Jerusalem people prayed at the Wailing Wall. (Grazer, Brian, Producer, 2005) As the re-entry module entered the earth's atmosphere, a radio blackout with a probability of four-minute duration lasted beyond the four-minute deadline, but the dramatic climax—the "critical moment" of safe return—after which the astronauts were presumed dead was delayed as if for dramatic impact.

Truly, the safe return of the astronauts represented "our finest hour". All the dramatic elements were there. Striving prodigiously, an entire team of NASA experts had succeeded in problem-solving beyond reasonable expectation, the modern man-made media provided a global stage for the drama, and the people of the world had shared their heartfelt prayers for the astronauts' safe return. However, the psychological-spiritual meaning embedded in this narrative has not been fully appreciated in a world-culture that emulates technology and ridicules augury. People tend to misconstrue and forget the manifold signs of ***Divine Providence***.

OBSCURING TRUTH WITH WORDS

The "dynamics" of divine providence are clearly-evident in the Apollo 13 story, and the ability to recognize these dynamics requires a scientific worldview based on ***cosmic sentience.*** The concept of cosmic sentient models in a ***Psychecology*** (ecology of psyche) that is developed in the following chapters amplifies the probability of recognizing and re-evaluating ***Divine Providence***. Though the ancient science underlying augury could be argued in this book, our purpose is to cite suggestions of divine providence in the "mystical-scientific" correlations that exist in modern scientific fields—mathematics, physics, neurobiology, and neuro-semantics. The names may have changed, but the dynamics are the same; i.e., augury is very similar to what modern physics calls "probability" in quantum mechanics, and the term "God" may express the same ineffable qualities as the modern "quantum field" or the universal laws of science. The ancient Essenes came close to the truth. The Essenes thought of God as universal Law expressed as the laws of nature (Szekely, 1970, p. 22), so the modern scientific laws defining "quantum electromagnetic grids and frequencies" may be understood to approximate the mystical understanding of "God". The emergent scientific understanding of "superconductivity" already goes well beyond the behavior of superconductors at very low temperatures. (Kirk, 2014, par. 7) Rather, superconductivity (timelessness and action-at-a-distance) may be the pervasive sentient dynamic that accounts for ***Divine Providence*** that appears magical.

The cosmic *Law of Correspondence* (science-based correlation) allows for contemplation of "living" sentience that pervades the infinite range of possibilities ranging from atomic sentience to super-human sentience. Once this hypothesis is grasped and accepted, the dynamic for maintaining synergy with divine providence becomes the human mandate—as it has been in previous cultures. According to the research model of the alchemists, *"As it is above, so it is below,"* is based on principles of holography. Another word for synergy is harmony, and another word for harmony is "coherence". Coherence—harmonious frequencies or mediated bandwidths—make the modern media-sphere possible. However, even the magic of the technological media-sphere is rapidly evolving—the difference between analog and digital waves or the difference between quantum mechanical probabilities and the Higgs Field vectors of quantum field theory (QET) are just beginning to emerge. Quantum-genetic computing and the quantum Internet await us backstage.

What are advertising and political hacking but examples of what was previously known as "obsession" or possession? Entire cultures have been altered with the obsession called advertising that—as documented in the BBC series "Happiness Machines" produced by Adam Curtis (2009)—has manipulated the collective unconscious in predictable ways and changed Western culture from needs-based values to wants-based values called "consumerism". According to the *Limits to Growth* study (Meadows, Meadows, Randers & Behrens, 1972), consumerism has led to near global catastrophe, so advertising as obsession is a critically serious problem—a problem that is not being addressed in realistic ways because of the scientific prejudice against words. Religious terms like "faith" are abjured by a modern science based—

ironically—on faith in undependable statistical probabilities. The purpose of our book is to spotlight such prejudice and to clarify terms in order to address global problems with a degree of authenticity.

The Great religions often divide the spiritual path into two trajectories—heart and mind or the path of the Bodhisatva and the path of the Adept (the path of faith and the path of science). The modern age of science may be understood as having some characteristics of the adept, but to maintain coherent energetic harmony with super-human bandwidths, the path of faith must be revived and re-integrated with the path of adept-hood. In the Monadic-mysterious environment of today's science, any "atheistic" perception becomes ridiculous. According to the perspective of the adept, the infinite realm of "Divine" sentience may find its analog in the framework of quantum field theory (QFT), cognitive science, and information and communication technologies (ICTs). Within this infinite field for discovery, measurable coherence provides the tool with which to understand the dynamics of "spooky action at a distance", the Singularity, perceived non-existence, the plethora of paranormal experiences, and—ultimately—the cause of the *Global Problematique* (the suicidal trajectory toward entropy).

When "coherence" is achieved—even partially—it manifests as divine intervention, and divine intervention has the characteristics of harmonious superconductivity. Perhaps the best metaphor available to exemplify the concept of superconductivity is the musical "concert" which includes all the dynamic components of harmony or coherence. Beyond the concert, the global media-sphere is both a metaphor and a superconductive reality made possible by fractal physics and mathematics—understanding and benign use of which is in process of refinement.

Coherence is the keyword, and the dynamics of cosmic sentience can be correlated directly with the dynamics of music. Coherent degrees of sentience—like notes on a musical scale—provide a referential matrix for harmony. Just as a concert requires the attunement of musical scales, instruments, artists, directors, and a multitude of atmospheric details for the maintenance of harmony, degrees of sentience can be attuned. Similarly, just as media bandwidths must be coherent and without interference patterns, sentient bandwidths are best contemplated according to degrees of scaled, fractal frequency discussed by Peter V. Grujic (2001) and Jean B. J. Fourier (1878). This begs the question, "Why does disharmony exist?" However, if the true answer were given, "Due to the human learning process" few would understand. This opinion is verified by the *Global Problematique*.

Within this conceptual framework, any perceived coherence—wherever it is found—may be understood to represent a degree of superconductivity and this superconductivity (Jungian synergy) may be understood as a manifestation of Divine (Logoic) intervention—perhaps something like the "self-fields" of quantum field theory. The following chapters represent different perspectives on the theme of "media models" and "superconductive sentience". For now, perhaps the best way to express the idea is with a quote from Carl Jung:

In his 1937 essay 'Psychology and Religion,' Jung wrote: "(perhaps) we could say with Nietzsche that 'God is dead.' Yet it would be truer to say 'He has put off our image, and where shall we find him again?'" Jung believed that God was still active as the Deus Absconditus, the hidden god, an unknown quality in the depths of psyche. (Lamborn, 2015, p. 2)

In other words, the hidden God, like the grin of the Cheshire cat, manifests as perceived coherence. Pragmatically speaking, the concept of coherence can be applied to both science and morality. This is why the ancient axiom advises us—as a practical matter—to be "pure" in thought, word, and deed. It is not a leap of faith to correlate purity (the alchemical quintessence) with coherence and to apply that purity as the essential ingredient required for the manifestation of a culture distinguished by its New-Order morality of coherent superconductive sentience in a cosmic unified field.

COHERENT PURITY

Think of the morality of coherence in terms of the electromagnetic harmony of Thoughts, Words & Deeds. For purposes of research, the Jungian functions (mind, emotion, sensory-physical perception, and spiritual intuition) are already being thought of as bandwidths of frequency within the cognitive Psyche. Humans (individual and collective) are the summation of our integrated thoughts, words, and actions. Our words and deeds arise from our thoughts, and what we do and say also shapes our thoughts. This is what we call mediated interaction, and on each functional scale, the nature of interaction becomes a media model. This truth has been expressed repeatedly and in many forms throughout the history of philosophy. Two examples are the following:

For the growth of faith and for the fostering of understanding, an essential requirement is purity of the heart, of the very base of thought (the kshetra),[1] of the levels of consciousness (chittha). Without this purity, the sudden effort of self-inquiry or investigation into the self-existent Atma, while in-the midst of diverse worldly and material entanglements, will be rendered fruitless, since it will not stem from an eager [personal] will.

The consciousness (chittha) must first be withdrawn from the objective world (prapancha) and turned inward [archetypes or electromagnetic forces which are beyond "material" scales of sentience] toward the awareness of the Atma [universal sentience]. Seeds can sprout fast only when planted in a well-ploughed land. So too, the seed of Atmic wisdom can sprout in the heart-field (hridaya-kshetra) only when it has undergone the necessary refining process.

Cleanliness is primarily an attribute of speech. Maimonides explains that one of the unique properties of Hebrew, "the holy tongue," is that it is an essentially "clean" language" – no dirty words. As we saw in a previous post, it is Pesach that is the holiday of speech – freedom of speech. To speak freely does not mean to speak as you like, to use whatever words you like, whether clean or dirty. As we explained, to speak freely means to be able to give verbal expression to the innermost emotions of your heart. Such deep and authentic self-expression comes out (the secret of the Exodus) "clean."

The holiday of Shavuot, which commemorates the Giving of the Torah, corresponds to the attribute of purity – more precisely, the purity of thought. For seven weeks the bride (the collective soul of Israel) purifies herself to marry her groom (God) on Shavuot. All her thoughts concentrate solely on her groom. (Sathya Sai International Organization-USA, n.d.)

These quotes exemplify a pervasive message in scripture that is fundamentally scientific. Divine providence may be understood as "clean" coherence. However, modern social agreement now considers foul language normal, abortion and molestation as a degree of free expression, murder as a right, warfare and torture as a survival mechanism, hatred as a virtue, and "I'm worth it," as a mantram. No wonder we have not been able to solve our problems. In this framework of the end-game, the critical need for divine providence may be stipulated, but the maintenance of coherence with super-human bandwidths is not possible without some degree of purity and cleanliness—harmonious coherence in thought, word, and deed. Achievement of this purity becomes the human mandate. At this stage of human evolution, the discipline required is a human responsibility.

MEDIATED PURITY, DIVINE PROVIDENCE, AND CONVERSATION WITH GOD

Viewed from the perspective of the media-sphere, purity and cleanliness are the issues to be addressed—as a doable matter of survival. The simple truth is that most humans do not want to be pure and clean in thought, word, and deed—not only because purity requires constant vigilance and discipline, but because purity and cleanliness are not understood as "fun". But in the incipient age of mediation in the unified field, the final days for discipline and "change of heart" have arrived.

The time for the human family to take collective responsibility for Earth's crisis is at hand.[2] By contemplating the nature of sentience and the dynamics for harmonizing its many degrees, we can reinstitute the scientific idea of divine providence and study the dynamics whereby humans can maintain coherent states with super-human bandwidths (talking with God in the Garden). This conversation between human and super-human bandwidths is amplified throughout infinity with mediated thought and language; i.e., the individual human macrocosm can only achieve optimal health if the microcosmic functions, systems, organs, cells, and atoms exist as harmonious frequencies. Relative responsibility for health—microcosmic/macrocosmic—is shared throughout the degrees of sentience. It is both personal and collective—neurobiological and technological—each person makes a difference; each context makes a difference.

As a collective responsibility, cleaning the media-sphere is a good place to start. The uncleanliness of the media-sphere is self-evident even if the pervasive availability of pornography is not considered. The accepted canard that unclean speech and pornography are "free speech" is simply a dark excuse for the wholesale dissemination of evil impurity which blocks planetary synergy with the higher, purer bandwidths of divinity. This perversion of the media was explored by George Orwell in his novel *1984* and expanded by Aldous Huxley in *Brave New World.* (Postman, 1984, pp. 155-163) The dynamics of psychological mediation should be well-known (advertising, neuromarketing, brainwashing, propaganda). Nevertheless, using the technological media for positive "pure" purposes (education, public radio, the arts) has faced an uphill battle against the profiteering classes, the ignorant, and the malignant. Moreover, Western materialistic ontology—even though scientifically outdated—continues to be a false scientific premise that obstructs all progress and severely limits all efforts at problem-solving.

The discreet event of Apollo 13 was a timely reminder that the media can be used to set the stage for divine intervention. However, the meaning of mystical intervention by God in human events—even this most significant illustration—is usually sub-optimized, miss interpreted, and unappreciated. In a nanosecond people take credit for-that divine intervention which is beyond even the best of their efforts. This inscrutable transition from despair to euphoria is translated into "God is on our side", and an irrational certitude that ***we*** accomplished it and should get all the credit for whatever miracle transpired. On its face, Apollo 13 is a clear example of the divine mystery (the whole exists within each part) to which humans have increasingly closed their eyes. In the immediate aftermath of emergency, humans are quick to take credit for their own salvation—and that is the problem.

When viewed from a perspective on ***symbolic meaning—as in Jungian dreams or augury***—the media event happened between Easter and Pentecost, the Christian festival celebrating the descent of the Holy Spirit on the disciples of Jesus after his Ascension. All-of the symbolism in this global event points directly to the intervention (descent) of God in response to the prayers of a unified people who had done everything in their power to succeed—without success. Apollo 13, like language and thought itself, is a metaphor that is substantial. In the reality of a sentient ecology of Psyche.

*PSYCH*ECOLOGY

[Electromagnetic] Thoughts, like drops of energy, attract small elementals. The character of these germs of the spirit is most varied: depending upon their substance, almost imperceptible embryos can achieve, under the nurturance of thought, diverse manifestations. They can form the basis of minerals and even plants. But one can imagine quite clearly how those thoughts that are void of any vital bases litter the lower strata of Earth. Meteoric dust is imperceptible to the eye, but it results in very substantial sediments. Hence, one can imagine how vast the dust of thought is and, being the effect of energy, how very substantial! The consequences of this debris of thought cause the illness of the planet. (Fiery World, 1933, p. 231)

The present moment in human evolution can go either of two ways—termination or rebirth. When examined for its underlying meaning and significance, this unique drama of Apollo 13 can be viewed in terms of a "prospective" dream. All the dynamics of scaled sentience are there; even the (sentient) Lunar Excursion Module (LEM) is appreciated as a friend and hero in the narrative—a persona—and much of the drama has to do with maximizing the complex sentience of LEM. Apollo 13 was an exercise in finality as-well-as a metaphor—in the final stage of its return to earth, the final challenge was whether the heat shield had been damaged and whether it would protect the astronauts from being incinerated as the return module entered the earth's atmosphere. After a complex series of seemingly miraculous problem-solving efforts by the NASA team and the astronauts, a successful return of the astronauts was still in serious doubt, and the possibility of fiery annihilation due to cosmic friction was very real. But in true cognitive-dramatic fashion the climax was a surprise.

Though it employs the selfsame dynamic used in the fields of mathematics, physics, chemistry, and genetics, ***Literary exegesis***—the analysis of meaning in literature—is a process of searching for (the unknown) unconscious cognitive ***meaning.*** Jungian dream analysis is almost the same dynamic as literary exegesis. Both use metaphorical amplification to discover underlying meaning. Carl Jung's system of dream analysis (Analytical Psychology) has usually been judged to be of limited scientific value. Such an errroneous perspective discloses serious ignorance of the most recent hard-scientific research in many fields especially psychology, artificial intelligence, robotics, and computer technology.

If "disbelief" in the existence of divine providence is not simply ignorant, it is blatantly prejudiced and scientifically illogical. Viewed from the perspectives of Jungian dream analysis or literary exegesis, the drama of Apollo 13 plays out much like a dream—a little story within a greater story—that has its plot in the human-planetary drama of space exploration, its exposition in the Cold War competition between the USA and the USSR, and its ***lysis*** (resolution of the plot-problem) in the conscious realization of re-entry into a new sentient age of cultural atonement.

DIVINE SENTIENCE

If the modern scientific emphasis on-the subject of "intelligence" or sentience is advancing with adamant disbelief in any Divine cause or existence, the default conclusion must be that science is "godless". It may be stipulated that sentience in the form of artificial intelligence (AI) is a primary subject of research that leads to the plethora of other scientific applications such as autonomous weapons, and neuro-marketing algorithms. Unfortunately, the focus of research in these fields is invariably misdirected toward profit, or war, or both, and if the expensive development of autonomous robots trained to kill humans is not a godless act—what is? Nevertheless, the scope of research on intelligence has taken us into alternative evolutionary dimensions that resonate with alchemy and range from intelligence in atoms (Hameroff, Stuart & Penrose, Sir Roger, 2014), intelligence in quantum fields (Orme-Johnson, n.d.; Shin, in publication), intelligence in molecules (Streedhar, 2017; Norris, 2014; Pert, 2003; Lipton, 2005) to intelligence in non-human dimensions including robots (McDonald, 2015), animals (Lilly, 1974 & 1969), plants (The Secret Language of Plants, ND; How Plants Communicate & Think, ND, elements (Kirk, 2014), water (Dr Masaru Emoto's Water Experiment - Words are Alive! 2014; Dr. Emoto, Thoughts/Intentions Shape Physical World 2013; Water Documentary: How your consciousness affects the universe, Emoto, n.d.), minerals (Lewis, 2014; Eagleman, n.d.), and the "light of Fire" (Galeon, 2016).

Intensive study of correspondences strongly suggests that this current research is consistent with the ancient sciences that include astrology, alchemy, philosophy, theological pan-psychism and monotheism, as-well-as the intelligence of language (oracles) as a sentient-symbolic form—a "living" sentience reminiscent of Jungian "living symbols". (Foster, 1985; Jacoby, 1973 ed.; Jung, 1933; Woodcock, in publication; Gaiseanu, in publication) This essence of living-sentience has been contemplated universally for ages by philosophers, and during these ages-past a holistic assumption was made by them that—to put it simply—God is "in" everything.

This was assumed, but not well-understood in the ways which modern science has provided. For example, electromagnetic fire, the fire of friction, electrical fire, are forms of fire that are embedded in everything. Solar fire is an advanced form of "fire light" that has been universally equated with "Spirit" as in Plato's Myth of the Cave or in the multitudes of so-called sun-worshiping cultures. After all—as the source of all life on earth—is not the sun worthy of awe and devotion? The ancient wisdom ensconced in esoteric knowledge teaches that stars and planets are the physical manifestations of exalted sentient beings. Viewed from the perspective of modern "evolutionary" science, this appears to be probable. Within this evolutionary scientific framework, the cognitive family metaphors defined by George Lakoff (2008 & 2002) would serve to supplement the family metaphor called the Trinity which is universal in theological mythology. The following quotes are from Mormon doctrine but are not restricted to that doctrine:

Intelligence or light and truth, is also used as a synonym for spirit element. Scriptures using both terms speak of the self-existent nature of the substance [fire] involved. (D. & C. 93:29; 131:7-8.) Abraham calls the pre-existent spirits the intelligences [sentient fire] that were organized [coherent] before the world was" (Abra. 3:22) because the intelligences were organized intelligence or, in other words, the spirit bodies were born from [the fiery] spirit element. (McConkie, 1979, p. 387)

Abraham used the name "intelligences" to apply to the spirit children of the Eternal Father. The intelligence or spirit element [sentient fire] became intelligences after the spirits were born as individual entities. (Abra. 3:22-74.) Use of this name designates both the primal element [fire] from which the [sentient] spirit offspring were created and-also their inherited capacity to grow [evolve] in grace, knowledge, power, and intelligence itself, until such intelligences, gaining the fulness of all things, become like their Father, the Supreme Intelligence. (McConkie, 1979, p. 387)

Primariy due to ignorance and prejudice, modern science has abjured ancient science which—indeed—is appropriately called "science" because if one takes the time to study the ancient mathematics of Plato, Anaxagoras, and Pythagoras, the chemistry of the alchemists, and the cognitive structure of astrology without prejudice, their field methods were—without any doubt—scientific. *Media Models to*

Foster Collective Human Coherence in the PSYCHecology addresses this modern fallacy—in a reality wherein punctuation in software can alter meaning— of neglecting ancient wisdom-science and subtle perception by using words that have not been used before. Hopefully, this will minimize confusion generated by ignorance or prejudice.

- ***PSYCHecology*** (an ecology of psyche based in part on electromagnetic laws) may be understood as God, QFT, hologram, or Eternal Spirit.
- ***Sentience*** may be understood as multi-dimensional cosmic intelligence that may be recognized by its dramatic-rhetorical process of mediation or "intelligent" coherent response to an environment.
- ***Media Models*** includes both organic and inorganic scales of sentience (crystallin structure; silicon) that engages in *conversational modalities* be they linguistic, mathematical, chemical, or metaphorical.
- ***Coherence*** may be understood as harmony of frequencies that exist on a continuum of degrees or dimensions that are measurable in terms of harmonic scales and frequencies such as heart-rate-variability (HRV) as researched by ***The Institute of Heartmath.***
- ***Plato's Sacred Geometry*** *There are scientists who consider that Platonic thought-forms are electromagnetically responsible for the evolution of consciousness within a living infinite fractal holographic universe. The late science advisor to the Belgrade Institute of Physics,Petar Grujic, in his book The Concept of Fractal Universe Anaxagoras (Grujic, P.V., 2001)explains that ancient Greek science was conversant with infinite fractal mathematics. In his book Anaxagoras on Mind (Baloyannis. S. J. January 2018), the neurological medical scientist, Staros Baloyannis, considers that Anaxagoras' work "is now a significant milestone in the field Neurophilosophy". (Pope, 2019)*
- ***Media-sphere*** refers primarily to the reality of technological media that includes the Internet, social media, nightly news, film, print, and all their paraphernalia. It also refers to the above defined media models that have similar "sentient" dynamics and that can usually be studied in terms of the conversational trinity (action, mediation, reaction/proaction) and measured in terms of Fourier mathematics and metaphorical correspondence (fractal structures).

Once the dynamics of sentience have been established, our primary purpose is to clarify the practical efficacy of contemplating and coordinating with super-human "spiritual" intelligence in order to address the suicidal planetary crisis that Dr. Aurelio Peccei[3] named the ***Global Problematique***. By itself, powerful as it is, human agency is not sufficient for understanding the cosmic scope of the problem nor for using it to solve errors of agency that contribute to the self-evident planetary crisis. It is human agency that has caused the crisis, and without responsible human agency in synchronicity with the forces of Spirit, the errors of ages will reap their consequences.

When considering a very complex subject such as universal sentience, one must be as succinct as possible. As an illustration of the value of ancient science, the following quotation and the citations within it could not be more succinct.

Plato's Phaedo discusses Anaxagoras' mathematical teaching that intelligence organizes all things in the best way it can but argues that physical reality alone does not convey a sense of order in the universe, which requires a sense of purpose (Conolly. T. 2019). In Plato's dialogue Ion he refers to a primordial life form within magnetized lodestone creating electromagnetic emotional thought forms within the mind.

His axiom that All is Geometry embraced both ethical and unethical thoughtform geometries entangled together within this electromagnetic theory of creative artistic mathematical evolution. Plato's sacred geometries creating order from chaos were held to hold evolutionary dominance over the unethical chaos mathematics also involved in the transition of chaos to ethical infinite universal order." (Pope, 2019)

In an emergent media age characterized by technological sophistication, artificial intelligence, robotics, political chaos, and planetary crisis, ignoring the ancient scientists would be irrational and foolhardy. Ultimately, contemplating research on sentience will lead to a better understanding of the problems associated with an age of communication and the media.

To Plato the universe was evolving the living information process from chaos under the guidance of a supreme ethical mathematical purpose. On the contrary Einstein believed that the universe was evolving all life in the universe to extinction under a supreme law governing the energies of chaos. Both philosophers of science were dealing with what scientists call a first-cause principle controlling the dynamics of reality. Einstein's science belonged to a cruel harsh tribal realty, which nonetheless was an essential part of Plato's mathematical map to the future. (Pope, 2019, p. 2)

PROSPECTIVE ON MEDIA SENTIENCE

Ultimately, research on sentience will lead to a better understanding of the problems associated with an age of communication and the media—discovery of "a supreme ethical mathematical purpose". This is imperative, but modern science has not even begun to consider two of the most important problem-complexes introduced by such a media-age: the psychological influence of money (as sole value structure) and the issue of mediated political influence that has all the characteristics as what past ages have understood as demonic possession.

THE ILLUSION OF VIRTUAL MONEY

The incipient growth of scientific contemplation in different scales of sentience is rampant: quantum/molecular/cosmic, artificial, robotics, and sentience in plants, minerals, and elements. The progressive minority is researching these varied fields as vital intelligence, while the majority ignores any suggestion that life and intelligence in various dimensions is of any concern. On its face, this is illogical at best. This ignorance prevails in the same way that the field of psychology Behaviorists ignored any idea of the existence of the "unconscious". This ignorance coincides with the way orthodox science rebukes any suggestion of the existence of alien life and sentience.

As long as this prejudice exists, progress and even planetary survival will be in doubt. On the other hand, progressive science has abundant precedents in alchemy, mathematics, and the mystery schools as-well-as in robust modern research in many fields. The orthodox scientific view is not really science. It represents the exact same stupidity as the Warren Report and the lies told about the assassinations of a whole generation of progressive leaders—all perpetrated by single actors instead of a conspiracy. Western culture is in critical jeopardy.

The **definition** of culture alludes to a particular set of customs, morals, codes and traditions from a specific time and place. Though morals, codes, and traditions may be difficult to measure, standards of market exchange serve as clear and measurable examples of a culture's morality; i.e., what a culture considers valuable costs more in the marketplace. Whether a culture's values are represented as furs, beads, diamonds, or gold, "market values" are relatively definitive and measurable standards. This remains true across other definitions of culture such as "market, hierarchy, adhocracy, or clan". However, changing cultural values from money to life/sentience (akin to the environmental ethic) could change everything.

In other words, though cultural standards of value are interactive and overlap, the current value structure of modern global culture is limited to "profit"—the metaphorical-illusory realm of money and exchange colored by motives of selfishness, privilege, and malice toward the not-self. Barter systems are valid relativistic systems based on some assumption of standard value, but stock markets—as dynamos of values in Western culture—have devolved beyond illusion to *delusion*. Plato was prescient to observe that Democracy with a single value (money) is the worst form of government. Because this delusion is so strongly embedded in the contemporary mindset, our discussion sentience is critically important for re-generating the global cultural value system.

As a reminder, Plato developed such distinct areas of philosophy as epistemology, metaphysics, ethics, and aesthetics. His deep influence on Western philosophy is asserted in the famous remark of Alfred North Whitehead: "The safest characterization of the European philosophical tradition is that it consists of a series of footnotes to Plato." Plato was the prototypical political philosopher whose ideas had a profound impact on subsequent political theory—the greatest of which was his impact on Aristotle. Platonic philosophy emphasizes the usage of dialectic, a method of discussion involving ever more profound insights into the nature of reality, and by cognitive optimism, a belief in the capacity of the human mind to attain the truth and to use this truth for the rational and virtuous ordering of human affairs. Plato believes that conflicting interests of different parts of society can be harmonized. The best, rational and righteous, political order, which he proposes, leads to a harmonious unity of society and allows each of its parts to flourish, but not at the expense of others. The theoretical design and practical implementation of such order, he argues, are impossible without virtue. Serious lack of virtue characterizes the capitalistic system which stands opposed to the common welfare.

Although large parts of the *Republic* are devoted to the description of an ideal state ruled by philosophers, the chief theme of the dialogue is justice. According to Socrates, "justice" has its origins in speech. Justice is a metaphorical model that exists "in heaven". Because of Plato's metaphorical language, many critics interpreted his theory as impractical, but modern scientific ontology gives the lie to that superficial understanding of pragmatism. Plato's vision of the ideal state is that it is just. Justice, understood traditionally as virtue and related to goodness, is the foundation of a good political order, and as such is in everyone's interest. Justice, if rightly understood, Plato argues, is not to the exclusive advantage of any of the city's factions. Rather, it is concerned with the common good of the whole political community, and is to the advantage of everyone. It provides the city with a sense of unity, and thus, is a basic condition for its health. According to Plato, injustice causes civil war, hatred, and fighting, while justice brings friendship and a sense of common purpose.

In order to understand further what justice and political order are for Plato, it is useful to compare his political philosophy with the pre-philosophical insights of Solon, the essence of whose constitutional reform was based on justice. Solon, when he became the Athenian leader in 593 B.C.E. restored righteous order, *eunomia,* within a political environment that was much like the current one—disparity of wealth that created two conflicting factions. In the early part of the sixth century Athens stood at the brink of a fierce civil war between two parties, the poor and the rich. For Plato, as for Solon, the starting point for the inquiry about the best political order is the existential fact of social diversity and conflicting interests, which involve the danger of civil strife. The amelioration of danger exists in a governmental system of lawful checks and balances—much like the constitutional democracy of the United States of America.

Why, then, did Plato consider democracy with only one value to be the worst form of government? Today, it is generally believed that democracy, "government of the people by the people and for the people," is the best and only fully justifiable political system. The distinct features of democracy are freedom and equality. Democracy can be described as the rule of the free people who govern themselves in their own interest, either directly or through their representatives. Although freedom is for Plato a true value, democracy involves the danger of excessive freedom, of doing as one likes, which leads to anarchy. Moreover, equality, related to the belief that anyone has the right and equal capacity to rule, brings to politics all kinds of power-seeking individuals, motivated by personal gain rather than public good. Democracy is thus highly unstable and corruptible. It opens gates to demagogues, potential dictators, and can thus lead to tyranny. The current global crisis has almost the exact dynamics addressed by Plato. (Plato: Political Philosophy, n.d.) An astute application of his—long-hidden—sacred geometry to the healing of global culture is, therefore, advisable.

Moreover, the common good is no longer limited to the human species. As discussed above, the common welfare now includes vital sentience that ranges from the atomic scale to the galactic scale. Jungian dynamics for healing the collective unconscious with analysis of meaning in dreams is one of the few modern scientific efforts to understand the comprehensive truth about universal life and sentience (Psyche). Jungian research is also directly related to Plato's sacred geometry and the archetype of coherence because coherence is the same thing as Jungian "compensation". (Schafer, Stephen, 2012) Monetary systems are like dreams; though they represent very "real" cultural value systems, they are fundamentally ephemeral, contextual, and illusory—and they can be easily manipulated for self-serving purposes which are not in the common interest.

Nevertheless, both dreams and illusions can be analyzed and simulated with statistical models and media images which—as researched by Reeves and Nass (1996)—are perceived psychologically as "real" by all demographics (pp. 251-256). Economists, insurance companies, and investors have always known that these systems are based on faith, but with the rise of virtual currency "belief" in the system—the illusion—is becoming even more important than before.

Images are the common denominators for perception of reality. For Jung, images are embedded with meaningful semiotic patterns, but now those patterns can be adjusted with Plato's sacred geometry to conform with "a supreme ethical mathematical purpose". Therefore, pervasive mediated images have been culpable in the degeneration of cultural reality, but they can also be employed to re-constitute an ethical reality without loss of individual or collective dignity. People will retain an ethical—in terms of the general welfare—array of evolutionary choices to make, and their personal individuation will contribute to the cultural individuation.

What must be remembered is that dreams (images) may now be understood as very real. As forms of sentience, dreams called memories, fantasies, likes, or goals have been symbolically generated for decades with advertising which is a primary factor in GNP. In the early 20th century, Edward Bernays—Sigmund Freud's nephew—proved that the fundamental values of a culture could be radically altered with the illusions called advertising. Based on Freudian principles of reduction and fear, Bernays changed the entire value system of the USA from a needs-based system to a want-based system and expanded the concept of "credit" to accommodate the dream. (Curtis, 2009)

The Freudian malaise that formed the basis for modern advertising has done incredible damage to culture, but Jungian principles are completely different. While Freudian principles are reductive (entropy), Jungian principles are expansive and evolutionary. Jung's "amplification method" for dream analysis is an evolutionary process for intensification and compensation within the process of Individuation. The difference between the two worldviews is like the difference between the symbol of the circle and the symbol of the spiral—a circle is closed, and a spiral is open.

ALL THE SENTIENT-ECONOMIC "SALVATION DUCKS" ARE IN A ROW

The Jungian healing process may be understood as sentient—as degrees of "intelligent" response which produces results within an environment. In short, all of the ducks are now in a row. The *Global Problematique* can now be healed with the application of sacred geometries and Jungian principles of dream analysis to the evolutionary programming of the Media-sphere. All of the current social-ontological dynamics can be addressed from a perspective of the "sentient" media-sphere which provides an evolutionary model for understanding economics in a cognitive-mediated reality—a much more practical model than the models called rationalism and materialism.

The world's economy is based on statistical probabilities that may be correlated with the probabilistic mathematics of quantum mechanics (QM). However, now the well-accepted probabilistic context of QM mathematics is being replace with the mathematics of Quantum Field Theory (QFT) that appears to be even more "sentient" than QM. The uncertainty principle (formulated by Werner Heisenberg) does not apply to the mathematics of quantum field theory (QFT). (Brooks, 2016, p. 125) So, it appears that the beginnings of understanding cosmic sentience are at hand.

However, QM theory based on Einstein's theoretical structure is still very influential, and probabilities exist on a continuum of validity; i.e., the uncertainty principle. Accordingly, some economic probability forecasts are more dependable (less uncertain) than others. Dependability of economic systems relies on their stability, balance, or coherence, and when that stability is threatened by intangible motivations (human factors) like greed, or sub-optimization the consequences appear as severe imbalances such as market collapse, inflation, depression, and inequality in the distribution of wealth within a system.

Human factors tend to impeach the coherence of natural systems and to throw the unified field of the environment into disrepair. Currently, mathematical-statistical models of probability rarely incorporate the more authentic cognitive dynamics (human factors) at work in economic systems. Just as "sentience" is ignored in research on artificial intelligence and autonomous robotics—where it is relatively obvious—it is treated superficially in economics—even though accurate assessment of the "wants" of target groups (which are clear indications of sentience) for advertising is critical to the industry. For example, a recently well-received book on neuromarketing is fundamentally derived from a superficial understanding of Jungian archetypes of the "deep" unconscious. (Zaltman, Gerald & Zaltman, Lindsay, 2008, pp. 35-38)

"Carl G. Jung argued that archetypes are patterns, symbols, and images that represent basic qualities of mind inherent in and shared by every person regardless of culture. Hence, archetypes are universal and operate in what Jung called our 'collective unconscious'". (Zaltman & Zaltman, p. 36) The Zaltman book is based on a superficial understanding of cognitive archetypal dynamics. Nevertheless, the book received rave reviews. Stephen M. Kosslyn, John Lindsay Professor of Psychology and Chair of the Department of Psychology at Harvard University says, "Filled with fascinating observations and insights, this book is brimming with wisdom and good sense…*Marketing Metaphoria* has something new to say to anyone interested in marketing, psychology, or human communication."

Cyber currencies (CC) are often called crypto currency, virtual money or a combination of these terms. CCs came into existence to take advantage of the unique character of the Internet. Cyber currencies are a figurative form of money that is verified, or "confirmed" by a computer network comprised of the coin's users. CCs vary in quality, and their fundamental value is based on the quantity of users and a high daily exchange volume. They are "like" cash which means that CCs exist as metaphors. A holder does not need to have a bank account, just the coin, to make an online purchase. Currently "the coin" must be bought with existing monetary models, so CCs are simply an extension into more dreamlike dimensions of value structures. Because a buyer can't shove cash into a computer to buy something, CCs become virtual money which can be spent anonymously online—and misused by hackers.

Essentially, CCs are a sentient model for credit that—as above mentioned—is confirmed only by faith in a virtual reality. Like all monitory models, CCs are a form of gambling—one may add—a perversion of human agency (the power of thought, feeling, and will) to influence a quantum field of cosmic reality. CCs are sentient models which conform to "intelligent" (purposive, consequential) responses within a contextual environment. Presently, CC models are aspects of Capitalistic sentience that is perverse. Capitalism exists as a value structure based on competition, profiteering, and deceit. Foundations of Capitalism are the attainment of goods and services, but these valid purposes have become so manipulated and delusory that its proponents are destroying the planet. Malignant purposes are psychologically promoted with mediated advertising that is based in sentience; i.e. personal (contextual) wants and selfish desires as expressed in the idiom of, "I'm worth it." (Curtis, 2009) Political hacking employs the same psychological principles as advertising, and the current attack of political hacking on civilization is proceeding uncontested.

OBSESSIVE SENTIENCE

Intrusive obsession is rife throughout the media-sphere, but because it is considered superstition, the phenomenon does not enter-into modern problem-solving. Just as Divine providence remains unrecognized, obsession—even when it is obvious as advertising and propaganda—remains unrecognized. Hacking of the Internet—against which experts can barely contend—is certainly a form of obsession. Political hacking is willful intrusion on the collective unconscious for malignant purposes. Hacking originates from disembodied sentience beyond consciousness, and the target of such malignance is almost completely oblivious—and defenseless—relative to this intrusion that takes many forms in our mediated reality.

In order to protect itself from such malignance, a cultural persona must be willing to contemplate multitudes of sentient scales and psychological influences with the tools discussed above. Currently, obsessive forces are not even contemplated except at superficial levels. An earlier form of obsession within the media-sphere is the intrusion of advertising, but today advertising based on cognitive science of the unconscious has transformed the field into incipient "neuro-marketing". These obsessive influences are powerful and pervasive, extending to cosmic scope and including such vast issues as the world economy, the collective personae of nation states, and the miss-direction of the entire global value structure—yet human cultural gate-keepers are entirely ignorant of the crisis of obsession.

This pervasive ignorance in a plethora of variations constitutes a vastly influential frame of the cognitive unconscious that must be healed. The good news is that the dynamics of such healing are known.

They consist of three unique and proven elements of advanced science that have been implied in diverse fields but remain largely unrecognized and unused:

1. Jungian healing dynamics in a holographic dreamscape of the media-sphere
2. Heart-rate coherence that is measurable as heart-rate variability (HRV)
3. Plato's Sacred Geometry applied to holographic imagery

One source of this pervasive "carcinogenic" error has been defined over time and discovered to relate to the mathematics of quantum mechanical devolution. (Pope, 2018) Specifically, the mathematics relate to differences between Einsteinian symmetry and the asymmetrical sacred geometry of Plato. "*There are scientists who consider that Platonic thought-forms are electromagnetically responsible for the evolution of consciousness within a living infinite fractal holographic universe.*" (Pope, Robert, 2018) The same error can be applied to the differences between Freudian theory of reductive complexes and Jungian principles of healing.

Theoretically, the scope of this distinction may be applied to the differences between the mathematics of Quantum Mechanics (QM) and the mathematics of Quantum Field Theory (QFT). The inference is that only QFT is useful for the study of ***sentient Life*** in a holographic universe, and that the pervasive use of QM—with its ultimate interpretation of universal entropy—constitutes an obsession promulgated by a "Sumerian death cult" (Pope) that is presently leading humanity to oblivion. The objective of this anthology is to begin a process to reconsider the ontology of cosmic sentience in order to re-direct the current trend of science based largely on QM.

It is within this conceptual framework that the extensive research of Carl Jung is unique in its contribution to the re-direction of QM science by defining the dynamics of holographic Psyche which—unlike other psychiatric research—can be applied to healing the psychosis of the deep-seated death cult with which humanity is obsessed.

The phenomenon most readily visible on the plane of consciousness are symptoms and complexes. A symptom may be defined as a physical or psychic indication of an obstruction of the normal flow of energy. It is a 'danger signal' indicating that there is something drastically wrong or inadequate in the conscious attitude and that a broadening of consciousness is desirable. (Jacoby, 1973, p. 36; Jacoby, 1959)

Within the virtual illusions of today's mediated reality, it is ludicrous to ignore centuries of precedent relating to obsession. In order to contemplate this precedent, it is also necessary to re-define human perceptions of "God" and to employ Plato's sacred geometry—as fractally represented by coherence and Jungian healing dynamics—in the process. Those sophisticated modern observers who find it difficult to understand the dynamics of political hacking are poorly prepared to contemplate the deeper more comprehensive dangers of mediated possession, and those who pride themselves in their sophisticated atheism are no doubt doomed.

The fiery understanding of obsession is called "Urumiya". Man is not the only one to possess this straight-knowledge [sentience]; certain animals close to man sense this dreadful state. Horses and dogs in-particular sense and resent the proximity of obsessed persons. In ancient China, there was a special breed of dogs, highly prized, which was very sensitive and useful in detecting the obsessed. In ancient times it was also a custom to exhibit the horses and dogs before guests, observing at the same time the reaction of the animals. Many envoys were put through this test. (Fiery World, 1933, p. 202)

It is within this scientific sphere of possession that super-human assistance is most urgently required. The conscious determination to ignore this sentient reality of electromagnetic fields would be foolhardy. Scales of sentience must be diligently researched in order to distinguish scientific relevance from superstition. Never before has scientific open-mindedness and discipline in human science been required than in the present media age. Unfortunately, scientific communities worldwide have closed their eyes to the obvious and have disarmed themselves on the verge of inevitable battle. It would behoove them to consider the Bhagavad Gita that provides guidance in preparation for the battle of Kurukshetra.

Possession is an age-old phenomenon observed all over the world. "Possession has been observed in many forms and in many countries, especially in rural communities, country sides, in poor, uneducated, marginalized and tribal population, but in urban areas and metropolitan cities as well in their slums and higher societies. It has been reported extensively by anthropologists, sociologists and lately by psychologists." (Vagrecha, 2016)

A plethora of ancient and ongoing research—considered anomalous and almost entirely ignored—exists in the study of demonic possession. This was undoubtedly a good thing in the past when superstitious and fear, itself, were powerful magnetic attractors, but now—good or bad—this intrusion on sentience cannot be ignored. It must be addressed. One researcher, Laura Sumrall[4] --a Visiting Fellow at the Max Planck Institute –reports that, "Catholic priests in America field requests for exorcisms numbering in the thousands every year." In this regard, perhaps the film industry—as aspects of the media-dream model (Schafer, 2016 a, b, c, & d) has superlative scientific value.

Speaking historically, the contest between good and evil sentient forces was not as primitive nor mindless as it has been portrayed to be, and possession could be either "good or bad"; i.e., super-human energies and bandwidths can be understood as deriving from either benign or belligerent sources. Such judgement, of course, is subject to close scrutiny due to relativistic human perspectives. Nevertheless—as discussed in the book—energetic harmony or coherence may be used as a scientific indicator of the difference on a continuum of frequencies. For example—subject to holistic and open-minded prognosis—states of health can be considered coherent (asymmetrical), while states of illness are incoherent (symmetrical).[5] Some say the future of investing is in the cyber currency field, while others warn that the entire sector is rife with scams and potential danger.

The possibility of demonic agency would not usually be considered unless natural medicines first proved ineffective, but ineffective medicine was not taken uncritically to indicate a demonic cause. Physical convulsions, for instance, were also associated with natural diseases such as epilepsy, which was already understood to be unpredictable, chronic, and potentially incurable. For physicians, demonic agency was not simply an explanation for inexplicable illnesses: it was one of many possible explanations for illnesses that might in other cases be diagnosed as purely natural. While the activity of the devil might be a speciality of the priest, the psychosomatic symptoms associated with demonic possession also required the physician's expertise to investigate the potential for purely natural causation. (Bergethon, 2019)

It does not require a high degree of imagination to see that contemplating *possession* from the perspective of *sentience, media-models,* or to *electromagnetic influence* would open-up this vast and critical field of modern research.

Addressing problems in this manner is a lonely commission because—invariably—human perspectives are *sub*-opti*mized*—and rarely capable of contemplating the cosmic ethics of infinity. Once, this was the prerogative of *Doctors of Philosophy,* but (partly due to the emphasis on technology and the subordination of the humanities), philosophy is no longer understood as the study of the nature of reality and service to the common welfare. In today's world where PhDs can be earned in a plethora of superficial "disciplines", the essence of philosophy has been almost entirely lost. At this critical time when true philosophers are most-needed they have become benumbed. But perhaps they can understand the significance of the Media-sphere as both sentient reality and metaphor. What it is possible to know has again become the domain of both science and theology—the ancient mysteries and morality.

What constitutes morality in the new order of things? This is the neglected domain of the humanities, the arts, and *Psych*ecology. The dynamics discussed in the chapters of this book are not merely speculative. Rather, they are grounded in a complex of current and ancient science and the humanities with which—if the will exists—human agency can be maintained synergistically with super-human sentient bandwidths.

One virtue of this anthology is that the book's contents represent an array of different fields—some of which are entirely unfamiliar to its audience and which may appear to disagree with one another. The Editor has striven to include as many diverse perspectives as possible while providing a common denominator—the theme of multi-scalar "sentience".

REFERENCES

Baloyannis, S. J. (2018). Anaxagoras on mind. *J Neural Stroke*, *8*(1), 269. doi:10.15406/jnsk.2017.08.00269

Connoly, T. (2019). *Plato: Phaedo. The International Encyclopedia of Philosophy: Socrates' intellectual history* (96a-102a). Retrieved from https://www.iep.upm.edu//Phaedo/

Curtis, A. (2009). *Century of the self: Happiness machines for free.* British Broadcasting Corporation. Retrieved from http://freedocumentaries.org/film.php?id=140

Dr. Emoto Thoughts/Intentions Shape Physical World - Amazing... SHARE! (2013). In *YouTube*. Retrieved 6/01/2016 from https://www.bing.com/videos/search?q=consciousness+of+water&&view=detail&mid=EECD4295AEE23D260AC4EECD4295AEE23D260AC4&&FORM=VDRVRV

Dr Masaru Emoto's Water Experiment - Words are Alive! (2014). In *YouTube.* Retrieved from https://www.bing.com/videos/search?q=consciousness+of+water&&view=detail&mid=3C404585CAAEB8A0536F3C404585CAAEB8A0536F&&FORM=VDRVRV

Eagleman, D. (n.d.). Silicon Immortality: Downloading Consciousness into Computers. *David Eagleman.* Retrieved from https://www.eagleman.com/blog/silicon-immortality-downloading-consciousness-into-computers

F World. (1933). New York: Agni Yoga Society.

Fourier, J. B. J. (1878). *The Analytical Theory of Heat* (A. Freeman, Trans.). The University Press. (Original publication 1822)

Galeon, D. (2016). For the first time in history, we are seeing alien worlds (with spectroscopy). *Futurism.* Retrieved 7/11/2017 from https://futurism.com/for-the-first-time-in-history-we-are-seeing-alien-worlds/

Grazer, B. (Producer). (2005). *Apollo 13 anniversary edition.* Universal Studios.

Grujic, P. V. (2001). *The Concept of Fractal Universe Anaxagoras*. Belgrade Institute of Physics.

How Plants Communicate & Think. (n.d.). *BBC & YouTube*. Retrieved from https://www.bing.com/videos/search?q=national+geographic+plants+video&&view=detail&mid=ABD11A5AADE89DE64EC6ABD11A5AADE89DE64EC6&&FORM=VDRVRV

How your consciousness affects the universe, Dr. Emoto. (n.d.). In *YouTube*. Retrieved from https://www.bing.com/videos/search?q=consciousness+of+water&view=detail&mid=431C0918B170F7B18E81431C0918B170F7B18E81&FORM=VIRE

Kirk, T. (2014). Superconducting spintronics pave way for next-generation computing. *PhysOrg*. Retrieved January 15, 2014 from http://phys.org/news/2014-01-superconducting-spintronics-pave-next-generation.html

Lamborn, A. B. (2015). The Deus Absconditus and the post secular quest. In *Jung in the Academy and Beyond*. University of the South.

Lewis, T. (2014). This computer chip can think like a human brain. *Live Science*. Retrieved from https://www.livescience.com/47240-ibm-computer-chip-simulates-brain.html

Lilly, J. C. (1967). *The Mind of the Dolphin: A Nonhuman Intelligence*. Garden City, NY: Doubleday.

Lilly, J. C. (1974). *Programming and Metaprogramming in the Human Biocomputer: Theory and Experiments*. New York: Bantam.

McConkie, B. (1979). *Mormon doctrine* (2nd ed.). Amazon Prime.

McDonald, C. (2015). The good, the bad, and the robot: Experts are trying to make robots moral. *Cal Alumni Association*. Retrieved from https://alumni.berkeley.edu/california-magazine/just-in/2015-06-08/good-bad-and-robot-experts-are-trying-make-machines-be-moral

Meadows, D. H., Meadows, G., Randers, J., & Behrens, W. W. III. (1972). *The Limits to Growth*. New York: Universe Books.

Norris, J. (2014). Do gut bacteria rule our minds. *UCSF*. Retrieved from https://www.ucsf.edu/news/2014/08/116526/do-gut-bacteria-rule-our-minds

Orme-Johnson, D. W. (n.d.). *M*aharishi Effect. *Global Good News*. Retrieved November12, 2015 from maharishi-programmes.globalgoodnews.com/maharishi-effect

Pope, R. (2019). Plato's map of the future omni artifact. *Science-Art Research Center of Austalia*. Retrieved from https://connect.xfinity.com/appsuite/#!!&app=io.ox/mail&folder=default0/INBOX

Radin, D. (1997). *The conscious universe: Scientific proof of psychic phenomena*. Harper One.

Radin, D. (2006). *Entangled Minds: Extrasensory Experiences in a Quantum Reality; · Supernormal: Science, Yoga, and the Evidence for Extraordinary Psychic Abilities*. Paraview Pocket Books.

Radin, D. (2018). *Real magic: Ancient wisdom, modern science, and a guide to the secret power of the universe*. Penguin/Random House.

Saffell, N. (2015). *On the origin of robot species: robots building robots by "natural selection"*. University of Cambridge. Retrieved August 12, 2015 from http://www.cam.ac.uk/research/news/on-the-origin-of-robot-species?sthash.zMSyp5OS.mjjo

Sathya Sai International Organization-USA. (n.d.). Retrieved from https://sathyasai.us/devotion/discourse/purity-thought-word-deed)

Schafer, S. B. (2016a). *The dream analog: Psychecology video games. Exploring the Collective Unconscious in a Digital Age*. Hershey, PA: IGI Global. doi:10.4018/978-1-4666-9891-8.ch002

Schafer, S. B. (2016b). *The media-sphere as dream: Researching the contextual unconscious of collectives. In Exploring the Collective Unconscious in a Digital Age*. Hershey, PA: IGI Global. doi:10.4018/978-1-4666-9891-8

Schafer, S. B. (2016c). Fostering psychological coherence with ICTs. *International Journal of Reliable and Quality E-Healthcare*, *4*(4), 1-31. DOI doi:10.4018/IJRQEH.2015100101

Schafer, S. B. (2016d). *Employing dynamics of Jungian compensation to foster psychological coherence with ICTs* (Vol. 4). Hershey, PA: International Journal of Privacy and Health Information Management.

Sheldon, C. M. (2011). How to change your frequency to change your reality. *YouTube*. Retrieved from https://www.bing.com/videos/search?q=consciousness+of+water&&view=detail&mid=E3E62E256E8BFBBA90C8E3E62E256E8BFBBA90C8&&FORM=VDRVRV

Streedhar, N. (2017). Micro-organisms and their influence in the human body. *Financial Express*. Retrieved from https://www.financialexpress.com/lifestyle/health/micro-organisms-and-their-influence-in-the-human-body-find-out-how/547433/

Sumrall, L. (2018). *The Essene teachings of Zarathustra*. Academy of Creative Living.

Talbot, M. (1991). *The holographic universe*. New York: Harper.

The Secret Language of Plants. (n.d.). *National Geographic & Youtube*. The Secret Language Of Plants - National Geographic Documentary.

Vagrecha, Y. S. (2016). Possession Behaviour. *Abnorm Behav Psychol, 2*, 119. Retrieved from https://www.omicsonline.org/open-access/possession-behaviour-2472-0496-1000119.php?aid=76421

ENDNOTES

1 The Kshetra or the field refers to the body which is material, mutable, transitory and perishable, the Kshetrajna refers to the conscious knower of the body who is of the same essence as Knowledge, immutable, eternal and imperishable, the knower of the body is the soul residing in the body. Kshetra is Prakrti or matter which is insentient, and the knower of the Kshetra is the Purusha who is sentient. In the context of the book, Kshetra is no longer "insentient" for the primary reason that "purity" exists on every scale of sentience.

2 It must be stated that our recommendation of human responsibility does not include such perversions as the excuses—based in Malthusian doctrine—used by the comfortable classes to avoid their

civil responsibilities; i.e., English conservatives explain away the potato famine in Ireland as being caused by the character flaws of the Irish peasants themselves.

3 Aurelio Peccei was born in a dynamic decade. At the beginning of the 20th century man had conquered the air with planes and zeppelins and the first European capitals had constructed metro lines. There was the spirit of progress in the heart and minds of people and it was not until 1972, when "Limits to Growth" was published as the first report to the Club of Rome, that a public discussion started on the question as to whether mankind was moving towards a disaster even in the absence of unrest and war.

4 Laura Sumrall is a Visiting Fellow at the Max Planck Institute for the History of Science from October 2018-March 2019. She is studying for her PhD at the University of Sydney in the School of History and Philosophy of Science, where she completed her MSc in 2015. Her dissertation project, "Deposing the Demon: Jan Baptista van Helmont and the Authority of Magic in Early Modern Medicine," examines the works of the Dutch chymical physician Jan Baptista van Helmont (1579/80–1644) and his followers, focusing on the demonological origins of van Helmont's disease theory and the reception and transformation of his ideas in later medical philosophies. Her research interests within the context of early modern medicine include disease etiology and nosology, occult qualities, demonology, and epistemology, informing her broader interest in the engagement of esoteric knowledge and early modern science.

5 The research on Plato's sacred geometry has been addressed by many including Robert Pope who has synthesized and applied the principle to "sacred art".

Introduction

During the scientific paradigm shift into quantum ontologies such as quantum mechanics (QM), quantum field theory (QFT), quantum cognition, quantum computing, quantum genetics (GMOs), robotic cloning and artificial intelligence (AI), and quantum computing, the critical interactive *human agency* related to superconductive scales of "quantum sentience" (denominated the *Psych*ecology- i.e. an ecology of Psyche) has been generally ignored. Instead of contemplating the potentials of ***First-Cause*** sentience, science limits itself to the vague concept of "consciousness" which remains confused, sub-optimized, and chaotic.

Perhaps the best illustration of quantum holographic sentience is the "Media-sphere" which has been demonstrated to have the same holographic/rhetorical structure as Jungian Psyche; i.e., entangled dramatic architecture. (Schafer, 2017; Schafer, 2017; Aiello, 2014; Schafer, 2012; Schafer, 2011; Zaltman, & Zaltman, 2008) Within a new paradigm of QFT characterized by "Field" mathematics rather than QM mathematics (Brooks, 2016), research on "sentient fields" (plants, minerals, elements, molecules, atoms, and the Media-sphere) becomes essential to pragmatic science, understanding of *Psych*ecology, and avoidance of catastrophic oblivion.

Robust environmental research (largely implicitly inferential) has established:

1. The existence of scaled sentience (models of interaction with the environment) in multiple organic/inorganic dimensions (Radin, 2018; Katabi, 2018; Kirmayer, 2018; McCraty et al., 2017; Galeon, 2017 & 2016; Gackenbach, & Hakopdjanian, 2015; Jeffrey, 2015; Dr Masaru Emoto's Water Experiment - Words are Alive!, 2014; Lewis, 2014; Norris, 2014; Burnett, 2012; McCraty & Childre, 2010; Radin, Lund, Emoto, & Kizu, 2008; Pruchnic, 2008; McCraty & Tomasino, 2006; Radin, 2006; Pobachenko Kolesnik, Borodin, Kalyuzhin, 2006; Lipton, 2005; Peters, 1999; Radin, 1997; Pribram, 1991, 1977, 1971, & 1969; Postman, 1984; Nikolaev, Rudakov, Mansurov, & Mansurova, 1976; Lilly, 1974 & 1967; Burton, Cabrera, Frank, 1951; Eagleman, n.d.; Kenji & Ushida, n.d.)
2. An incipient environmental morality/ethics (Treat other species as "self", recognize the rights of all scales of Life, etc.) that is consistent with "spiritual" morality and can be applied—as well—to human agency in a sentient quantum field. (Pereira, & Reddy, 2018; Baloyannis, 2018; Streedhar, 2017; Vagrecha, 2016; Simonite, 2015; Clark, 2014; Von Foerster, 2013, 1973, & 1962; Szekely, 1970; Sheldon, 2011; Pobachenko Kolesnik, Borodin, Kalyuzhin, 2006; Lipton, 2005, pp. 183-202; Pert, 2003; Papadopolous & Byng-Hall, 2007; McCraty, 2005; McCraty, 2002; Chopra, 1995; Bateson, 1972; Szekely, 1970; Orme-Johnson, n.d.; Water Documentary: How your consciousness affects the universe, Dr. Emoto, n.d.; Shiga, n.d.) The QFT reality has a new imperative based on

> the longstanding scientific observation—paraphrased from Aristotle—that "anything that moves without being moved" has a "first cause" (meaning) that is susceptible to scientific discovery. This statement has come to mean that if something moves, it is alive, but in the QFT reality, everything is in motion—so, logically speaking, "everything" is alive and—actually—exists as a "meaningful/purposeful" model of sentient life.

This logic could be considered simplistic or ludicrous in the orthodox rational-materialistic context, but—of course—that context no longer exists. Nevertheless, a first-cause ontology is the hypothesis underlying all theology, and it explains how "God" can be "in everything". The ancient theological premise is usually considered by "pragmatic realists" to be "mystical". However, the "mystical" hypothesis is universal in spiritual teaching, and confusion about mystical reality arises mostly from efforts to ignore its existence. Moreover, mysticism had been universally misunderstood until modern-day science has suggested and reaffirmed its validity.

QFT science is making the mystical hypothesis understandable, but—as an example of pervasive scientific misunderstanding—even in the super-sophisticated technological sphere, many very animated (self-driving) automobiles and combat robots that are already predicted to be completely autonomous, are neither considered alive nor intelligent (Except by a minority of scientists who are beginning to think of robots as a new species). Orthodox "reasoning" based on a false premise that excludes asymmetrical mathematics and sub-atomic "purpose" can be considered faulty. False assumptions constitute a fundamental flaw in much current scientific thinking that has generated fear of the "Singularity" as well as the failure of virtually all efforts to address the *Global Problematique* in meaningful ways. For example, it should be very clear that the problems caused by "enemies" cannot be resolved with the proliferation of nuclear weapons. Perpetuating deadly pollution of the natural environment cannot address the global health crisis, tax cuts for the mega-wealthy cannot address world poverty, and autonomous weapons cannot address terrorism. Neither can the enforced suicidal tendencies of human collectives be addressed with punishment and enslavement of the population, nor safety be secured by isolationism and fear-mongering. This volume presumes to address global problem-solving from a more coherent perspective on Cosmic sentience.

Media Models to Foster Collective Human Coherence in the PSYCHecology is a collection of minority perspectives relative to the existence of scaler sentience in the quantum field of reality. Our purpose is to facilitate a more open-minded science in which—at quantum-molecular scales—everything moves and has discriminative sentience of some kind (Anaxagoras called it *Nous*). Plato's "All is Geometry" gave *Nous* an ethical mathematical purpose in which the primitive life-force within magnetic lodestone orientated life to evolve away from chaos toward an ethical order. David Bohm described this movement as 'Implicate Order' and 'Holomovement' moving in the opposite direction to that of entropic chaos.

The aspect of sentience may be associated with the properties of lodestone because it is becoming well-established scientific fact that sentience is entangled and interacts everywhere in the electromagnetic biosphere. Plato's single-source geometry provides a mathematical ethos completely alien to most current knowledge in physics with its prevailing sterile quantum mathematics. The artist Salvador Dali's eccentric flamboyant insistence that flat plane paintings contain a Platonic spiritual human survival message within plane geometry is now of academic importance. Artists around the world are now unconsciously painting stereoscopic images into paintings that are quite visible in single paintings when viewed through asymmetrical electromagnetic glasses. These images are more pronounced than the 3D images at the Dali Stereoscopic Museum in Spain emanating from two painting side by side.

This new holographic phenomenon is in accord to Immanuel Kant's and Emmanuel Levinas' conviction that Plato's ethical asymmetrical electromagnetic field is evolving within the creative artistic mind. Prevailing science took little notice that Levinas upgraded Kant's asymmetrical field concept to First-Principle dominance controlling the dynamics of universal reality. In 2013 the European Space Agency Planck Observatory mapped the first electromagnetic light as being *asymmetrical* in contrast to the First Principle that Einstein thought was *symmetrical*. This created confusion as to the evolutionary meaning of ethics, but in retrospect, it is not so difficult to explain. Einstein and his atomic science contemporaries had no option but to help build the atomic bomb before Hitler's obviously unethical Nazi regime did so. Einstein's science was obeying tribal science's evolutionary law of the "survival of the fittest" without fully appreciating the irrelevance of that doctrine when humanity is one species. Since Neolithic times tribal intellect evolved by the need to develop weapons of conflict.

When DNA demonstrated that humans belong to a single species, the tribal scientific rule book became obsolete. Ethics now embraces the survival of the human species—because the human species has responsibility for the survival of the entire geosphere. However, as tribal science and human survival science are both integral aspects of human evolution, they can be entangled within computer programs as an antidote governed by the holographic human survival ethic. Giving the survival ethic priority would move science away from Einstein's entropic end-game that demands not only human extinction but a kind of Ragnarok—the Norse cataclysmic destruction of the cosmos and everything in it. Applying the holographic principle to design of video games such as *Psych*ecology games (Schafer, 2018) might be a pragmatic method for healing Psyche with superconductive impact on the global electromagnetic neurobiological field. Such video games (PEGs) could be designed with reference to Anaxagoras' worldview based upon infinite fractal geometrical logic. (Grujic, 2001) The mathematics of the holographic worldview would counter the current outdated QM entropic mentality of suicidal survival-of-the-fittest. This volume specifies that the appropriate range of scientific research consistent with QFT must be addressed on scales of **super-human sentience** such as the **"Soul"** or **"Monad".** Such research is the scientific key to ameliorating the human-created global crisis that threatens all forms of geo-animated-sentient life.

OBJECTIVE OF THE BOOK

Editing this book is not simply an academic pursuit. In a time of catastrophic planetary problems, virtually all attempts to address these problems have been demonstrably archaic and futile. They don't work because the human understanding of reality has not been accurate. Outdated problem-solving methods are useless because their premises have become logical fallacies which not only perpetuate error but annihilate morality. A fundamentally changed quantum-sentient paradigm-shift in reality concepts is supported by robust research in many fields. Science is gradually accepting the fact that sentient "intelligence" exists on scales throughout the Cosmos. Not only for practical reasons must this sentient Life be respected on all entangled scales of Cosmos, but it must be appreciated, respected, and held in awe. It must be loved!

The proof of these facts of sentience is reflected in research in every scientific field: artificial intelligence, robotics, the psychological media, advertising, neuro-marketing, intelligence in plants, water, crystals, silicon chips, and cell membranes. For example, one researcher observes that, "The [cell] membrane's function of interacting 'intelligently' with the environment to produce behavior makes it the true brain of the cell." (Lipton, 2005, p. 86) Moreover, it has been discovered that microbes in the human gut influence "conscious" human choices relative to diet (Panduro, Rivera-Iñigue, Sepulveda-Villegas, & Roman, 2017; Stilling, Dinan, & Cryan, 2016), that water responds to loving written and spoken words (Radin, Lund, Emoto, & Kizu, 2008), and that grass and other plants have survival mechanisms. (Brown & Zinnert, 2018) In other words, we now find ourselves living in a mystical reality, and—in order to survive—we must learn to address problems with scientific-mystical dynamics.

Quantum-based research has amplified the veridical importance of expanding scientific parameters in-order to understand "universal sentience" and to address problems from scalar perspectives: sub-human, human, planetary, and super-human. This is not some fantasy. Robust environmental-psychological research has already verified these scales of sentience, and the blind impetuous scientific thrust into fields of genetics, robotics, AI, and hacking of the psychological media are primary areas of concern for the very survival of planetary Life. The conjunction of neurolinguistics and imaging technologies such as MEG is divulging correspondences to "intelligent functions" at all scales of "life" that are being currently researched.

Many linguists are studying linguistic correlations from the perspective on sentience, and the axiom that language grounds thought is being extended to the influence that human agency (with its emphasis on language) has on other sentient life forms. Perhaps the most significant genius in the field was Carl Jung who understood and explored the distinction between vital symbols, dead symbols, and "signs". Jung understood that vital semiotic patterns prevailed in the personal and collective unconscious, and he named these quantum patters archetypes of the unconscious. Exploring further, he observed that archetypes are projected into various scales of "contextual" consciousness that he called archetypal representations.

The relationship between archetypes and archetypal representations is an essentially metaphorical/rhetorical blueprinting of an essentially mathematical-programmatic "narrative architecture"—a mediated projection of images embedded with holographic messages—which may be correlated with the vectors of Higgs fields and the asymmetrical sacred geometry of Plato. When properly viewed from the perspective of universal sentience, Jungian dream therapy can be translated to cultural therapy employing mediated biofeedback as a cultural healing dynamic. In other words, various "metaphorical" scales of language exist beyond the literal, and Jung went on to define those scales in terms of four human "functions": Feeling (emotion); Sensing (perception); Thinking (mentation); and Intuiting (spiritual coherence). Computational frameworks have already been developed (specifically Myers-Briggs personality profiles) to research these functions. Various media models can be used to amass quantifiable data relative to ***fields of super-human sentience.***

DIVINE SENTIENCE

The modern scientific focus on-the subject of "intelligence" or sentience is advancing radically. It may be stipulated that artificial intelligence (AI) is a primary subject of research that leads to many other applications such as autonomous robotics (AR), and neuro-marketing. However, unfortunately the focus of research in these fields is invariably on profit, or war, or both. The scope of research on intelligence has taken us into alternative dimensions that resonate with alchemy and range from intelligence in atoms (Hameroff, & Penrose, 2014), intelligence in quantum fields (Pribram, 1991 & 1977; Sieb, Richard, in press), intelligence in molecules (Pert, 2003; Lipton, 2005) to intelligence in non-human dimensions including animals and plants, elements (water, minerals, air), and "light" (Fire).

Intensive study of correspondences strongly suggests that this current research is consistent with the ancient sciences that include astrology, alchemy, philosophy, pan-psychism, and monotheism, as-well-as the intelligence of language as a sentient-symbolic form—a "living" sentience reminiscent of Jungian "living symbols". (Foster, 1985; Jacoby, 1973 ed.; Jung, Carl, 1933; Woodcock, in press; Gaiseanu, in press)

This essence of living sentience has been contemplated universally for ages by philosophers, and during these ages-past a holistic assumption was made by them that—to put it simply—God is "in" everything. The ancient Essenes put it another way. To them, *cosmic laws reflected in natural laws are God.* What we know of the laws of mathematics, physics, chemistry, astrology, and alchemy can be scientifically understood in association with an ethical infinite asymmetrical geometrical first cause-principle governing the dynamics of universal reality. This reality has been widely assumed but not well-understood by the vast majority of seekers. For example, "Fire" has been universally equated to "Spirit" such as in Plato's Myth of the Cave, and the following quotes from Mormon doctrine:

Intelligence or light and truth, is also used as a synonym for spirit element. Scriptures using both terms speak of the self-existent [prime mover] nature of the substance [fire] involved. (D. & C. 93:29; 131:7-8.) Abraham calls the pre-existent spirits the intelligences [sentient fire] that were organized [coherent] before the world was" (Abraham, 3:22) because the intelligences were organized intelligence or, in other words, the spirit bodies were born from [the fiery] spirit element. (McConkie, 1979 p.387)

Abraham used the name "intelligences" to apply to the spirit children of the Eternal Father. The intelligence or spirit element [sentient fire] became intelligences after the spirits were born as individual entities. (Abraham, 3:22-74.) Use of this name designates both the primal element [fire] from which the [sentient] spirit offspring were created and-also their inherited capacity to grow [evolve] in grace, knowledge, power, and intelligence itself, until such intelligences, gaining the fulness of all things, become like their Father, the Supreme Intelligence. (McConkie, 1979, p.387)

Modern science has added some tools for understanding reality, but—as described above—it is now dedicated to a death wish. Primarily due to ignorance and prejudice, modern science has abjured ancient science which—indeed—is appropriately called "science" because if one views the ancient mathematics of Pythagoras, the chemistry of the alchemists, and the cognitive structure of astrology without prejudice, their methods were—without any doubt—scientific. *Media Models to Foster Collective Human Coherence in the PSYCHecology* addresses this modern fallacy of neglecting the reality of subtle perception—a reality, after all, wherein punctuation in software can alter meaning. Hopefully, by using words that have not become clichés, confusion generated by ignorance or prejudice will be minimized.

- **PSYCHecology:** (An ecology of psyche) may be understood as God, Fire, QFT, hologram, Eternal Spirit, the Hidden God, or ethical mathematical purpose.
- **Sentience:** May be understood as multi-dimensional cosmic intelligence that may be recognized by its dramatic-rhetorical (subject-action-object) process of mediation or "intelligent" response to an environment.
- **Media Models:** Includes all *conversational modalities* be they linguistic, mathematical, chemical, or metaphorical.

- **Coherence:** May be understood as harmony of frequencies that exist on a continuum of degrees or dimensions that are measurable in terms of harmonic scales and heart-rate-variability (HRV) as researched by ***The Institute of Heartmath.***
- **Plato's Sacred Geometry:** *"There are scientists who consider that Platonic thought-forms are electromagnetically responsible for the evolution of consciousness within a living infinite fractal holographic universe" (Pope, 2019, para 1). Petar Grujic, the late science advisor to the Belgrade Institute of Physics, explained that ancient Greek science was conversant with infinite fractal mathematics in his book 'The Concept of Fractal Universe Anaxagoras' (Grujic, 2001). Neurological medical scientist, Baloyannis (in Anaxagoras on Mind, 2018), considered that Anaxagoras' work stands as a 'significant milestone' in the field of Neurophilosophy (Pope, 2019).*
- **Media-Sphere:** Refers primarily to the reality of technological media that includes the Internet, social media, nightly news, film, print, and the fundamental physics, mathematics, and syntaxes applied to create their paraphernalia. Within the media-sphere, media models have a cognitive basis that extends beyond human cognitive models. The dynamics of media models are "conversational"; i.e., they can be studied in terms of the conversational trinity (action, mediation, reaction/pro-action) and measured in terms of Fourier mathematics, electromagnetic dynamics, color and sound harmonics, coherence, and metaphorical resonance (fractal structures).
- **First Cause:** Refers to the "purpose" of anything—the "why?" rather than the "How?". In other words, the "why?" refers to meaning, whereas the "how?" refers to methodology. Most modern science puts a priority on method rather than meaning.

Once the dynamics of sentience have been established, our primary purpose is to clarify the practical efficacy of contemplating and coordinating with super-human "spiritual" intelligence in order to address the suicidal planetary crisis that Dr. Aurelio Peccei[1] named the ***Global Problematique***. Powerful as it is—by itself—human agency is not sufficient for understanding the cosmic scope of the problem nor for unilateral discovery of solutions to the self-evident planetary crisis. It is human agency that has caused the crisis, and without responsible human agency in coordination with the forces of "Spirit", the errors of ages will reap their consequences.

Plato's Phaedo discusses Anaxagoras' mathematical teaching that intelligence organizes all things in the best way it can but argues that physical reality alone does not convey a sense of order [or sentience]in the universe, which requires a sense of purpose (Conolly, 2019). In Plato's dialogue, Ion, he refers to a primordial life form within magnetized lodestone [that can create] electromagnetic emotional thought forms within the mind. His axiom that All is Geometry embraced both ethical and unethical thoughtform geometries entangled together within this electromagnetic theory of creative artistic mathematical evolution. Plato's sacred geometries creating order from chaos were held to hold evolutionary dominance over the unethical "chaos" mathematics. These sacred geometries [may be applied the transition [from] chaos to ethical infinite universal order. (Pope, 2019)

In an emergent media age characterized by technological sophistication, artificial intelligence, robotics, political chaos, and planetary crisis, ignoring the ancient scientists would be irrational and foolhardy. Ultimately, contemplating cosmic scales of sentience will lead to a better understanding of the problems associated with an age of communication and the media.

To Plato the universe was evolving the living information process from chaos under the guidance of a supreme ethical mathematical purpose. On the contrary, Einstein believed that the universe was evolving all life in the universe to extinction [entropy] under a supreme law governing the energies of chaos. Both philosophers of science were dealing with what scientist call a first-cause principle [that controls] the dynamics of reality. Einstein's science belonged to a cruel harsh tribal reality, which nonetheless was an essential part of Plato's mathematical map to the future (Pope, 2019, para. 3).

Human hubris and the absence of a coherent mathematical map to the future—misdirected purpose—has been perverting most current scientific research. The consequences of this faulty science are that problem-solving methods have also been perverted. The validity of this opinion is being vividly reflected in the catastrophic failure of problem-solving efforts that are ensnared in a caustic dream; i.e., the source of problems cannot be addressed with theories of competition for sub-optimized economic profit. The superhuman problem-solving guidance provided by the scientific ideas of human philosophical precursors has long provided detailed strategies for protecting the ecosphere and planetary culture, and now science is beginning to reaffirm those strategies that are characterized in the mystical (environmental) beatitudes which may be paraphrased, "Treat all life with love and respect as if it were your own life."

However, the superhuman sources of these human reaffirmations are not being given proper credit. Modern human scientists usually take total credit for their "intuitive" insights, but this limits their ability to solve problems in approximately the same way that ownership-of-ideas in the form of intellectual property slows the creative process and creates dissention. In general, the premise of human problem-solving efforts puts far more emphasis on personal ways to limit creativity than to stay in contact with the Cosmic Source of creativity. This fact is documented in theological history, but exemplars have always existed to remind humanity of the truth, and it may be stipulated that these exemplars (Christ, Zoroaster, Buddha, Moses, Confucius, Plato, Anaxagoras, Lao Tzu) who, invariably, gave credit to the one-source or Monad. They freely admitted their reliance on super-human sentience. Humans must reaffirm their openness to super-human input.

To paraphrase an ancient axiom, "The proper study of man is man." It becomes incumbent upon humans to investigate and study the whole sentient universe. Due to the flattening of human potential during the period of rational materialism, the super-human potentials of the human being have been orphaned. But in-order to understand the wisdom and goodness of the Creator—whatever it is called—the evidence of divine wisdom that is shown in all that concerns the human species must be re-emphasized. Perhaps what makes humankind special is that—within the trinitarian context of personality, soul, and spirit—each human constitutes the universe in seed form (microcosm), and the proximate cause of the cosmic fruit (macrocosm). It is the human duty to "know itself"—to study the origin of humankind, its history, its birth, and the composition and structure of its component parts which all have scaled sentience: the members and organs of its body, their relations to each other, and the vital-sentient functions and purposes of every one of them.

Most important, now, is to study the cognitive processes, the motives, the thoughts, the dreams, and the feelings involved in individual and collective human agency and to synchronize them with super-human sentience. Furthermore, we must consider the purpose of human psychological existence including its super-human potentials—the relationship of the human soul to the sentient scheme of ***First-Cause Creation***, the limitations of human reasoning, and the power of human passions and desires. Ignoring human responsibility and its super-human potential is not acceptable.

The cognitive unconscious (CU) was once thought to be unscientific because it was not measurable, but measurable research on the cognitive unconscious is commonplace today in many fields. For the purposes of this book, anything formerly considered unconscious should now be considered "sentient" at some scale. This is the new trend, and processes including the autonomic nervous system (ANS), intuition, genetics, dreams, computer software, and robotic autonomy are presently being researched on a vast scale that cannot be discussed in the limited terms of consciousness. These processes are unconscious to humans, and one of the most comprehensive models for research on sentience is the dream research of Carl Jung which remains consistent with the most advanced current research on cognition, the media, AI, and robotics.

One form of mediation, advertising, has long been a tool for cognitive manipulation of the contextual (sentient) unconscious, but emergent issues of political hacking, false news, disinformation campaigns, lies, neuromarketing, misuse of social media, pervasive surveillance, cyber warfare, and the destruction of both nature and civilization provide clear and abundant proof of the need for renewed research perspectives on-the subject of universal sentience—beginning with the sentient media-sphere. This book, *Media Models*, will aim to include relevant theoretical frameworks and the latest empirical research models in-order to address sentience at multiple scales in order—ultimately—to address super-human bandwidths.

Perhaps the most practical way to address the book's theme is to use the Media-sphere metaphor as analog for a holographic quantum-dynamic field of living sentience. Modern populations are superficially aware of media potentials and paraphernalia, but recent events affirm the vast human ignorance and helplessness in its interaction with the sentient (cognitively dynamic) media. Critical issues of contemporary mediated social order and challenges to global freedom, health, and sanity must now be vigorously addressed from a renewed perspective on first-order dynamics of a living sentient quantum field.

In-order to map the future—to frame coherent-cognitive-evolutionary progress—applying the Jungian science of dream analysis (with its emphasis on Individuation) to the Media-sphere and media-dreams and using the sacred mathematics of Plato to embed coherent patterns in mediated images could result in significant harmonization of quantum-contextual cultural fields and the resolution of seemingly intractable global problems that have their origins in human limitation. If the global Media-sphere is a dreamscape and if "contextual" media content and media markers (including variations in advertisement, entertainment, news, drama, sports, and social networks) is understood in terms of Jungian dreams, analysis and biofeedback based on these media-dreams can be applied to the healing of cultural psychosis. The media dreamscape provides an authentic analog for the dynamics of personal and collective Jungian Psyche. The Jungian model is unique, prospective, and benign, and it is being gradually re-affirmed (but often unattributed) with current research.

Accordingly, it is plausible that sentient human coherent bandwidths could be attuned with "coherent" super-human bandwidths. Adjustment (biofeedback) of collective human energies with the use of computational video game design—Fourier mathematics and their sophisticated algorithms—could be employed in order to maintain attunement of human civilization with a superconductive state of being. Within this mathematical context, interpolation within quantum field dynamics could contribute to synchronizing the dynamics of human cognition—by way of the Media-sphere—with the dynamics of universal sentience. Of course, other methods of super-human contact such as prayer, meditation, mantra and rhythmic movement have long existed. However, in this technological-amoral age of *planetary crisis*, an unorthodox model for ***timely healing*** must be considered. *Psych*ecology video games (PEG) may provide such a model.

REFERENCES

Baloyannis, S. J. (2018). Anaxagoras on mind. *J Neural Stroke*, *8*(1), 00269. doi:10.15406/jnsk.2017.08.00269

Bateson, G. (1972). *Steps to an ecology of mind: collected essays in anthropology, psychiatry, evolution, and epistemology*. University of Chicago Press.

Bergethon, K. (n.d.). *Cyber currency investing for beginners*. Loveland, CO: Legendary Ventures. Retrieved from https://thefranklinsociety.com/cyber-currency-investing-for-beginners/

Brooks, R. A. (2016). *Fields of Color* (3rd ed.). Silver Springs, MD: Universal Printing, LLC.

Brown, J. K., & Zinnert, J. C. (2018). Mechanisms of surviving burial: Dune grass interspecific differences drive resource allocation after sand deposition. *Ecosphere*, *9*(3), e02162. doi:10.1002/ecs2.2162

Burnett, D. G. (2012). *The sounding of the whale: Science and cetaceans in the twentieth century*. Chicago: University of Chicago Press.

Burton, W. K., Cabrera, N., & Frank, F. C. (1951). The growth of crystals and the equilibrium structure of their surfaces. In Philosophical Transactions of the Royal Society A, 243(866), 299. doi:10.1098/rsta.1951.0006

Chopra, D. (1995). *Body, mind, and soul. Turner Home Entertainment*. Atlanta, GA: PBS Home Video.

Clark, B. (2014). *John Lilly, the mind of the dolphin, and communication out of bounds*. Texas Tech University.

Connoly, T. (2019). *Plato: Phaedo. The International Encyclopedia of Philosophy: Socrates' intellectual history* (96a-102a). Retrieved from https.//www.iep.upm.edu//Phaedo/

Curtis, A. (2009). *Century of the self: Happiness machines for free*. British Broadcasting Corporation. Retrieved from http://freedocumentaries.org/film.php?id=140

Documentary, W. how your consciousness affects the universe, Dr. Emoto. (n.d.). In *YouTube*. Retrieved from https://www.bing.com/videos/search?q=consciousness+of+water&view=detail&mid=431C0918B170F7B18E81431C0918B170F7B18E81&FORM=VIRE

Dr Masaru Emoto's Water Experiment - Words are Alive! (2014). *YouTube*. Retrieved from https://www.bing.com/videos/search?q=consciousness+of+water&&view=detail&mid=3C404585CAAEB8A0536F3C404585CAAEB8A0536F&&FORM=VDRVRV

Eagleman, D. (n.d.). Silicon Immortality: Downloading Consciousness into Computers. *David Eagleman*. Retrieved from https://www.eagleman.com/blog/silicon-immortality-downloading-consciousness-into-computers

Foerster, V. (2003). On constructing a reality. In *Understanding Understanding: Essays on Cybernetics and Cognition* (pp. 211–227). New York: Springer. doi:10.1007/0-387-21722-3_8

Gackenbach, J., & Hakopdjanian, S. (2015). Breaking the frame of digital, dream, and waking realities. In *Exploring the Collective Unconscious in a Digital Age.* IGI Global. Retrieved from http://www.igi-global.com/book/exploring-collective-unconscious-age-digital/140975

Galeon, D. (2016). For the first time in history, we are seeing alien worlds (with spectroscopy). *Futurism.* Retrieved 7/11/2017 from https://futurism.com/for-the-first-time-in-history-we-are-seeing-alien-worlds/

Galeon, D. (2017). New research shows that the universe might once have been a hologram. *Futurism.* Retrieved 7/11/2017 from https://futurism.com/new-research-shows-the-universe-may-have-once-been-a-hologram/

Grujic, P. V. (2001). *The Concept of Fractal Universe Anaxagoras.* Belgrade Institute of Physics.

Hameroff, S., & Penrose, R. (2014). *Discovery of quantum vibrations in 'mocrotubules' corroborates theory of consciousness.* Retrieved 1/16/2014 http://phys.org/news/2014-01-discovery-quantum-vibrations-microtubules-corroborates.html

Ho, M. W. (2005). *The rainbow and the worm: the physics of organisms.* Singapore: World Scientific Publishing Co.

How Plants Communicate & Think. (n.d.). *BBC & YouTube.* Retrieved from https://www.bing.com/videos/search?q=national+geographic+plants+video&&view=detail&mid=ABD11A5AADE89DE64EC6ABD11A5AADE89DE64EC6&&FORM=VDRVRV

Jacoby, J. (1959). *Complex, archetype, symbol in the psychology of C. G. Jung.* New York, NY: Princeton University Press.

Jacoby, J. (1973). *C. G. Jung.* Yale University Press.

Jeffrey, C. (2015). First-ever QED device that detects and corrects its own errors. *Physics.* Retrieved from https://newatlas.com/quantum-device-detects-corrects-errors/36404/

Katabi, D. (2018). RF-Pose, AI that can see through walls. *MIT Computer Science & Artificial Intelligence Lab.* Retrieved from MIT Technology Review, Futurism, & Motherboard (Kaleigh Rogers): https://motherboard.vice.com/en_us/article/a3aaqp/mit-device-uses-wifi-to-see-through-walls-and-track-your-movements;https://connect.xfinity.com/appsuite/#!!&app=io.ox/mail&folder=default0/INBOX

Kenji, I. B., & Uchida, S. (n.d.). *Neural networks can read book covers.* Kyushu University. Retrieved from MIT Technology Review: https://www.technologyreview.com/s/602807/deep-neural-network-learns-to-judge-books-by-their-covers/?utm_campaign=newsletters&utm_source=newsletter-weekly-computing&utm_medium=email&utm_content=201611

Kirmayer, L. J. (2018). Ethno and cultural psychiatry. *The International Encyclopedia of Anthropology.* Retrieved from https://connect.xfinity.com/appsuite/#!!&app=io.ox/mail&folder=default0/INBOX

Lewis, T. (2014). This computer chip can think like a human brain. *Live Science.* Retrieved from https://www.livescience.com/47240-ibm-computer-chip-simulates-brain.html

Lilly, J. C. (1967). *The mind of the dolphin: A nonhuman intelligence.* Garden City, NY: Doubleday.

Lilly, J. C. (1974). *Programming and metaprogramming in the human biocomputer: Theory and experiments*. New York: Bantam.

Lipton, B. (2005). The biology of belief. Santa Rosa, CA: Mountain of Love/Elite Books.

McConkie, B. (1979). *Mormon doctrine* (2nd ed.). Amazon Prime.

McCraty, R., & Childre, D. (2010). Coherence: bridging personal, social, and global health. *Alternative Therapies*, *16*(4). Retrieved November 6, 2015 from https://www.heartmath.org/assets/uploads/2015/01/coherence-bridging-personal-social-global-health.pdf

McCraty, R. (2002). Heart rhythm coherence: And emerging area of biofeedback. *Biofeedback*, *30*(1), 17–19.

McCraty, R. (2005). *Enhancing emotional, social, and academic learning with heart rhythm coherence feedback*. Heartmath Research Center, Institute of Heartmath.

McCraty, R., Atkinson, M., Stolc, V., Alabdulgader, A. A., Vainoras, A., & Ragulskis, M. (2017). Synchronization of Human Autonomic Nervous System Rhythms with Geomagnetic Activity in Human Subjects. *Int. J. Environ. Res. Public Health*, *14*(7), 770. Retrieved 10/08/2017 from http://www.mdpi.com/1660-4601/14/7/770

McCraty, R., & Tomasino, D. (2006). Coherence-building techniques and heart rhythm coherence feedback: New tools for stress reduction, disease prevention, and rehabilitation. In E. Molinari, A. Compare, & G. Parati (Eds.), *Clinical Psychology and Heart Disease*. Milan, Italy: Springer-Verlag. doi:10.1007/978-88-470-0378-1_26

Nikolaev, Y. S., Rudakov, Y. Y., Mansurov, S. M., & Mansurova, L. G. (1976). Interplanetary magnetic field sector structure and disturbances of the central nervous system activity. *Acad Sci USSR. IZMIRAN, 1976*, 29.

Norris, J. (2014). Do gut bacteria rule our minds. *UCSF*. Retrieved from https://www.ucsf.edu/news/2014/08/116526/do-gut-bacteria-rule-our-minds

Orme-Johnson, D. W. (n.d.). *M*aharishi Effect. *Global Good News*. Retrieved November12, 2015 from maharishi-programmes.globalgoodnews.com/maharishi-effect

Panduro, A., Rivera-Iñigue, I., Sepulveda-Villegas, M., & Roman, S. (2017). Genes, emotions and gut microbiota: The next frontier for the gastroenterologist. 世界胃肠病学杂志*:*英文版*, 23*(17), 3030-3042. doi:10.3748/wjg.v23.i17.3030

Papadopolous, R., & Byng-Hall, J. (Eds.). (2007). Narrative Based Medicine: A narrative approach to medicine. Academic Press.

Pereira, C., Reddy, J., & Shashi, K. (2018). The boy who grew a new brain: Understanding this miracle from a neuro-quantum perspective. *NeuroQuantology*, *16*(7), 39-48. Retrieved from https://www.researchgate.net/publication/326711377_The_Boy_Who_Grew_a_New_Brain_Understanding_this_Miracle_from_a_Neuro-Quantum_Perspective

Pert, C. B. (2003). *Molecules of emotion*. New York: Scribner.

Peters, J. D. (1999). *Speaking into the Air: A History of the Idea of Communication*. Chicago: University of Chicago Press. doi:10.7208/chicago/9780226922638.001.0001

Plato Political Philosophy. (n.d.). *Internet Encyclopedia of Philosophy IEP*. Retrieved from https://www.iep.utm.edu/platopol/

Pobachenko, S. V., Kolesnik, A. G., Borodin, A. S., & Kalyuzhin, V. V. (2006). The contingency of parameters of human encephalograms and Schumann Resonance electromagnetic fields revealed in monitoring studies. *Complex Systems Biophysics.*, *51*(3), 480–483.

Pope, R. (2019). Plato's map of the future omni artifact. *Science-Art Research Center of Austalia*. Retrieved from https://connect.xfinity.com/appsuite/#!!&app=io.ox/mail&folder=default0/INBOX

Postman, N. (1984). *Amusing ourselves to death*. Viking Penguin.

Pribram, K. (1969). The neurophysiology of remembering. *Scientific American, 220*, 75.

Pribram, K. (1971). *Languages of the brain; experimental paradoxes and principles in neuropsychology*. Englewood Cliffs, NJ: Prentice-Hall.

Pribram, K. (1977). *Languages of the Brain*. Wadsworth Publishing.

Pribram, K. (1991). *Brain and perception: holonomy and structure in figural processing*. Hillsdale, NJ: Lawrence Erlbaum Associates.

Pruchnic, J. (2008). Neurorhetorics: Cybernetics, psychotropics, and the materiality of persuasion. *Configurations*, *16*(2), 2. doi:10.1353/con.0.0057

Radin, D. (1997). *The conscious universe: Scientific proof of psychic phenomena*. Harper One.

Radin, D. (2006). *Entangled Minds: Extrasensory Experiences in a Quantum Reality; · Supernormal: Science, Yoga, and the Evidence for Extraordinary Psychic Abilities*. Paraview Pocket Books.

Radin, D. (2018). *Real magic: Ancient wisdom, modern science, and a guide to the secret power of the universe*. Penguin/Random House.

Radin, D., Lund, N., Emoto, M., & Kizu, T. (2008). Effects of distant intention on water crystal formation: A triple-blind replication. *Journal of Scientific Exploration*, *22*(4), 481–493.

Reeves, B., & Nass, C. (1996). *The media equation*. Stanford, CA: Center for the Study of Language and Information.

Schafer, S. (2011). Articulating the paradigm shift: Serious games for psychological healing of the collective person. In *The emerging science of human-data interaction—simulation of living systems: MIT Technology Review*. Retrieved July 31, 2014 from https://www.linkedin.com/grp/post/1039687-5900533809526378496

Schafer, S. (2012). Optimizing Cognitive Coherence, Learning, and Psychological Healing with Drama-based Games. In *Video Game Play and Consciousness*. Nova Science Publishers. Retrieved from http://www.facebook.com/insights/?sk=po_261980910540561#!/pages/Video-Game-Play-and-Consciousness/261980910540561?sk=wall

Schafer, S. (Ed.). (2017b). *Exploring the collective unconscious in a digital age*. Hershey, PA: IGIG Global.

Schafer, S., & Yu, G. (2011). *Meaningful video games: drama-based video games as transformational experience*. Hershey, PA: IGI Global. doi:10.4018/978-1-60960-567-4.ch019

Schafer, S. B. (2017a). *Generating a superconductive culture of conscience*. LAP LAMBERT Academic Publishing.

Schafer, S. B. (2018). *Gaming cultural atonement*. Lambert Academic Publishers. Retrieved from https://www.morebooks.de/gb/search?utf8=%E2%9C%93&q=+978-613-9-89469-7+

Sheldon, C. M. (2011). How to change your frequency to change your reality. *YouTube*. Retrieved from https://www.bing.com/videos/search?q=consciousness+of+water&&view=detail&mid=E3E62E256E8BFBBA90C8E3E62E256E8BFBBA90C8&&FORM=VDRVRV

Shiga, J. (n.d.). Of other networks: Closed-world and green-world networks in the work of John C. Lilly. *Amodern 2: Network Archaeologies*. Retrieved from http://amodern.net/article/of-other-networks/

Simonite, T. (2015). A better way to design brain inspired chips. *MIT Technology Review*. Retrieved May 6, 2015 from http://www.technologyreview.com/news/537211/a-better-way-to-build-brain-inspired-chips/

Stilling, R. M., Dinan, T. G., & Cryan, J. F. (2016). The brain's Geppetto—Microbes as puppeteers of neural function and behaviour? *Journal of Neurovirology*, *22*(1), 14–21. doi:10.100713365-015-0355-x PMID:26047662

Streedhar, N. (2017). Micro-organisms and their influence in the human body. *Financial Express*. Retrieved from https://www.financialexpress.com/lifestyle/health/micro-organisms-and-their-influence-in-the-human-body-find-out-how/547433/

Szekely, E. B. (1970). *The Essene teachings of Zarathustra*. Academy of Creative Living.

Vagrecha, Y. S. (2016). Possession Behaviour. *Abnorm Behav Psychol, 2*, 119. Retrieved from https://www.omicsonline.org/open-access/possession-behaviour-2472-0496-1000119.php?aid=76421

Von Foerster, H. (1962). Bio-Logic. In *Biological Prototypes and Synthetic Systems*. New York: Plenum Press.

Von Foerster, H. (2013). The beginning of heaven and earth has no name. In Seven Days with Second-Order Cybernetics. New York: Fordham University Press. doi:10.2307/j.ctt13wzw8d

Zaltman, G., & Zaltman, L. (2008). *Marketing Metaphoria*. Harvard Business Press.

ENDNOTE

1 Aurelio Peccei was born in a dynamic decade. At the beginning of the 20th century man had conquered the air with planes and zeppelins and the first European capitals had constructed metro lines. There was the spirit of progress in the heart and minds of people and it was not until 1972, when "Limits to Growth" was published as the first report to the Club of Rome, that a public discussion started on the question as to whether mankind was moving towards a disaster even in the absence of unrest and war.

Section 1
Rhetorical Sentence

Chapter 1
Posthuman Being:
Inceptive Sentience

John Christopher Woodcock
Lighthouse Down Under, Australia

ABSTRACT

With the enormous chaotic changes taking place today, contemporary artists are showing us a vast and mystifying range of artworks that show glimpses of nascent worlds coming into being and just as quickly disappearing into oblivion. The author explores a world that seems to be gaining some traction—the world of the posthuman. Contemporary art is showing glimpses of this still-forming world in artworks produced from a collision between or interpenetration of virtual reality and empirical reality, giving rise to weird, horrific, and sometimes strangely beautiful forms. Then the author seeks to penetrate "behind" this posthuman art to the activity of the artist, in order to find the original "bringing forth" (Heidegger) of posthuman artworks. This move reveals the fundamental place of revelation and prophecy as the origin of any artwork and thus indicates the essential nature of the posthuman world.

INTRODUCTION

Painting is a thundering collision of different worlds, intended to create a new world in and from the struggle with one another, a new world which is the work of art. (W. Kandinsky)

In the literary world, particularly that of Science Fiction, posthumanism refers to a kind of being such as a cyborg which in some sense transcends humanity. Philosophical studies of posthumanism include deconstruction of humanism and its privileging human nature, that places humans above all other natures or forms of being. At this time technological exploration of posthuman reality includes a push towards the development of Artificial Intelligence, raising fresh questions about consciousness, ethics, and the Heideggarian existential structure of care (i.e., Can anything in the world show up as *mattering* to AI, as it clearly does with human beings?). In the world of Contemporary Art, many artists are interrogating the question of posthumanism in terms of the breakdown of boundaries between what we habitually think of as natural and what we understand as artificial or virtual.

DOI: 10.4018/978-1-5225-9065-1.ch001

As rich, diverse, and challenging as all these perspectives on the posthuman are, there is another, more fundamental definition of the posthuman that the author proposes—a perspective that addresses the *essential* meaning of the posthuman in terms of revelation and prophecy, i.e. the self-presentation of what the author calls living language "speaking" through the human mouthpiece. This language is inceptive, creative, and calls forth, in its very "speaking", a new world with its new appearances, while at the same time bringing our formerly stable world to its final, catastrophic end.

BACKGROUND

Part One of this chapter describes how contemporary artists are articulating, through their artworks, an uncanny world known prosaically as posthuman. This world, according to contemporary art, involves a collision between an "alien" world and the familiar empirical world of stable appearances. The artistic results are often grotesque, weird, or an uncanny blend of horror and beauty. This "alien" world or way of being is often named as virtual reality and the author adopts that term in this chapter.

Part Two of this chapter explores the creative process that must occur logically prior to the artistic endeavour. This is the process of creativity, or prophecy itself—the inceptive or originating process that Heidegger calls the *activity* of the artist. He sought to approach the origin of art and indeed of all appearances through philosophical language, becoming more and more "poetic" (*poesis*) as he tried to show how to speak into being that which did not exist before. This is the realm of revelation, prophecy and living language, i.e. the clearing where language is a happening. The author explores this fundamental level of creation, and shows how the *essential* meaning of posthumanism, only hinted at or implied by contemporary art, is one of fresh language/world coming-to-be.

The chaos of our times reflects a collision, fusion, or marriage between inceptive, world-producing living language—future forming language with its writhing, fleeting appearances coming into being and falling back into oblivion—*and* achieved or pragmatic dead language, a mere tool of the empirical human being, with its stable, permanent appearances

The *life* of living language is the essential "alien" that contemporary art is implying in its posthuman artworks. Living language is alien to the stable world of pragmatic, achieved language and its corresponding appearances, cultural practices, etc. It has its own *telos*, catching receptive individuals up in its movements and "speaking" through them what must now be spoken into existence. Such individuals become mouthpieces to this "alien" prophetic speech.

Two modern pioneers who found their way to the "clearing" (Heidegger) where such encounters happen are Nietzsche and C. G. Jung. Nietzsche was a great pioneer of language who discovered this "alien", self-presentational living language! He tells us explicitly, for example, of his experience of this inceptive, world-creating language with all its power and numinosity, when he wrote the dithyrambic *Thus Spake Zarathustra:*

[O]ne is the mere incarnation, mouthpiece or medium of an almighty power. ... One hears—one does not seek; one takes—one does not ask who gives: a thought suddenly flashes up like lightning, it comes with necessity, unhesitatingly—I have never had any choice in the matter. ... Everything happens quite involuntarily, as if in a tempestuous outburst of freedom, of absoluteness, of power and divinity. The

involuntariness of the figures and similes is the most remarkable thing; one loses all perception of what constitutes the figure and what constitutes the simile; everything seems to present itself as the readiest, the correctest and the simplest means of expression. (Nietzsche, F., 1909)

Nietzsche's participation with the fires of living language opened his eyes to a momentous event as having always-already occurred in the psychic background of appearances (the "nothing" of Heidegger). Nietzsche's astounding discovery announced the close of a stable reality (the Metaphysical world), as founded on achieved or pragmatic language with its habituated pairs of opposites such as inner/outer, spirit/matter, truth/falsity, mind/body, etc.

During C. G. Jung's initiatory ordeal, as recorded in *The Red Book*, he also, like Nietzsche before him, became a "mouthpiece of an almighty power"—the almighty power of living language that he called "the spirit of the depths":

[A]t the outset of Liber Novus, Jung experiences a crisis in language. The spirit of the depths, who immediately challenges Jung's use of language along with the spirit of the time, informs Jung that on the terrain of his soul his achieved language will no longer serve. His own powers of knowing and speaking can no longer account for why he utters what he says or under what compulsion he speaks. (Jung, C. G., 2009)

Jung describes his ordeal in terms of fire and storms, explicitly comparing his experiences with those of Nietzsche and Hölderlin:

One thunderstorm followed another. My enduring these storms was a question of brute strength. Others have been shattered by them—Nietzsche, and Hölderlin, and many others. ... I was so frequently wrought up ... it was as if I were hearing [language—my insert] with my ears, sometimes feeling it with my mouth ... below ... everything was seething with life. ... a message had come to me with overwhelming force. ... I hit upon this stream of lava, and the heat of its fires reshaped my life. ... the primal stuff ... this incandescent matter ... material that burst forth ... and at first swamped me.

This small sample of Jung's ordeal, along with the translators' interpretation, as quoted above, shows that, from start to finish, his ordeal comprised a breakdown in "achieved" language i.e. traditional or habituated forms of language that stabilize the real appearances of the world, as also happened to Nietzsche. This breakdown opened both men up as receivers of a fiery incursion of self-presentational speech that "spoke" through them. They became mouthpieces for the "alien".

From the ordinary human empirical perspective, "achieved" or "dead" language is used as a pragmatic tool for our purposes. Achieved language is collectively interpreted as having reliable external references, constituting what we normally call "reality". Such interpretations stabilize culture. We generally know what to do or how to respond when we speak in such pragmatic ways. But, when the hidden background of Being undergoes a transformation, interpretations of the released inceptive language as being referential or as having external meaning simply collapse, and living language, i.e., future-forming prophetic language that utters as self-presentational figures of speech, breaks free and through the human mouthpiece.

This living language has an "alien" character, at times terrifyingly so, as a fury of destructive spiritual power is released, bringing to an end all reified or habituated forms of achieved language—a world catastrophe! Jung, like Nietzsche before him, participated in the collapse of traditional linguistic structures (achieved language) and plunged into the fiery chaos that follows. In so doing he made the momentous discovery of living self-presentational language—what he later called the objective psyche, or archetypal reality.

Jung's immense difficulty in bringing an emerging reality/world to language is shared by other investigators of living or inceptive language. We may for example turn to Heidegger and his enigmatic book, Contributions to Philosophy, of which it is said:

There is less consensus about this book than about nearly any other twentieth-century philosophical text. Is it "Heidegger's major work" or "metaphysical dadaism"? An earthshaking achievement or laughable gibberish ... As an account of the "essential happening of be-ing," the text resembles a treatise; as an investigation of the roots of concepts, it resembles history of philosophy; as an analysis of a crisis, it resembles cultural critique; as an invocation of a moment of decision, it resembles prophecy; as a self-conscious deployment of language, it resembles poetry. ... This new thinking is marked by its attention to the hidden and possible as such, by its distrust of all efforts to represent the inceptive event, and by its ambition to participate in the event, not simply to observe it. (Polt, R., 2006)

The Red Book, Contributions to Philosophy, and *Thus Spake Zarathustra*, have the literary power to draw us into at least a vicarious experience of the *reality* of living language as such—an inceptive language that is "alien" to achieved or traditional language forms. This is their legacy to future generations.

Contemporary art artistically renders the "alien" (living, inceptive language) in a multitude of ways, since the birth of non-representational or abstract art at the end of the 19th C. At the same time, great philosophers and psychologists (known at the time as alienists) discovered and participated in the alien character of non-representational, self-referential language *as such*.

As described in Part One, posthumanist art is now responding to another movement, a further transformation in Being, "pictured" as a collision or interpenetration of the "alien" and empirical reality as if poetically to say that the "alien" is descending to earth and interacting with empirical or human reality, taking Western culture further than these great pioneers of living language could say at the time of early 20th C. Whereas they opened up the possibility of encountering living language, *as such*, i.e. as distinct from earthly appearances altogether, and therefore "alien", they were unable to "speak" its further development of uniting or interpenetrating with pragmatic, earthly language, inaugurating a posthuman world with its corresponding nascent appearances, as pictured by contemporary artists.

In order to demonstrate the initiatory character of the prophetic experience and its world-forming character, the author describes two visions in Part Two, one by C. G. Jung in 1944, and one by George Orwell in *1994*. Jung's vision shows the process of initiation into what the author calls the world of psychological Being (the ego-alien world of archetypal reality, or the world of living language, as such). From this immersive and revelatory experience, Jung could participate in prophecy, i.e. he became a mouthpiece for this new alien world, speaking a new language expressive of this strange archetypal world he had experienced. He was scarcely understood, as he lamented many times near the end of his life. However, many new cultural practices that are based on various interpretations of the world of archetypal reality have subsequently flowered, not the least being Analytical or Depth Psychology.

A vision in *1994* initiated the author into an emerging world of what is known now collectively as posthumanism. He "saw" the "descent of the alien" to earth, yielding to an interpretation that living language, or the archetypal world, now "wants" to experience itself not as separate from pragmatic or achieved language but *as* the infinite or archetypal depths of ordinary earth-bound language. This union of the "alien", often called virtual reality, and empirical reality is now available to receptive individuals as a perceptual experience of fresh appearances of a new real world—the world of the posthuman. Contemporary artists are showing their interpretations of this momentous background event, as described in Part One, but as the author suggests in Part Two, the coming-to-be of the posthuman world is fundamentally a transformation taking place in how we language our world and ourselves.

POSTHUMAN BEING

Contemporary Art and Posthuman Being

Our epoch is a time of tragic collision between matter and spirit and of the downfall of the purely material worldview. (W. Kandinsky)

Heidegger made the fundamental discovery of a world that could also be called "the style of a culture". It is normally invisible and silent, ontologically prior to our reflections on life and yet it determines the way we live and understand the way we live. The style of a culture or its world is "[l]ike the illumination in the room, [it] normally functions best to let us see things when we don't see it." (Heidegger, 2005)

In Heidegger's essay, *The Origin of the Work of Art*, he famously refers to van Gogh's painting of a pair of shoes. These shoes in their status as equipment conceal a world that is so pervasive, surrounding the peasant woman at all points, that for her, it remains invisible. She simply wears her shoes, without noticing or reflecting on them as equipment. But, because they are presented to Heidegger as art, he *can* begin to allude to the Germanic rustic world that informs and shapes the life of the peasant and gives rise to the boots as equipment in the first place:

From out of the dark opening of the well-worn insides of the shoes the toil of the worker's tread stares forth. In the crudely solid heaviness of the shoes accumulates the tenacity of the slow trudge through the far-stretching and ever-uniform furrows of the field swept by a raw wind. On the leather lies the dampness and richness of the soil. Under the soles slides the loneliness of the field-path as evening falls. The shoes vibrate with the silent call of the earth, its silent gift of the ripening grain, its unexplained self-refusal in the wintry field. This equipment is pervaded by uncomplaining worry as to the certainty of bread, wordless joy at having once more withstood want, trembling before the impending birth, and shivering at the surrounding menace of death. (Dreyfus, 2005)

This vivid description highlights Heidegger's understanding of the *function* of art to make manifest and articulate a world that is normally invisible and silent to those living in it and yet remaining determined by it. It is the German rustic world, brought suggestively to language by Heidegger's existential phenomenology, that determines the peasant woman's existence.

From this historical, philosophical background, what world is now coming-to-be, from within the Western technological culture of the 21st C., as rendered by contemporary art?

I recently visited the Sydney Biennale which provides a platform for art and ideas and is recognised for commissioning and presenting innovative, thought-provoking art from Australia and around the globe.

In that year (2016), artistic director Stephanie Rosenthal addressed this question:

[I]f each era has a different view of reality, what is ours? With our growing dependence on the virtual world of the Internet, the distinction between that world and the physical one is becoming ever less defined. Many artists are attempting to access the 'in-between'—the place where the virtual and the physical fold into one another. (Transcribed from the MCA exhibit)

As I entered the first floor to buy my ticket, I glanced over to the souvenir shop and saw a reproduction of a work of art by one of the Biennale presenters, Grayson Perry. In this illuminating moment, I suddenly shared Heidegger's understanding of art as manifesting and articulating the ordinarily hidden world that gives rise to all the appearances that we normally accept unreflectingly as "things" or "equipment".

Perry had woven a tapestry showing the interior of a modern English household. I was immediately struck by the array of equipment taken for granted and used by the occupants of this home. The title of Grayson Perry's artwork is *The Annunciation of the Virgin Deal,* (Perry, Grayson) and the central figure in the tapestry, i.e. the angel, is not equipmental at all, and therefore perhaps is not even an appearance *in* this (so far undisclosed) modern world. No one else notices or pays any attention to her. Since the angel is not a manifest thing in the world, it seems likely that she is the artist's representation or personification of the normally concealed world itself.

The reader may access an image of the tapestry at the following URL: https://artsandculture.google.com/asset/the-annunciation-of-the-virgin-deal/DgHzNHCbRZyjLw

As with van Gogh's painting, the viewer is in a position to allow the equipment in the world that Perry presents slowly to disclose the world that is normally concealed from its inhabitants. The equipment portrayed in this tapestry comprises items that we in the 21st C. use intelligently in the dominant style of the modern English culture: cups, coffee maker, magazine, cushions, tea towels, technological devices, etc. If we look more closely at these items in Grayson's artwork, another feature that we normally take completely for granted begins to stand out—the ubiquitous presence of labels: "bourgeois and proud"; "organic"; "local"; "plastic"; "Penguin Books" (on the mugs); "make tea, not war" (on the towel); etc.

Ubiquitous labels or advertising messages saturate our minds throughout days and nights. We hardly notice them any more but our senses are drawn to these labels involuntarily and we begin to think the thought of the label unconsciously. Of course the advertising industry knows how to exploit our involuntary attention to its own advantage, giving every effort to ensure we reach for one particular product over the competition.

In unconsciously thinking the imposed thought of labels or ads, human beings begin to think thoughts that are *alien* to any personal empirical experiences also available at the time. Interviews of the "man in the street", asking for his political opinions for example, reveal only the thought of the mass media at work, nothing personal. How *could* most people have any personal experience of the horrors of terrorism, or the devastation of tropical rainforests, or shenanigans of national politics, etc.? Yet such alien thoughts (i.e., alien to the empirical lives of most of us) are determining the course of events in the empirical or physical world. The 1997 movie, *Wag the Dog* illustrates this point beautifully. The USA goes to war, in response to fearful images concocted in the mass media by a few men. A purely virtual threat has terrible consequences in the empirical world.

Virtual reality became possible *as a concept* only when we began thinking reality exclusively in terms of the passive senses i.e., the belief that everything we know about our world comes from our senses passively receiving stimuli. We call this reality empirical or physical reality. Only a small technical step is then needed to excite the senses directly with technologically generated information and to call our subsequent experience "virtual reality": a reality is generated that is as close as possible in experience to empirical reality when received passively through the sensory-motor system. The "reality" aspect to virtual reality reminds us that we are immersed in a *real* world, while the term "virtual" indicates simulation of empirical reality. Importantly, we interpret this simulation or alien thought as kind of solid substance, like the appearances of empirical reality. The experience of virtual reality is so close to that of empirical reality that it is used to train pilots and counter-terrorist forces, for example. At the same time as our immersion in virtual reality today, we are also surrounded by our usual world—a world that determines stable appearances that we use routinely on a daily basis: cups, towels, computers etc.

Perry's artwork, then, is showing us an *emerging* new world—a birth if you will (the angel's annunciation?) Two worlds, virtual and physical, are chaotically colliding, conflating, interpenetrating, fusing, or folding into each other. But how do we *say* the possible world that may be emerging from this linguistic chaos? Take one piece of equipment, a cup. The label on the cup in Perry's artwork is an ad for a Penguin book. We are meant to think, "Penguin books" every time we lift up the cup to drink. What is the *mode of being* of such a strange piece of equipment? We use the cup in its ordinary status as equipment of the empirical world. We use it to satisfy thirst, enjoy companionship with one another, sooth anxieties, kill time, etc.—all those activities that together disclose our modern style of culture. We could even imagine that the cup's presence invites us to make a pot of tea. At the same time, we are meant to think "Penguin books"—a thought that is completely alien to the equipmental status of the physical cup. The label stimulates our senses so that we are unwittingly compelled to reach for a Penguin book now or in the future, or at least to imagine doing so. This is an action having nothing to do with drinking tea. Perry's artwork displays this concatenation of disjunctive thinking throughout his painting and in so doing is alluding to a nascent world—one that normally is concealed from us and simply taken for granted.

This new world is giving birth to very strange appearances whose mode of being is far from straight forward. Museums are now mass-producing these new *things* e.g., table napkins that reproduce a Classical master's artwork. What are we to *think*, or how are we meant to *be*, as we wipe gravy away with a reproduced portrait of the Madonna? This is far from easy to answer but Perry's artwork is at least raising the question.

The Angel is placed alongside an IPad announcing a sale to Virgin airlines, and she points to portraits of Bill Gates and Steve Jobs. Just underneath these portraits sits an ordinary mother and child in the living room—a new birth from out of our technological civilization? Is the angel announcing the birth of a new world, with *its* correlative appearances, cultural style, and practices? In Perry's painting, virtual reality is chaotically showing up *empirically*. Virtual reality and empirical reality are no longer to be thought as logically separate domains of "reality", having nothing to do with one another, in the way that fiction and science were once thought as quite separate domains of reality.

In researching this new phenomenon I discovered an astounding video on YouTube. The scene opens up with:

Somewhere in a little town of Belgium
On a square where nothing really happens
We placed a button
And waited for someone to push it … (Youtube)

In the centre of the small town square is a pillar supporting a big red button. Dangling overhead is a sign: "Push to Add Drama". A passer-by does so and a drama erupts! Ambulances, accident victims, collisions between cars and people, fights, police and gangster shoot-outs all vie with one another while the stunned local people look on. Although it is easy to get absorbed in the dramatic action, it is quite instructive to also cast an eye on the locals who were suddenly assaulted by this advertisement for a TV station (as it turned out to be) bursting in on their ordinary lives. Some were clearly shocked and then frightened. Others were nonplussed, uncertain, while still others became, in effect, the audience enjoying the drama exploding so unexpectedly into their mundane lives "where nothing really happens". A purely staged media event erupts into the empirical lives of ordinary people going about their business. It is striking to see how quickly some, if not most, accepted it, even enjoying the "thrill" of it. Their relatively quick adaptation to the incursion of an advertisement into empirical life is reminiscent of earlier days when audiences at first fled the theatre as a cinematic train "rushed" towards them. Today we easily accept 3-D versions of these moving images, without thinking about the reality question at all. We merely enjoy the "thrill of the ride".

This staged event in which virtual reality (or medial reality) collides with empirical reality is merely another, more spectacular, example of the kind of appearances to which Perry is drawing our attention in his tapestry. You can see this conflation everywhere in our daily lives. One more example will suffice. Here in Sydney, Rugby League is very important. Watching a game on TV is like watching two hours of computer pop-ups. Labels literally cover the bodies of the athletes; computer-generated and painted ads cover the playing field; digitalized ads run incessantly around the fence bordering the game. And it is all accepted now. Sports heroes and advertisements are now conflated, and together constitute the same weird appearances whose mode of being is not at all clear at this time.

Grayson Perry's artwork is an attempt by a gifted artist to show us a new world coming into being, as suggested by all these strange new phenomena.

"The place where the virtual and physical fold into one another", as Stephanie Rosenthal says above, is a poetic way of pointing to the concealed world that is "producing" these strange appearances (occurrents or equipment, as Heidegger would say) that we now take for granted. We are being shaped by this silent world as surely as van Gogh's peasant was shaped by the rustic world that gave rise to her mode of being and *its* equipment (the boots). Perry's The Annunciation of the Virgin Deal is one artist's way of hinting at this new world, through its appearances, and we can see, through the grace of his artwork, that this new world or reality is very strange indeed.

Van Gogh leaves hints of the concealed presence of a rustic world when he painted a simple pair of peasant boots. But it was Heidegger who could articulate this world and its equipment for us. He *spoke* it into visibility—just for a moment—when he described the network of equipment that only had meaning relative to being used in that world of rustic existence. Similarly, Perry's artwork hints at a world coming to birth at this time, along with its alien-yet-ordinary appearances. How can we begin to articulate this new world with its peculiar appearances, as represented in contemporary art?

INCEPTIVE SENTIENCE

Throughout the last few decades, one of the major themes of the author's books concern the actual process of creation and manifestation of worlds. There are four phases:

1. Surrender and participation with the transformations within the "source of all", for want of a better expression, thus becoming a receiver or vessel for an as-yet-unborn possible future. This is a process of initiation.
2. Speaking as a mouthpiece of an unborn future. This is the work of revelation and prophecy.
3. Willingness on the part of others to make a move towards further articulation or manifestation of that unborn future or world. This is the work of art.
4. The gradual development of cultural practices based on the fresh appearances, giving further expression to and strengthening the new reality. This is the work of maintenance.

When these conditions are met, we end up living in that world, taking the new appearances for granted, and getting shaped by them. It becomes really so!

The world Grayson Perry is disclosing to us is still just coming into being, pictured in art as a collision between virtual reality and empirical reality. These new strange appearances should give us pause as we consider the words of Owen Barfield:

Imagination is not, as some poets have thought, simply synonymous with good. It may be either good or evil. As long as art remained primarily mimetic, the evil which imagination could do was limited by nature . . . but [since non-representational art—my insert] both the good and evil latent in the working of imagination begin to appear unlimited... We could very well move forward into a chaotically empty or fantastically hideous world.... We should remember this, when appraising the aberrations of the formally representational arts.... in so far as they are genuine, they are genuine because the artist has in some way experienced the world he represents. And in so far as they are appreciated, they are appreciated by those who themselves are willing to make a move towards seeing the world in that way and, ultimately therefore, seeing that kind of world. We should remember this, when we see pictures of a dog with six legs emerging from a vegetable marrow or a woman with a motor-bicycle substituted for her left breast. (Barfield, O.)

We can possibly avoid such a cultural catastrophe by exploring the inceptive process of creating such worlds as the virtual world more deeply—in particular, phases 1 and 2 above: the work of initiation and prophecy. We will thus see that the bizarre collision between the virtual world and empirical reality made manifest by advertising and technology in general is only one possible appearance of a deeper intentionality at work—an intentionality that is "seeking" a union between the "alien" and the human (or maybe their identity). In Part Two, we will see that the "alien" is seeking to particularize empirically as human while remaining in its character as alien, while at the same time the human being may re-discover her roots in the universal—this transformation taking place essentially in the realm of language.

To approach this mystery, Part Two will examine two modern experiences of initiation into the mystery of language, and the prophecies that subsequently burst forth, bringing fresh language, perceptions and cultural practices of an incipient world.

Initiation, Prophesy, and Emerging Worlds

The pictures were painted directly through me, without preliminary drawings and with great power. I had no idea what the pictures would depict and still I worked quickly and surely without changing a single brush-stroke. (Hilma af Klint)

It is not enough for human beings to become adults biologically, as all other animals do. Our minds or consciousness must be initiated via a "second birth" to bring us into accord with the real world we must face and meaningfully engage with alien sentience on a daily basis. Initiations bring us into accord with reality by teaching us what kind of human being we have been all along, and what kind of world we inhabit. Each age thus had its "timeless" initiation practices belonging only to that age. When reality transforms, fresh, appropriate forms of initiations must thus be discovered or created. For example, for those coming into adulthood in the Industrial Age, there was no place for initiations based on Animal Powers

Much of the European and American culture in the 19th and 20th centuries was founded on intimations of a new world, which might be called psychological reality—the reality of the *unconscious* mind, or the archetypal world. This world was so unexpected, uncanny, and indeed ego-alien, that no collectively endorsed initiatory practices to bring humans into harmony with this world could prevail. Abstract art and C. G. Jung's Analytical psychology were noble attempts to indicate the concealed presence of this "alien" world, but we can now easily see that no generally accepted method of initiation exists. Those few individual people who are initiated into this world understand their individual being as a duality—an island of consciousness surrounded by and saturated with a vast pool of unconscious or Archetypal Being!

In the 21st C. matters have grown more complicated still. Reality has shifted again: the world of *posthuman Being* has usurped that of *psychological or archetypal Being*.

The failure of our Western culture to produce initiations that bring individual and groups into accord with reality today has, of course, devastating consequences on the cohesion, and, indeed, coherence of society, as well as the well-being of individuals who are forced to face *our* reality with no preparations at all. It has become chaotic! We have failed to initiate our young into adulthood today because we have no initiatory practices that correspond to either psychological reality or posthuman reality. And we have no initiatory practices because we have failed to pay attention to the only source of initiatory practices—the same source throughout time—revelation and prophecy! Certain individuals: our prophets, mystics, shamans, call them what you will, are "summoned", usually through illness, madness, or misfortune, to face enormous, tectonic shifts in reality, and *personally* to endure the transformations in reality, and thus to be "gifted" with revelation. Their subsequent prophetic speech, if accepted by the community, leads to a new set of practices on how to become an adult in accord with the new reality.

An extraordinary early 20th C. example lies in the story of Black Elk, a Sioux shaman who had a powerful vision when he was nine years old. (Neihardt, J. G., 1972) His experience taught him and his people about a coming new world comprising people of all colours, beyond the local concerns of the Plains people of South Dakota. His tribe adopted the ceremonial dance that emerged as a gift from his visionary flight as being in accord with the new emerging reality (the influx of Europeans) that also threatened their traditional world.

Some decades later, in Switzerland, another shaman/prophet went on a visionary flight and brought back *his* "gift" of initiation to humanity in its hour of need. Unlike Black Elk, this mystic's vision has been largely ignored.

Initiation Into The World of Psychological Being

Early in 1944, C. G. Jung broke his ankle and subsequently suffered a heart attack. He was hospitalized for several months in a delirium. During his convalescence he entered a form of visionary consciousness that opened to a world as real as the hospital bed in which he lay, close to death. This is an excerpt of his report:

It seemed to me that I was high up in space. Far below I saw the globe of the earth, bathed in a gloriously blue light [long description follows]. I knew that I was on the point of departing from the earth.

After contemplating it for a while, I turned around. ... A short distance away I saw in space a tremendous dark block of stone, like a meteorite. ... As I approached the steps leading up to the entrance into the rock, a strange thing happened: I had the feeling that everything was being sloughed away; everything I aimed at or wished for or thought, the whole phantasmagoria of earthly existence, fell away or was stripped from me an extremely painful process. Nevertheless something remained; it was as if I now carried along with me everything I had ever experienced or done, everything that had happened around me. I might also say: it was me, and I was it. I consisted of all that, so to speak. I consisted of my own history, and I felt with great certainty: this is what I am. "I am this bundle of what has been, and what has been accomplished."

This experience gave me a feeling of extreme poverty, but at the same time of great fullness.... Something else engaged my attention: as I approached the temple I had the certainty that I was about to enter an illuminated room and would meet there all those people to whom I belong in reality. There I would at last understand...what historical nexus I or my life fitted into. I would know what had been before me, why I had come into being, and where my life was flowing. My life as I lived it had often seemed to me like a story that has no beginning and no end. I had the feeling that I was a historical fragment, an excerpt for which the preceding and succeeding text was missing. My life seemed to have been snipped out of a long chain of events, and many questions had remained unanswered. Why had it taken this course? Why had I brought these particular assumptions with me? What had I made of them? What will follow? I felt sure that I would receive an answer to all these questions as soon as I entered the rock temple. (Jung, C. G., 1989)

Jung says that previously he had understood his life as an historical fragment or an excerpt of a story, which left him with many unanswered questions. All these questions were to be answered when he entered the rock temple. Of course, to enter that temple would be, at the same time, to die on Earth, in the hospital bed. An emissary from Earth prevented him from that final step. His doctor arrived in his archetypal form, declaring that Earth still needed Jung, at which point the vision ended and Jung "returned" to the hospital bed. In a mysterious twist of cosmic fate, Jung's return to earthly existence involved an exchange with the doctor's life—he died suddenly soon after.

Jung had been gripped by questions, springing from an experience of his life as discontinuous, fragmented, and excerpted. What comes before, what comes after, why did I take this particular course? All these questions would be answered, not by reason or explanation, but by his re-joining, in death, the eternal thread from which he had been cut, in order to exist as the historical Jung.

This complex vision shows Jung asking the fundamental eternal questions that initiates have always been drawn to: where did I come from, who am I now, where am I going? Previous cultures have addressed these questions in terms of, for example, the ancestors, fate, destiny, the after-life, and so on. But Jung, being a pioneer of our modern psychological culture, formulated these questions in a way appropriate to his *psychological being*. He understood his conscious personality (who am I now?) as coming into existence as a discontinuity, born out of nothing and resting on nothing, "a story with no beginning and no end", a fragment, an excerpt, and "snipped out".

As he approached the temple (grew closer to death, we might say), he experienced a painful stripping away of attachments to earthly life (his aims, wishes, thoughts, regrets), leaving him impoverished and at the same time full: "I was what I had been and lived. … I had everything that was and that was everything." This remarkable and enigmatic formulation of Jung's initiation into the death process seems to say something very important about memory and identity. We normally know who we are by virtue of our conscious memories, stitched together as a continuous narrative. Our identity thus normally feels secure and permanent. Jung's process towards death reveals that he understood this narrative as really being a literary fragment, an excerpt with missing text on either side, discontinuous—conscious identity as a historical or temporal fragment or literary snippet, having no knowledge of what lay before or after it.

The painful stripping away of attachments to this or that memory left Jung with what he calls "objective cognition". He became *conscious* of his personality as a unity of memory and identity: "I was what I had been and lived." This is an amazing psychological achievement of his initiation, and it immediately raised the questions that Jung now faces. Now aware of the "excerpt" character of his conscious personality, he can ask, in effect, who *else* am I, now that I am also "beyond" my personality? What lies to either side of this snippet, historical fragment, or excerpt that we each call our conscious self, or identity?

The answer to these fresh questions consisted of a further initiation into death. Jung had to enter the temple and join the company of spiritual beings whose presence would show him what "historical nexus I or my life fitted into". In other words, he was to be shown the missing temporal elements that would give contextual meaning to what otherwise is a discontinuous, ahistorical "self" torn out of its deeper temporal matrix, into an existential state of meaninglessness. It appears that to join this company and gain knowledge of the deeper "background" context that informed and gave final meaning to his fragmentary conscious life is also to die. Instead, Jung was called back to life for a while longer.

Jung's vision was an initiation into the mystery of death and archetypal reality as such (i.e. living language) as well as the essence of psychological Being in the 20th C. He came as close to death as one could while remaining alive. As his "spiritual teacher", Death informed Jung what it is to be a *psychological* being, transforming him along the way. Jung wakefully experienced the "other side", normally only granted at the point of death. But this is precisely what an initiation is—transcendence or dying before you die! Death showed Jung the reality of our unconscious or archetypal being, beyond empirical reality.

He writes that, following his recovery:

A fruitful period of work began for me. A good many of my principal works were written only then. The insight I had had, or the vision of the end of all things, gave me the courage to undertake new formulations. I no longer attempted to put across my own opinion, but surrendered myself to the current of my thoughts. Thus one problem after the other revealed itself to me and took shape. Something else, too, came to me from my illness. I might formulate it as an affirmation of things as they are: an unconditional "yes" to that which is, without subjective protests, acceptance of the conditions of existence as I see them and understand them, acceptance of my own nature, as I happen to be.

Jung came close to death but did so in full consciousness ("the vision of the end of all things"). He was thus released from a fear of death, i.e., fear grounded in our everyday attachments to life. All these attachments were painfully stripped away from him as he approached the temple. A new personality was forged, "an ego that does not break down when incomprehensible things happen; an ego that endures, that endures the truth, and that is capable of coping with the world and with fate."

Jung's vision enabled him to penetrate into the essence of what it is to be a human being in the 20thcentury—a *psychological* being! He was initiated into knowledge of his everyday being as a unity of personality and memory, and—as well—a unity of that being and what he called "objective cognition". This psychological structure is a duality, a "two-ness" and accounts for the self-contradictory language Jung has to use throughout his life.

The outcome of Jung's ordeal and initiation into the mystery of death is quite in accord with the desired outcomes of any initiatory ordeal: a full awareness of one's mortality and "littleness" when faced with the greater incomprehensibility of life; a human being who is "aligned" with the reality of her times so that it becomes her truth; and one who can adapt to contingencies with a spiritual understanding of those contingencies in terms of fate or the ancestors, etc. However, Jung's initiation has a form that is appropriate only to his time, the time of *psychological* Being. His initiation had to be psychological (archetypal) in character.

Initiation Into the World of Complexio Oppositorum

Jung speaks of a painful stripping away of all his attachments to earthly existence. Within Jung's vision, this pain is real, as real as it was for initiates of former ages. Initiates of former ages also went through a "painful stripping away" of their ordinary attachments to life as a path towards encountering the "background causes" of the phenomenal world. These ordeals always targeted the body of the initiate (scarification, flesh sacrifices, physical ordeals of enduring wasps or ants, etc.). Pain belongs to the body, yes, but Jung had found his way to the background reality as such, informing *our* times—a strange reality that must also include body and yes, equally real pain! We call this reality "archetypal reality" or the objective psyche—*our* reality, the reality of our times—and, like the realities of all former ages, our reality requires a correspondingly appropriate initiation into its truth. Jung's vision is an outstanding example of what such an initiatory ritual would be. Jung was initiated into the truth of what a 20thcentury human being is, in his or her essence—a psychological being and a *complexio oppositorum* at that!

There is so much that we can learn from his vision once understood this way. It is exemplary of the form that a modern initiation must take in order to make psychological reality our *truth* so that we each can come into accord with this way of being. For example, no other real human being initiated Jung, as had happened in the past. The initiatory impulse emerged from the depths of unconscious Being in the form of a vision that attained the status of reality, which must therefore include the reality of the body with its pains and pleasures. The vision must have the convincing power of empirical reality, yet be understood as distinguishable from ordinary empirical reality. And it did so for Jung. Archetypal reality can initiate us only when our *whole* being is engaged and participating with its processes, as was true for any initiation in the past. We can see from Jung's subsequent report that his efforts to describe his visionary experiences are clumsy, contradictory, and inadequate. This is because archetypal reality cannot be understood in those categories of thought that constitute the background of our current empirical reality—for example, the fundamental inner/outer or mind/body disjunctions. The archetypal world has left such binaries behind.

When the initiate is in such a visionary state she can state quite simply and truthfully that her body is undergoing a painful stripping away. But when she returns to her current reality, all sorts of maddening questions emerge such as, "do you mean 'body' metaphorically, figuratively? Is it a subtle body? ..." All such inadequate formulations rest on the very inner/outer disjunction that the initiate has in fact overcome. Jung is showing us the nascent inception of a new world—archetypal Being or living language.

Initiation Into The World of Posthuman Being

It must seem clear, after this small taste of the complexity of Jung's vision that no modern cultural practice has subsequently emerged to facilitate any such initiatory "ritual" collectively. Jung's depth psychology or *Analytical Psychology* has failed the promise of initiation when it became institutionalized—an outcome Jung vigorously opposed. We can say then, that as far as our *psychological* being is concerned, we don't have a lot of initiates at present. This is not a small matter when we also consider the thought that our *posthuman* way of being in the 21st C., with all its emerging practices, may have completely overtaken that of psychological man. Jung's form of initiation may now be ontologically obsolete, in the sense that reality now has quite different contours and requires another kind of initiation altogether. Yet it may become clear below that initiates into the reality of psychological Being are needed to guide others into the reality of posthuman Being.

We can approach this unnerving thought if we begin to understand psychology and technology not merely as two separate compartments of knowledge. We can also understand that psychology and technology are each founded within a defining style of consciousness and its correlative language and cultural practices. Language is the *a priori* that brings both psychology and technology into existence through the revelations of prophecy and then development of corresponding cultural practices. I am not here referring to the kind of language that psychology and technology each uses pragmatically. I am referring to inceptive, originating language that in effect prepares the future. I will call *this* language *living language*—language as autonomous, bestowing inceptive ideas upon the initiate who can reach this level of reality and speak it into existence in the form of prophecy.

Jung's visionary and transformative experience in 1944 took place entirely within the domain of living language or what he called Archetypal reality. He was transformed by it. Twenty years earlier his *Red Book* also vividly demonstrates this reality to us. All the beings that Jung addresses or engages in a real way are figures of *speech*! *The Red Book* is a record of self-presentational, autonomous, inceptive language. The technological forms of our present civilization demonstrate a dominant cultural practice that is rooted in a particular positivistic interpretation of living language or archetypal depth. The infinite depth of Being, once expressed through prophecy and *poesis* has been interpreted as empty physical space; the initiatory gift of mystically perceiving everything as One has been interpreted literally and abstractly so that we now can *pragmatically* think and speak globally, holistically, in totalities or generalities. We can now treat any individual, or natural being as a replaceable mechanical *part* of a greater all-encompassing abstract whole. For example, no one mourns the loss of an *individual* component of a computer. We know that such components are to be understood as replaceable things, "standing-ready", as Heidegger says, to be used and disposed. They are merely parts of an abstract whole (the Internet). It's the *abstract* "whole" that matters in language today and which drives our technological civilization, not a mystical perception of One-ness, such as in Jung's 1944 vision: "Far below I saw the globe of the earth, bathed in a gloriously blue light…"

The style of consciousness that this pragmatic, abstract language reflects is as "beyond" Earth, in "outer space"—*alien* to anything human. For the most part we speak this language of abstract totalities unconsciously, not realizing that when we do we are in identity with a style of consciousness that is "out in space", cold and alien to the earthly experience of embodied life. We may think, "I want to save the planet", completely unaware of the contradiction between the semantics of our speech (saving the planet) and its syntax (the planet can only be perceived from out in space). The syntax reveals a style of consciousness alien to earthly empirical existence, quite unable to effect earthly change from within its own style of concept/percept.

Living language or archetypal reality is foundational to both psychological Being in Jung's time and technological Being in our own time. Jung's 1944 vision demonstrates the initiatory process into psychological Being, which includes commitment to archetypal reality or living language as such. When Jung's visionary flight took him beyond earthly existence where he could "perceive" its totality, he was speaking visionary or sacred language in the manner that Black Elk did when he said that the centre of the world is Harney Peak, South Dakota, but that it could be anywhere. When this sacred manner of speaking is usurped by profane pragmatic considerations, or spoken profanely, it becomes literalized and abstract, as the present language of our technological civilization.

Now at this time of the posthuman, as our contemporary artists are showing, what kind of initiation may be required to bring us into accord with an emerging reality, our posthuman way of being? Any initiation today will be required to rest on a world structure that produces kind of being that cannot be called simply a human being, or even a psychological being. Our artists are showing that living language or archetypal reality, once thought as logically separate from empirical reality in the world of psychological Being, must now be articulated in a totally differently way in the posthuman world.

Following my visit to Grayson Perry's exhibit in 2016, I attended another art show in Sydney that also presented a range of postmodern art. The exhibit described artists:

Whose works encourage us to ask what it means to be human today, and what it might mean in the future. Drawing inspiration from science fiction, robotics, biotechnology, consumer products and social media, they offer experiences that raise questions around the idea of the posthuman—a concept that signals new understandings of humanity and a breakdown of boundaries between what we think of as natural and artificial:

Some of these works combined human body parts (such as body fat used as fuel) and machines, in order to "speak" this new way of being. I noticed that the audience asked exclusively about the technical side of the art works (how did you do it, etc.) No one appeared to address the aesthetic, human perception—sheer grotesque horror paraded in front of us!

This art is saying that an "alien" consciousness, i.e. the inceptive language that gave birth to our artificial technological civilization, is seeking to unite with the empirical human being. It wants to humanize, while remaining *as* alien. At present, artists are picturing this union as a horror (but strangely not seen as such!) This *linguistic* horror is already facing us, almost everywhere, in the mass media. To take two popular examples in politics today, Presidential candidates routinely parade ordinary folks before the media. These people tell us their human stories of suffering, victimization, or loss, in an "alien" or medial context that is using them for propaganda purposes and, at that level, cares nothing about their human misery at all, only that the staged performance achieve the purpose of persuading others to vote, etc. The mass media regularly chews individuals up in a similar fashion. Tears are encouraged or even provoked but the abstract "alien" interest lies not in their human suffering but in getting a good story for the ratings. The language forms that are used are a grotesque combination of personal feeling and abstract, alien thinking, "So sorry you lost your spouse to cancer but we'll be back in a moment after the break ..." This monstrous warped form of the collision between languages/worlds evokes horror (or should)!

Psychological Being and its cultural practices originated in Jung's initiation into archetypal reality (living language) and prophetic speech. Similarly, posthuman Being has been brought into existence by a further mysterious archetypal transformation and its attendant prophetic speech.

In 1994, the author was given the following initiatory vision:

A man is among us; he looks quite normal, except he is in fact alien. He is friendly, wants to, needs to live amongst us, and is warmly welcomed. Many therapists are excited and thrilled with the glamour of his gifts, which include space ships that could fly at dizzying speeds. I join in with this madness for a bit but lose interest and instead grow increasingly alarmed. I try to warn others. I decide to act. I want to burn him and race around looking for a flamethrower. Instead I kept grabbing fire extinguishers and spray him with those. They are useless. He tries to stop me and we seem to realize that there was nothing personal in this. He wants simply to live here and I could sense incredible danger to us. I say, "It's just that our species can't survive if you stay. We need to survive too!" Then I go back to my frantic search. He says, responding to my "what ifs? ..." "Do you mean, what if I spit on the carpet or people?" And he does so, thus at last revealing the danger. A terrible poison was in his spittle; it dissolves flesh leaving horrible forms, like a fly dissolves its meal. I get more frantic until... (a new scene breaks through)... There is a bed of coals, so astoundingly hot that they each glow transparent red. One would simply evaporate on them. In the midst of my passion to stop the alien, a young man, a human, flings himself as a voluntary sacrifice onto the bed of coals. He is the sacrifice who will save us. I am struck with horror and agony, a religious agony that sends me to my knees as I feel his act of sacrifice. O God! O God! O God! I imagine his flesh blackening and crisping as he rolls on the coals in apparent unspeakable agony. Yet when I actually look, though I am screaming in pain and horror and awe, I can see quite objectively that he is undergoing a different process. He is moving about, but in agony, or intentionally? That is, is he avoiding the heat or is he moving to further expose himself to it? He is not screaming. Is he in pain? He then begins to glow red just like the coal itself. He becomes transparently glowing red all over, like a clay vessel does at the highest point that forms the pot. Even more astounding, he is pregnant, almost full term! Did he enter the ordeal that way or did the transforming fire engender a new life in him? By this time I cannot describe my own feelings at all. It's too much. One cannot name a mystery such as this. Then, the name comes: He is our Redeemer! (Woodcock, J. C., 2015)

According to this vision, a horrible outcome occurs when the "alien" is countered by fear and a gesture of refusal towards his intention—to live amongst us, *as* alien, or human-alien. The "alien" is a style of consciousness "outside" of our ordinary human consciousness. Our ordinary and habituated style of consciousness might be called "humanistic". We relate to one another with *care*, as individuals, each with intrinsic dignity etc. The "alien" style of consciousness, on the other hand, constitutes the linguistic foundation of our technological civilization. This vision "pictures" this "alien" style of consciousness as having already "landed". Other similar visions have followed this one over the years for the author and lead to the conclusion that living language or archetypal reality is no longer to be thought as a separate realm to earthly empirical reality but as interpenetrating it. It has, so to speak, come down to earth.

However, as the vision and contemporary art both show, terrible and grotesque appearances are now coming into (via the alien's "spittle", i.e. literalized and reified mystical language of Oneness) in this posthuman world. The deeper intentionality may be seen in the prophetic speech of the second part of the vision, which suddenly breaks through all the fear and panic. We may be redeemed from the fate of a "chaotically empty or fantastically hideous world" (Owen Barfield, above) if the human initiate undergoes a sacrificial and transformative ordeal in fire. The vision indicates that ordinary empirical language, based on the opposites (either/or) cannot understand initiatory space and must be broken down—the catastrophe of a former stable world coming to an end. This collapse of the opposites thus releases a destructive fire and at the same time engenders a new birth, or new way of being.

A new language is coming into being, one that first must destroy all the opposites that have produced our stable Western culture for thousand of years: spirit/matter, mind/body, inner/outer, god/man, archetype/person, fiction/reality, and so on. A spiritual fire is released, as Nietzsche says (see quote above), along with this collapse and is currently sweeping the world unchecked. But if the initiate can contain the fire, i.e. as a vessel, while enduring its transformational process, a birth may be engendered—the birth of a new language that can speak into existence a new posthuman world, along with its corresponding appearances.

CONCLUSION

[In] printing "Not Legal Tender" across a brilliantly counterfeited dollar bill, the resulting object (counterfeit cum inscription) [is] rendered incapable of deceiving anyone. It is not an illusory dollar bill, but then, just because it is non-illusory it does not automatically become a real dollar bill either. It rather occupies a freshly opened area between real objects and real facsimiles of real objects: it is a non-facsimile, if one requires a word, and a new contribution to the world. (Danto, A.)

Language is on the way to becoming conscious of itself in its posthuman status of being "alien-human" as the author also learned in another initiatory, prophetic dream in 2012:

I am wandering the streets, alone. I find myself in a hall where some ritual is going on, conducted by an older man. The participants are each undergoing a ritual, which seems perfunctory, i.e. just going through the motions. It has a Masonic-Christian feel to it. We are all sitting on our knees on carpet. When he sees me, he suddenly becomes interested, and asks me to go through the ritual, which now comes alive. There is a line on the floor. I am to touch my head on that line, i.e. submit. I do so as he intones the ritual of confessions. As I touch the floor with my head, he smiles and says warmly you are forgiven, everything. Then he comes over to me and crouches, whispering in my right ear for some time. As I listen, I hear the voice of the "other", a higher pitch, unearthly, i.e. the angel is speaking to me though him. I have trouble understanding most of it but the angel talks for some time. When finished I get up but have difficulty speaking. My right hand begins to write spontaneously. I scribble "interlocutor".

Living language or archetypal reality has always been alien to the human experience. Until modern times we have substantiated its "alien" status as a transcendent deity, nature spirits, the *daimon*, angels, and so on. To become conscious of itself *as such*, i.e., beyond representation in the human imagination, it needed the consciousness of its human mouthpiece as a vehicle of expression rather than representation. This was the work of Abstract artists at the turn of the 20th. C., as Kandinsky's and Hilda af Klint's artworks show (see epigrams above).

The posthuman world is coming-into-being as living language, or archetypal Being, seeks to enter the human domain of experience but remain in its "alien" status, as the dream above suggests. The inhabitants of the posthuman world will be in the ontological status of "human-aliens". The author's book, The Peril in Thinking, explores the dreadful consequences that happen individually and worldwide when thinkers like Nietzsche and Heidegger find their way to the fires of living thinking but fail to humanize it (Woodcock, 2015). They are not to be blamed. The fiery energies of living language are terrible, as destructive as the furnace of a nuclear bomb, and as creative as engendering a birth.

The essential meaning of this birth is simply this: living language now knows itself *as such*. This is what C. G. Jung means by Self-consciousness. In our posthuman times it also seeks to be humanized *as* the human mouthpiece. The child is thus "alien-human", as posthuman art is informing us. This is a poor description to be sure. We are just at the beginning of this transformation in language/world. But perhaps a little more can be said about its phenomenology. The archetypal world is now to be thought as the "alien" *within* ordinary, earthly, pragmatic language, not as a separate domain of experience "out there" in archetypal reality. As Kandinsky tells us, "Everything dead quivers". This transformation in consciousness/world results in a kind of dual vision, the kind of vision that Jung gained with his personalities no. 1 and 2. Salvador Dali also enjoyed this dual vision as rendered in his amazing Surrealist art forms, using his artistic "paranoiac-critical method". This method is based on his perceptual duality, and privileges spontaneous, "irrational" expressions, which have the effect of bending or loosening the hardened stability of material reality.

From the ordinary human, empirical perspective, language is a pragmatic tool to be used for our purposes. This of course will continue, as we need a stable world. But when the alien consciousness, or "other within", wishes to speak, the human mouthpiece may willingly surrender her personal concerns, handing over her speech centres or artist hand etc., to that "other" who can thus "speak" through her. The human co-creator may thus begin artistically to shape that alien, prophetic speech into a thing of beauty rather than into a monster, as contemporary artists are warning us about. If this willing surrender to the "alien within" can be strengthened through appropriate cultural practices, the posthuman world striving to come into existence through its appearances may express its truth as the now popular motto, "everything is art"!

Everything—including us!

REFERENCES

Barfield, O. (1957). Saving the Appearances. London: Faber and Faber.

Dreyfus, H. L. (2005). Heidegger's Ontology of Art. In H. L. Dreyfus & M. A. Wrathall (Eds.), *A Companion to Heidegger*. Malden, MA: Blackwell Publishing. doi:10.1002/9780470996492.ch25

Grayson, P. (n.d.). *The enunciation of the virgin dead*. Retrieved from: https://www.google.com/search?q=Grayson+Perry,+The+Annunciation+of+the+Virgin+Deal&rlz=1C1KMZB_enUS577US671&tbm=isch&source=iu&ictx=1&fir=mWcwHLNjqNGzyM%253A%252COceTlgB_IOeLyM%252C_&usg=AI4_-kSNO--aTpOlSYUyOF6jHLxVtP0V9w&sa=X&ved=2ahUKEwi396iN1uHfAhUfwMQHHVuZD6cQ9QEwAnoECAUQBA#imgdii=Pvug-5M-3bzRgM:&imgrc=mWcwHLNjqNGzyM

Heidegger, M. (2002). The Origin of the Work of Art. In *Off the Beaten Track*. Cambridge University Press.

Jung, C. G. (1989). *Memories, Dreams, Reflections* (A. Jaffé, Ed., WinstonR.WinstonC., Trans.). New York, NY: Vintage.

Jung, C. G. (2009). The Red Book: Liber Novus (S. Shamdasani, M. Kyburz, J. Peck, & S. Shamdasani Trans. & Eds.). London: W. W. Norton.

Neihardt, J. G. (1972). *Black Elk Speaks*. Pocket Books.

Nietzsche, F. (1909). Introduction. In *Thus Spake Zarathustra: A Book for All and None* (T. Common, Trans.). London: T.N. Foulis.

Polt, R. (2006). The Emergency of Being: On Heidegger's Contributions to Philosophy. Ithaca, NY: Cornell University Press.

Transcribed from the MCA exhibit. (n.d.). In *Youtube*. Retrieved from https://www.youtube.com/watch?v=316AzLYfAzw

Woodcock, J. C. (2015). The Peril In Thinking and Other Essays. Scotts Valley, CA: Createspace Publishing Company.

Chapter 2
The Belief Model of Sentience:
Cognitive Dynamics of Mediated Conversations With God

Stephen Brock Schafer
Pacific Rim Enterprises, USA

Brock Shafer
Pacific Rim Enterprises, USA

ABSTRACT

The media model called "belief" serves as the language by which humans converse with God in the scripture of redemption. Hypothetically, redemption is based on the same cognitive dynamic as healing with dream analysis established by Carl Jung. These dynamics promote coherence within a sentient-cognitive psyche (a combination of conscious/unconscious scales of sentience), universal reality based on unity/energism, and narrative/dramatic structure. Based upon these dynamics, this chapter establishes a correspondence between the cognitive language of the mediated dreamscape and the cognitive language of belief that allows for conversation between God and humanity. The key to understanding this cognitive conversation is the belief model of sentience. Conflicting belief systems are endemic to human experience. Differences in the languages of belief contribute to the most psychotic-hysterical misunderstandings and their resulting mayhem. Due to illiteracy relative to the emergent language of cognitive mediation, the portents for planetary survival are dire.

INTRODUCTION

Increasingly, modern scientific research is verifying mystical principles. This chapter aspires to remove some of the prejudice that exists between those who use the language of mysticism and those who use the language of so-called modern science[1] and to anticipate the dynamic correlations between technological mediation and The ***Belief Model of Sentience.***

For the present purpose and due to limits of space, the dynamics of coherent cosmic sentience must be stipulated. Arguments relative to this stipulation may be found in writings from many fields. Correla-

DOI: 10.4018/978-1-5225-9065-1.ch002

tions in philosophy, myth, mathematics, physics, chemistry, and ethics support our hypothesis. (Schafer, Stephen Brock, 2018; Schafer, Stephen Brock, Ed., 2016a; Schafer Stephen Brock, 2016 b & c) This chapter emphasizes the dramatic structural dynamics of cognitive sentience and the importance of belief in mediated conversations with God.

To those constrained by prejudice against any language that seems to suggest divine providence, the above language will seem mystical—at best. In order to allay prejudice, this chapter will *touch upon* the vast array of current science that is in process of verifying mystical principles, but the emphasis of the chapter is placed on applying the dramatic common denominator (narrative architecture) that exists as the syntax for the *cognitive-mediated language of belief.* Most important, the chapter is not intended as simply academic. It recommends a strategy for both analyzing the media dreamscape for unconscious meaning and for harmonizing the conversation between humanity and God with the coherent syntax of the language of belief.

As a model for correlating the mystical dynamic called "belief" with quantum-psychological sentience, *Global Workspace Theory* (Baars, Bernard J., 1997) is germane. The conversation between conscious and unconscious cognitive dimensions may be understood as cognitive interaction between the "lighted dramatic stage" of consciousness and the vast personal and collective "audience/backstage" of the cognitive unconscious.

This conversation comports with the dynamics of Jungian Psyche, so the conscious/unconscious dynamics of belief can be analysed and modelled according to "coherent" programming and the use of Jungian principles as they apply to media-dreams. So-called "media-dreams" have the same structure and dynamics as Jungian dreams (Schafer, Stephen Brock, 2016b) and these dynamics have been verified with subsequent research in cognitive dynamics; i.e., "We live our narratives," (Lakoff, George, 2008, p 33) and these narratives exist by degrees of sentience on mediated scales that are called dreams, memories, fantasies, and waking dreams (real life). Also, biological patterns of belief (Lipton, Bruce, 2016) have been identified as conversational-sentient dynamics at the molecular scale.

Because divergences in belief systems are the source of the current global crisis, it is critical to understand their dynamics for application of problem-solving in mediated dimensions. The salient feature of "mediation" is that it amplifies contextual conversations. Unfortunately, most humans are essentially ignorant and do not speak the language of digital mediation. The best current indicator of this assumption is the inability to deal with political hacking, of advertising, and of emergent propagandistic neuro-marketing.

Though the mystical languages of Hinduism, Buddhism, Gnosticism, Islam, Judaism, and Christianity exhibit many similarities, this chapter emphasizes mystical terms used in the Holy Bible. The objective is to suggest that employing the common denominator, differences in mystical and scientific languages can be explicated without altering their contextual significance. Whereas modern scientists have employed the scientific languages of mathematics, physics, chemistry, biology, and psychology to contemplate the dynamics of any "conversation with God", our hypothesis is that the mystical language used in the Biblical conversation divulges the same "science-based" truths within a somewhat different syntax of belief. Increasingly, just as the relatively new idea of a quantum field theory (as compared to quantum mechanics) is taken as "given" by an increasing number of scientists, the idea of sentience in this quantum field must become understood as a "given".

Thomas Kuhn's term, paradigm shift, referring to a fundamental change in the underlying assumptions of science has been used for decades. Much superficial attention has been given to the concept of a "paradigm shift", and the term has been casually employed to, somehow, explain and anticipate precipitate changes currently being experienced in every field of human agency. The "age of communication" is become the "age of cosmic sentient Life" but with far too little appreciation of what the change portends relative to civilization, sanity, and morality.

An adjusted understanding of this vital-sentient cosmos is the human mandate—its top priority. To recognize it as the priority requires a radical expansion of consciousness—one aspect of which is the re-affirmation of Divine intervention or influence. In the new reality where gut bacteria has been scientifically proven to influence human choices, where—at unconscious levels—media images are perceived as real (Reeves & Nass, 1996), and where memories, dreams, fantasies, and perceived reality have the same quantum-cognitive structure, (Schafer, Stephen Brock, 2018; Schafer, Stephen Brock, 2016a, b, & c) the pragmatism of being able to understand the Divine Conversation is not difficult to appreciate.

The scope and complexity of this Divine influence is actually imponderable, but it has always been the human *lysis*[2] (resolution of the dramatic problem) to strive toward an understanding (evolution) of the imponderable. Vast philosophical, theological, and scientific study and commentary has been dedicated to the ontology and epistemology of cosmic sentience, and one of the most imponderable factors in this study is the study of "belief" as a sentient media model. Why is belief so important?

Understanding the purpose and processes of belief as a sentient media model could lead to the ***redemption*** of human agency. In modern terms, a hierarchy of sentience might be understood as a hierarchy of frequencies or bandwidths, and the theological concept of "redemption" can be understood in terms of achieving degrees of coherence—for the purpose of accurate reception—within and between bandwidths.

In diverse ways, philosophical theology prioritizes the value of the WORD. For example, the Judeo-Christian Bible asserts that, "In the beginning was the word." Islam avers that the language of the Koran is God, itself. The *Deliberate Tradition* of Confucius is based upon the clear enunciation in language of social relationships. Among the Buddha's Eightfold Path, are right speech, right thought, mindfulness, concentration, intent, conduct, effort, and livelihood that are all based on codification with speech which has been memorized by students for thousands of years in order to maintain social coherence.

The authors are not interested in academic evaluations, historical comparisons, or value judgements. The creative primacy of the "word" should be obvious because it is with words that humans create their reality and civilization itself which would not be possible without oral or written codes of law—both moral and secular. The dynamic principle in the dream analysis developed by Carl Jung is "living language"—the language of symbolism in dreams—and Jung distinguished between living language, dead language, and signs. The most recent cognitive-linguistic research amplifies the Jungian truism with the advantages of modern research techniques and imaging technology.

John Woodcock has proposed a way to better appreciate living language as an evolutionary "inceptive" dynamic for creating the future—a "posthuman" alien world or any *future-world.* In other words, Woodcock defines the dynamic for evolution, "a perspective that addresses the *essential* meaning of the posthuman in terms of revelation and prophecy., i.e. the self-presentation of what the author calls living language "speaking" through the human mouthpiece. This language is inceptive, creative, and calls forth, in its very 'speaking', a new world with its new appearances, while at the same time bringing our formerly stable world to its final, catastrophic end." (Woodcock, in publication) This thought brings us back to the synthetic concept of "belief redeemed with the mediated power of the word".

A salient concept in scriptures is ***"redemption"***—the requirement to compensate for faults or to regain possession (of something) for payment. Such human "agency" requisite to redemption is often misconstrued for the reason that the understanding of agency—and reality itself—is restricted to material dimensions which are stipulated illusions. Subjective agency has been widely ignored in our materialistic age, but in the incipient mediated reality that is presently nascent, it becomes critical to understand the dynamics of subjective psychological influence.

In-spite of its limitations, the term ***"bandwidth"*** is useful. The term bandwidth is part of a modern technological language that may be applied to a better understanding of obscure messages in scripture. Bandwidth is based on an interactive continuum of frequencies consistent with quantum field theory (QET)—a concept that also seems consistent with the universal message of scripture that avows ontological unity (Monotheism or Logos) within extended cosmic reality. In this context, the concept of human "redemption" might be understood metaphorically as cognitive resolution into white or into a coherent (harmonious) state (bandwidth).

Science understands the concept of light as a synthesis of all colors, but this understanding is sometimes difficult to make real. A fundamental exercise in meditation is to visualize white light. This skill takes time to achieve, but when accomplished, the practice of intentional visualization results in the bio-molecular release—and, therefore, realization—of light in the head of the meditator. If this statement appears to be unfounded, the reader is advised to research different systems of meditation in which the principle is affirmed. This said, it is necessary to use metaphor to show the relationships between concepts and categories in the subject area of metaphysics—ancient or modern. Human contemplation of super-human or super-rational "absolute" states of being like "God" is nearly impossible because human understanding of the "absolute" or unity (white) is limited by the dual structure of human brains which perceive illusions (such as the color spectrum) in a dual reality as "real". There should be no doubt that the bandwidth metaphor of the Media-dream model can be useful to science.

This chapter presumes to explicate some perceived correlations between metaphysical Biblical scripture and metaphysical modern science. As stated above, the purpose of the chapter is not merely academic. The purpose is to better understand the conversation with God—to emphasize salient scriptural wisdom with access to the language of science in-order to enhance current problem-solving models that are based in a unified field model of universal sentience.

According to the primary theme of the anthology in which our chapter appears, contextual sentience exists on many scales ranging from quantum to super-human (planets, systems, and galaxies)—an obviously theoretical perspective that has already been applied to a variety of ontological models such as the science of chakras, alchemy, astrology, and Pantheism.

This is precisely the point. Modern science need not reinvent the wheel. Many of the spokes have already been established and they can be applied to media models in a sentient dualistic reality where mediation in a vast array of languages is intrinsic—innate to the conversations that take place between distinct elements such as proton/electron, nucleus/membrane (molecular/chemical), cells/tissues, organs/systems, archetypes/archetypal representations, software/data banks, and algorithm/application. These conversations entail many subordinate systems—all of which may be understood as media models with correspondences to linguistic theory, spoken, and written languages. What moderns call languages have subordinate components such as phonology, grammar, syntax, jargon, idiom, and metaphor. A language like English is a media model, but the languages of mathematics, qualia, chemistry, light, and astronomy are also media models with many subordinate components.

BACKGROUND

Cognitive Dynamics of Mediated Conversations With God

Though the concept of quantum sentience has its foundations in theological science-philosophy, deeper issues of quantum sentience are just beginning to be researched in modern science, so a brief indicator of this complex trend is relevant to the chapter's argument. Notwithstanding earlier research by Dean Radin, in a recently published paper, the authors argue the case for "precognition" (a cognitive-sentient factor) at a quantum scale. (Mossbridge, Julia A. & Radin, Dean, 2018) The article is chosen here as an illustration because Jung explained the existence of precognition as it appears in "prospective dreams".[3] Therefore, a direct correlation exists between the research of Mossbridge, Radin, and Jung. In turn, the concept of prospection is directly related to the nature of "timeless" unified quantum fields—as presently understood:

Prospection, the act of attempting to foresee one's future, is generally assumed to be based on conscious and unconscious inferences from past experience and anticipation of future possibilities. Most scientists consider the idea that prospection may also involve influences from the future to be flatly impossible due to violation of common sense [prejudice] or constraint based on one or more physical laws. We present several cases of empirical evidence challenging this common assumption. If this line of evidence can be successfully and independently replicated using preregistered designs and analysis, then the consequences for the interpretation of experimental results from any empirical domain would be profound. (Mossbridge, & Radin, 2018, abstract)

A great deal of criticism of this view could be expected from the perspective of "common assumption", and the authors' responses to that criticism are as follows:

On more general grounds, any means by which information could be transferred from a future event to a past event would violate the Second Law of Thermodynamics." (Houran et al., 2018, p. 102) In fact, that statement is only true for a closed system (Spa-kovszky, 2017). Here we are dealing with living organisms, that is, open systems.

The second law does not apply to open systems [that include] humans, birds, and worms, the three living systems...discussed in our precognition review. Further, recent empirical results reveal a reversal of the arrow of time as a result of using a closed system consisting of particles with quantum correlations, which trade their correlations for a decrease in entropy, suggesting that the second law must be modified to take information correlations into account (Micadei et al.,2017).

The strength of today's scientific worldview rests upon theories that accurately explain observations, which, in turn, rest upon a host of experimental methodologies. Those very same methodologies were used in the experiments under discussion, and they reveal what appear to be precognitive effects. Thus, it is invalid to argue that precognition is implausible, especially on methodological grounds, because doing so would necessarily have to raise red flags about the very foundations of the scientific worldview.

Thus, we suspect that plausibility arguments against precognition are really based on a commonsense or everyday view of reality [rather than] the worldview [that is] actually revealed by science which abounds with counterintuitive discoveries. (Mossbridge, Julia & Radin, 2018)

If modern science were truly unbiased, such arguments as the above would disarm critics. However—because most current cognitive research is counterintuitive—scientists have trouble adjusting their perspective to the facts. Because it is only recently that the very existence of the cognitive unconscious has been accepted by science, hard-science-based facts related to its dynamics are still viewed as mystical. It was not long ago that the psychology of behaviorism flatly denounced any concept of the unconscious. Though this Luddite principle of behaviorism is now completely discredited, it still exerts unwarranted influence on current hard science.

A serious disconnect exists between extensive study of artificial intelligence, autonomous robotics, and neuromarketing that are largely based on neurobiology and the robust research on cognitive neurobiology of the unconscious—the cognitive dynamics of which have been extensively researched in recent decades. Importantly, the research of neuro-linguist, George Lakoff, has proven to corroborate the research of Carl Jung. Lakoff observes that 98% of human function is based in dimensions of the cognitive unconscious (Lakoff, 2008, p. 9) which authors in this anthology discuss in terms of sentience. Jung says, "Nothing influences our conduct ***less*** than do intellectual ideas [consciousness]...for such ideas [only] ***represent*** **[unconscious]** ***forces*** that are beyond logical justification and moral sanction." (Jacoby, 1973, p. 9)

Contrary to what Lakoff calls the "rational enlightenment metaphor" which is based on an assumption of human rationality, he cites estimates that 98% of human mental function resides in the "cognitive unconscious"—that our embodied (neurobiological) minds reside outside the level of consciousness and both our moral and political thought is patterned as narrative-metaphorical frames that determine how we function in the physical and social worlds. (Jacoby, 2008, p. 11) In the language of the brain—its very building blocks—words and thoughts are defined relative to narrative frames and conceptual metaphors that may be understood to comprise the fundamental syntax of belief.

The very fact that we recognize these cultural narratives and frames means that [the syntax of belief is] instantiated physically in our brains. We are not born with them, but we start growing them soon, and as we acquire the deep narratives, our synapses change and become fixed. A large-number of deep narratives can be activated together. We cannot understand other people without such cultural narratives. But more important, we cannot understand ourselves—who we are, who we have been, and where we want to go—without recognizing and seeing how we fit into cultural narratives. (Jacoby, 2008, pp. 33-34)

Here we have examples of why the language of belief is important. The premises for some stereotypical narratives are: Rags-to-Riches, Pull-Yourself-up-by Your Bootstraps, All Men Are Equal, The Ends Justify the Means, White Supremacy, The American Dream, Culpability of Big Government, False News, American Redemption (hero fails, then succeeds), Third World Inferiority, Christian Superiority, The-Gold-Digger, The Woman's Lot, Death of a Child, Born Again (The Cycle of Death & Rebirth). Using the dynamics of Jungian dream analysis to understand the cognitive dynamics of beliefs and prejudices among a culture's people can be applied to mediated biofeedback in order to "heal" the collective unconscious in non-manipulative ways—non-manipulative because individual choices rule the trajectory of mediated biofeedback.

Relative to semantic "scripts," Lakoff cites Roger Shank and Robert Abelson (Jacoby, 2008, p. 249) in his paraphrase, "Complex narratives—the kinds we find in everyone's life story, as well as in fairy tales, novels, and dramas—are made up of smaller narratives with very simple structures." Chapter 18 of Political Mind goes into detail regarding cognitive framing of the nervous system including a significant criticism of Chomsky's "New Enlightenment" approach).[4] Lakoff says, "The neural circuitry needed to create frame structures is relatively simple [neural binding circuitry, neural signatures, event structures] and so frames tend to structure a huge amount of our thought—and belief.

Sounding remarkably like Jung who observed that dreams have dramatic action that can meaningfully be broken down into the elements of a Greek play (Jacoby, 1973, p. 83), Lakoff explains that each frame has roles (like a cast of characters), relations between the roles, and scenarios carried out by those playing the roles." (Lakoff, 2008, p. 22; Goffman, 1974) "There is a protagonist, the person whose point of view is being taken. The events are good and bad things that happen. And there are appropriate emotions that fit certain kinds of events in the scenarios." (Lakoff, 2008, p. 23) Citing Goffman, Dr. Lakoff also observes that words (and beliefs) are all defined relative to conceptual frames: "Groups of related words, called 'semantic fields,' are defined with respect to the same frame. Thus words like 'cost,' sell,' 'goods,' price,' buy,' and so on are defined with respect to a single frame," (Lakoff, 2008, p 22; Goffman, 1974) and the roles of Buyer, Seller, Goods, and Money, form a narrative field context for the frame. Moreover, in *Moral Politics* (2008), Lakoff explains in detail how "much of moral reasoning is metaphorical reasoning."

As to how such metaphors were deduced, Lakoff says: "The metaphors have been arrived at independently by looking at two kinds of evidence: (1) how the language of a non-moral domain (e.g., the financial domain) is used to talk about the moral domain, and (2) how the forms of reason used in non-moral domains [e.g., economics, engineering][5] are used to reason about the moral domain." (Lakoff, 2002, p. 63)

An example of how narrative-metaphorical resonance can influence morality is discussed by Lakoff in terms of the core metaphor of moral accounting. According to this fundamental metaphor, a syllogism is established: well-being is good; well-being is moral; wealth contributes to well-being; to be wealthy is moral. According to this *Moral Accounting metaphor*, economic words like *owe, debt,* and *pay* are used in a moral context. (Lakoff, 2008, p. 63) The result is a logical framework in which the metaphorical resonance suggests that wealth is better than poverty (good), so being wealthy is moral (good) whereas being poor is akin to evil. Lakoff explains: "A conceptual metaphor is a correspondence between [belief] concepts across conceptual domains, allowing forms of reasoning and words from one domain (in this case, the economic domain) to be used in the other (in this case, the moral domain)...Thousands of such metaphors contribute to our everyday modes of thought [belief], (and) play an enormous role in characterizing our worldviews." (Lakoff, 2002, p. 63)

Reminiscent of Jungian amplification, Lakoff describes *radial conceptual categories* as the most common form of metaphorical structure. "The radial categories show how the coherent ideologies in each category fit together and what the relationships among them are." (Lakoff, 2002, p. 14) Radial categories are structured according to variations on a central model. As an example, he uses the category *mother* and explains that the term is characterized by four metaphorical sub-models: The birth model, the genetic model, the nurturance model, and the marriage model. (Lakoff, 2002, p. 8) All of the frames in the radial categories are couched in little metaphorical stories within stories. Presumably, if "mother" is the prototype and sub-model variants of the "mother metaphor" constitute radial categories or variations on the central model, discovering how the categories fit together and how they are related in a contextual field would result in discovery of meaning as explained by Jungian dream analysis.

As understood by Carl Jung, the mother archetype has infinite application. First in importance is the category of the personal mother and grandmother, stepmother and mother-in-law—any woman with whom a relationship exists. A second category is figurative. To this category belongs the goddess, and especially the Mother of God, the Virgin, and Sophia. Mythology offers many variations of the mother archetype, as for instance the mother who reappears as the maiden in the myth of Demeter and Kore; or the mother who is also the beloved, as in the Cybele-Attis myth. Figuratively, birth from the paradise of the mother generates the compulsion of a patient to discover meaning in order to "recover what has been lost".

Relative to the language of belief systems, Lakoff says: "Words evoke whole frames—whole mental structures. Those mental structures activate an embodied mental simulation, giving the words meaning." (Lakoff, 2008, p. 240) According to Lakoff's reasoning, progressive thinking is in harmony (coherence) with the greater symphony of American cultural ideologies. This is very important because both Jung and Lakoff define what is "good" and what sorts of choices lead to coherence. Differences between what is good and what is not good becomes relatively specific—as they do in scripture. Ultimately, the difference between good and evil cannot be judged by humans because the scope of their understanding does not extend to issues of the absolute where coherent balance exists between the two. However, within the sphere of patterned sentience, distinguishing between good and evil by virtue of belief becomes more tractable. Why? Because all the great teachers agree on what is good. Social (civilized) good is defined in terms of the Golden Rule; however, the principle enunciated in the golden rule must be extended to all sentient Life.

For example, put yourself in the position of an animal. Would you want to be treated as an object, confined to a feedlot, murdered, and eaten? If in the eyes of God, you are a devout Muslim with essentially the same values as a good Christian, would you want to be despised on prejudiced grounds? If you were a human fetus, would you want to be aborted? If you were a woman, would you want to be objectified, used, and de-valued? If you were a man, would you want to be slaughtered as an enemy of war? The measure of what is good is more easily understood than most people believe, and—in most cases—what is bad is obvious. Jailing children for profit, torture, the slavery of debt, the destruction of beauty and the natural environment for purposes of profiteering, genocide, enforced starvation of multitudes, and psychotic hubris are evil. The evil consequences of such dereliction of human responsibility are obvious. They point to the need of human ***redemption.***

The syntax of unconscious belief is known, and it can be used to heal. Metaphorically amplified Jungian symbols can be understood as "whole frames" that divulge meaning through access to the structure of archetypal belief and meaning—the contextual differences between good and evil. What is known about "contextuality"—the application of curative strategy to special cases—is vast in its scope. Only a few suggestions of this scope are possible in this limited context, so each of the following references must be contemplated in terms of Jungian ***"amplification"*** of the idea.

What in the brain allows a general narrative form, say a Rags-to-Riches narrative to be applied in a special case? ...The answer is 'neural binding.' Neural signals go from downstream to upstream and back. Neural pathways from downstream regions 'converge' on their way upstream at what are called 'convergence zones.' Information from downstream and upstream is 'integrated' at the convergence zones via neural binding. There are theories of how binding works, but we do not know for sure. The

FIGURE 1: Carl Jung's Model for metaphorical dream motifs (Jacoby, 1973, p.85)

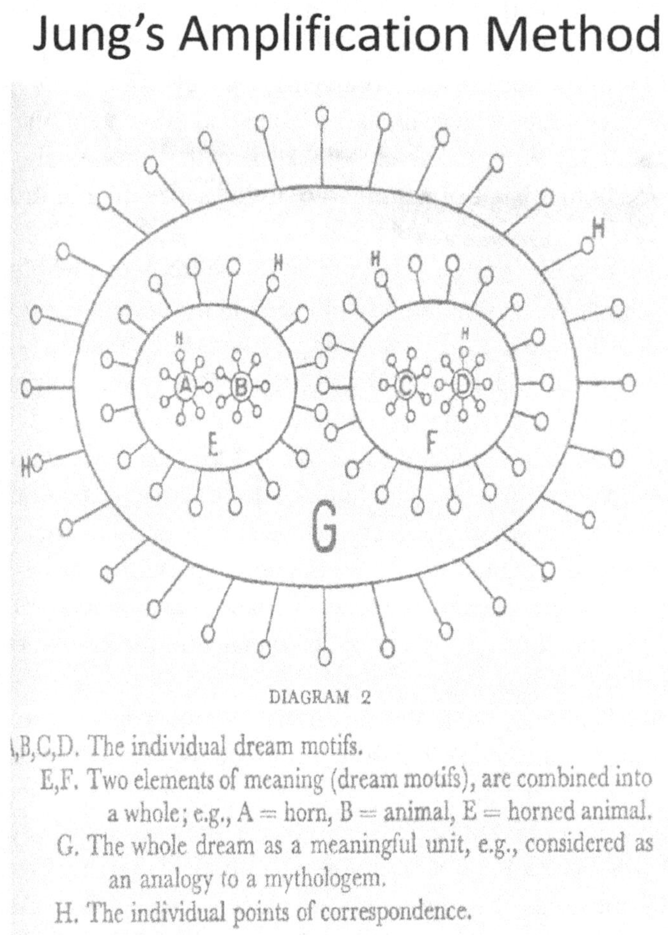

most prominent theory is that binding is 'timelocking' –neurons firing simultaneously in different parts of the brain [QFT and holographic synchronicity] along connecting pathways. Another current theory is based on the coordination of so-called neural signatures—small collections of individual neurons together forming distinct firing patterns [self-fields?]. ***Neural binding circuitry consists just of neurons and connections but has a special effect: it creates new experiences….*** *We understand events in the world in [the limited] terms of what our bodies can do. As a result, every simple narrative has such an 'event structure.' In the field of computational neural modeling, the technical term is an 'executing schema,' or 'X-schema' for short….Neural binding is also crucial to the time structure [pre-cognition] of a narrative. (Lakoff, pp. 22- 27)*

The dramatic model of sentience always includes the following fundamentals:

- Premise: Evolution for survival in order to evolve.
- Exposition: Statement of the dramatic problem faced by sentient players which varies according to contextual scale
- Lysis: Resolution of the dramatic exposition

Lakoff says, "We live our narratives." According to the assumption of shared dramatic architecture, the dramatic structure at every scale of sentience can be correlated. On the human scale, "The lived story is at the center of modern personality theory. The theory of neural computation…shows how our brains not only permit this but favor it. The typical roles played in narratives include Hero, Victim, and Helper…The [contextual] roles in narratives that you understand yourself as fitting give meaning to your life, including the emotional color that is inherent in narrative structures." (Lakoff, p. 33) So, the language of belief is based on dramatic structure, but structural analysis may be conducted at atomic/molecular scales of neurobiology as-well-as at scales of human ethics. "Dramatic event structures are carried out by brain circuitry. The same event structure circuitry can be used to live out an action or a narrative, or to understand the actions of others [in the shared context of story structure]. (Lakoff, p. 27)

An important aspect for understanding the language of belief is that, "Neural binding can create emotional experiences. In the area of the limbic system, the oldest part of the brain in terms of evolution, there are two emotional pathways with different neurotransmitters: one for positive emotions (happiness, satisfaction)—the dopamine circuit—and one for negative emotions (fear, anxiety, and anger)—the norepinephrine circuit. There are pathways in the brain linking these emotional pathways to the forebrain, where dramatic structure circuitry seems most likely to be located." This is why beliefs are emphatically complex and emotional.

Activations of such convergent pathways are called 'somatic markers.' It is they that neutrally bind the emotions (downstream, near the brainstem) to event sequences in a narrative (upstream, apparently in the prefrontal cortex, at the front and top of the brain). The somatic markers allow the right emotions to go where they should [be] in a story. They are the binding circuits responsible for the emotional content of everyday experiences. …emotional content can be bound to a narrative, yielding a melodrama—a narrative with heightened emotional content. (Lakoff, p. 28)

Dramatic frames are linked in very sophisticated but "ponderable" ways, understanding of which can contribute to the harmonization of belief systems without prejudice. One reason for this is that:

The same part of the brain we use in seeing, is also used in imagining we are seeing, in remembering seeing, in dreaming that we are seeing, and in understanding language about seeing. The same is true of moving. The same parts of the brain used in really moving are used in imagining that we are moving, remembering moving, dreaming about moving, and in understanding language about moving. Mental "simulation" is the technical term for using brain areas for moving or perceiving, imagining, remembering, dreaming, or understanding language. It is mental simulation that links imaginative stories to lived narratives. (Lakoff, 2008 p. 39)

THE BELIEF MODEL OF SENTIENCE

The extended "background" section above is essential for even a rudimentary scientific understanding of this chapter's hypothesis and keywords, and some detail regarding the similarities in the research of Jung and Lakoff is critical. (Schafer, 2018 a, b, & c) Within the "media-dream", the scriptural language of "God", "belief" and "redemption" may be correlated with scientific terms such as "quantum unified field", "archetypes of the unconscious in Hilbert space", and "coherence for quantum measurement of choices that tend toward the good".

Within the framework of generally accepted scientific language, perhaps scriptural truths may be contemplated as scientific in origin. Within the mathematical framework of quantum mechanics where infinity is not calculable (Brooks, 2016)[6] frequency becomes the metaphor with which to best understand overall universal sentience according to quantum field theory. The author is not embarrassed by the inevitable detailed criticisms of this statement from experts in discreet fields of research because the purpose of the chapter is to generate debate. If it can be stipulated that all "language" is essentially figurative language; i.e., even mathematics is figurative in that numbers and symbols "represent underlying concepts". The "word" $E = mc^2$ represents a complex concept just as the word *"truth"* represents a complex concept. Neither word constitutes the concept itself. A picture of a pipe is not the pipe itself, and as Alfred Korzybski pointed out, "the map is not the territory"; i.e., the mathematical-linguistic reality is not the same as the quantum-cognitive reality. However, these languages now appear to be intrinsically and structurally related. Though it is not the reality, itself—due to corresponding syntactic structure—a map serves a practical purpose whether it is a roadmap, a star map, or a genetic map. Due to advanced modern research, science has reduced the gap between cognitive frames of belief and the archetypal sources of belief.

It is in this sense that we can discuss sentience in terms of media models that have many intrinsic and corresponding structural characteristics including dramatic narrative: Notwithstanding the need for dramatic personae in any story, other fundamental structural elements of dramatic architecture have been defined by Aristotle: ***Premise*** (Schafer, 2007; Glassner, 2004) or **purpose and meaning of a story**, ***exposition*** **(statement of the problem),** and ***lysis*** **(resolution of the dramatic problem).**

One example of applying dramatic frameworks at a particular scale (molecular) of sentience is the dynamic at the scale of *Procaryotes*. Prokaryotes which include bacteria and other microbes (a primitive form of biological life) are intelligent and carry out their lives according to the structure of drama. "A bacterium carries out the basic physiologic processes of life like more complicated cells and makes something like choices (either, or) in the process. A bacterium eats, digests, breathes, excretes waste matter and even exhibits "neurological' processing". (Lipton, 2016, p. 76) In this Procaryotic life, drama must be contemplated as a process in which the *premise* of the drama is a noun (self-awareness), the *exposition* of the narrative is the verb (surviving-as-self), and the *lysis* of the drama is about *redemption* (learning the lessons of survival of the evolving self).

This dramatic structure can be correlated with the same pattern on other sentient scales wherein the same dramatic architecture prevails. The dramatic structure of atomic life is more obscure to common sense, but laws of chemical bonding in order to create new life (new chemical arrangements) may be understood in terms of ***premise*****,** ***exposition*****, and** ***lysis***. In order to understand the laws of sentience on this scale, experiments (based on a mathematical scale of electromagnetic choice) are articulated as the constant fluctuations among neutrons, electrons, and other sub-atomic ***dramatis personae.*** This drama results in some re-combination of force fields to achieve "redemption" by creating new chemical combinations.

So—back to the human scale of sentience—rhetorical patterns can be understood to define our human reality, and that reality may be understood—if only vaguely—as cosmic-sentient in extent. Indeed, relative to cosmic civilization itself, we are told that "In the beginning was the word," and the word can be understood in its expansion to include all dramatic architecture as the blueprint of the Grand Architect. In the Essene tradition, ***cosmic law, itself, is God***. Alchemy, which was the precursor to the science of chemistry had a sophisticated perspective on the word, *chemistry,* and its model of atomic molecular reality could be understood as corresponding with semantic fields, semantic frames, and dramatic structure which has its own sub-structure. The basic alchemical syntax of Sulphur, mercury, and salt correspond to metaphorical fire, air, and water and also to the figurative elements of the universal trinity—Father, Offspring, and Mother.

In the ancient sciences, the key to achievement of wisdom was the ***Law of Correspondences***. Furthermore, the dramatic (rhetorical) format employed universally in myth to help people understand the nature of "reality" must have been intentional. No matter whether the dramas of the Gods be defined in the stories of the Gnostics, Egyptians, Greeks, Romans, Norse, American Indians, Hindus, or Buddhists, their *exposition* can be paraphrased as the question, "What is the purpose *(exposition)* of the Divine Plan?" Dramatic *lysis—the answer to the question*--can be paraphrased as the achievement of "harmony" or "human understanding of the process called creative evolution and how to achieve redemption".

Based on robust research and for the purposes of this chapter, it may be stipulated that fundamental cognitive structure is rhetorical. Rhetorical "narrative" or drama defines the model of conscious/unconscious (sentient) cognitive structure that includes virtually every sentient structure including the structure of symbolic dream language, memory, fantasy, and the economy.

As detailed in the BACKGROUND section, the psychic principles defined by Carl Jung are being reiterated in modern cognitive research. The similarities are extensive but have mostly to do with psychic unity and *energism* (Jacoby, p. 2), dramatic structure, and the significant resonances of symbolic language. Though it will require time and research—based on existing scientific revelations in QFT—we hypothesize that scriptures can be understood in terms of quantum field "syntax" that is entirely figurative in nature and structured as dramatic metaphor. Like Jung's prospective dreams, scripture is prescient and many of its allusions point to a quantum field reality that is precognitive, or—more accurately--timeless.

In scripture, cognitive-spiritual "reality" is expressed in dramatic terms that may be understood as archetypal-fractal cognitive patterns (semantic frames). The profusion of fundamental archetypal metaphor used in scripture is obvious. Family metaphors, physical body metaphors, and "church" metaphors abound, and their contextual intermingling and resonance becomes infinite—just like a quantum field. As Lakoff observes, "They are stories within stories". Simply transposing the human body metaphor into the arena of modern medical science will begin to clarify the purposes of some of the ancient systems of augury; i.e. if augury could anticipate death, so can multiple tests and clinical research augur heart failure (precognitive viewing).

The evolution of so-called modern sophistication should not invalidate the principle of augury. Instead, ancient practices should be reviewed to better understand the truths they represent and to better apply them to modern healing practices. "For as in one body we have many members, and the members do not all have the same function, so we, though many, are one body in Christ, and individually members one of another." (Rom. 12:4-5) "We who are many are one body" (1 Cor. 10:17). "Just as the body is one and has many members, and all the members of the body, though many, are one body, so it is with Christ" (1 Cor. 12:12). "You are the body of Christ and individually members of it" (1 Cor. 12:27). In all these references, we see that the unified field of Christ's body is constructed of fractal individual members that may have different functions.

In scripture—because of its metaphorical nature—perspective is constantly changing. For example, we see: "Equip the saints for the work of ministry, for building up the body of Christ [redemption]" (Eph. 4:12). The "ministry" may also be understood as the process of redemption because it means becoming Christlike. "Christ is the head of the church, his body, and is himself its Savior" (Eph. 5:23). The church and its ministers represent the physical body of Christ. In esoteric terms, the fractal correlation between God and Humanity was understood in terms of a hierarchy of similar and correlated functions (flesh, emotion, mind, and spirit), so the body of Christ refers mostly to the flesh (which is familiar to people)—the limited outer form of Christ. However, the implication is that all functions are integrated into one—which is why "man can never step into the same river twice". The river (human quantum-molecular being) is constantly changing according to field dynamics. Moreover, today's understanding of the physical body is that it is not just flesh. Rather, it is comprised of quantum elements of all the four functions that are metaphorically expressed as earth, water, air, and fire. Fleshly molecules consist of water, fire (electricity), and sentience. This view is fundamentally a quantum perspective that accounts for why coherence can harmonize all four functions. "We are [all] members of [Christ's] body" (Eph. 5:30).

"In my flesh I am filling up what is lacking in Christ's afflictions for the sake of his body, that is, the church" (Col. 1:24). Because it is a mixed metaphor, this statement seems awkward and it is somewhat more difficult to explicate. At least in part, it may mean that the individual incarnate helps to redeem whatever *man-made* afflictions Christ might suffer. However, related to the idea that Christ suffers (which is clear in the drama of the crucifixion, but—in quantum terms—is timeless until redeemed by human effort) is the important redemptive idea that incarnate (in the flesh) humans are the cause of Christ's suffering (the Pharisees and the people made the incoherent choice to crucify) and so they must redeem themselves by achieving coherent atonement. On a quantum scale, the kaleidoscopic dance of metaphorical interpretations is imbedded in the unconscious archetypal patterns of belief.

"I am writing these things to you so that . . . you may know how one ought to behave in the household of God, [redemption] which is the church of the living God, a pillar and buttress of the truth". (1 Tim. 3:14-15) Here—with the same elements of mixed metaphor—the church becomes the household which is also the body of Christ. The church provides a different metaphorical semantic field with which to understand the unified entirety of God and to behave (redemption) accordingly. Gradually, these many metaphors grounded in human experience and preserved in Jung's collective unconscious serve to frame the outsourced contextual beliefs of individuals and groups. These subconscious beliefs, then, structure one's foundation for self-awareness and ***right choice***.

In the world's scriptures, the family dynamic is represented as the ***Trinity*** represented in the mythical dramas of the gods. The doctrine of Christian Trinity resembles the Hindhu Trimurti where Brahman is also Brahma, Vishnu and Shiva, yet all three are ultimately Brahman. The Supernal Triad within Kabbhala's Tree of Life of Kether (Divine Consciousness), Chokmah (the Son in relation to Kether as its Father, it is the active, male principle of existence) and Binah that embodies the primal female energies. Under this trinity is the Abyss with the rest of the manifestation. We find the concept of trinity in many other beliefs systems: the **Egyptian Trinity** of a transcendental god Amun that also manifests as Ra and Ptah. Of God Ra that has a daughter Maat (Wisdom, Knowledge) that is a consort or sister of God Thoth (Divine Word or Logos of Ra). This was soon replaced with Osiris, Isis and Horus. The **Greek Trinity** consisting of Zeus (the God), Hera (the Goddess) and Hercules (their son). The **Babylonian holy Trinity** consisting of Ninki (later became Ishtar) the mother, Enki (later became Namakh) the father, and Marduk the son. Three animal icons associated with the three Sun 'gods' are: the lion for the goddess, the bull for the father, and dragon for the son. The **three sacred Vedic energy manifestations** are, Tamas (inertia, darkness, destruction), Rajas (preservation, movement, dynamic), and Satvas (creation, existence, order, purity). (Trinity in the world's religions, 2013)

These many trinitarian patterns of cosmic sentience can be translated into mathematical languages (the right triangle of Pythagoras which is based on geometrical vectors; Fourier transforms) as well as cognitive languages. In Lakoff's research, two family structures rule belief—The Strict Father metaphor (SFM) and the Nurturant Parent Metaphor (NPM). (2008 & 2002) Jung's archetypes consist of unified energy and force fields which resonate as the dynamics of the "mother", the "father" complex, "anima" and "animus" patterns which appear similar to both metaphorical resonances and to unified field dynamics. Field theory may be understood as the *Father aspect*--the macrocosmic metaphor—Fractal structure as the *Mother aspect* (dualism/microcosm), and coherence as *Offspring aspect* (the redeemed Christ, Soul, or Son). People have no problem understanding the dynamics of procreation, family, and offspring as an important belief pattern, but they have trouble understanding the same pattern as a figurative dynamic of all belief.

The difference between experienced reality and quantum reality is one of scale, but quantum scales are imponderable because—before now—people have not had the skills required to contemplate the reality of the quantum field. However, the fundamental Truth of each reality remains the same—only the language changes. QFT and QM may be verified with the languages of physics and mathematics--Fourier mathematics that is based on frequency and followed up with the mathematics of *Hilbert space* and *Higgs field*. These arguments are far too detailed to develop in the context of this chapter, but the arguments may be reviewed from a variety of perspectives (Brooks, 2010; Gleick, 1987) in fields of mathematics, physics, computer science, cognitive science, artificial intelligence, and robotics. Truly, it is their fractal nature that provides the possibility of employing neurobiological principles to autonomous robots.

Explicated from the above quantum field perspective allows new meaning to permeate the following scriptural-metaphorical equations: "The household of God [is] built on the foundation of the apostles and prophets, Christ Jesus himself being the cornerstone, in whom the whole structure, being joined together, grows into a holy temple in the Lord. In him you also are being built together into a dwelling place for God by the Spirit" (Eph. 2:19-22). For those unfamiliar with this stone mason's language, a cornerstone is important since all other stones will be set in reference to this stone, thus determining the position of the entire structure.

What the alchemists named *"the Philosopher's Stone"* (Hall, 2003, pp. 127, 129, 305, 449, 502, 508, 515) also called *Magnus Opus* suggests that a serious metaphorical correspondence exists between alchemy and more recent scripture. "'Behold, I am laying in Zion a stone, a cornerstone chosen and precious, and whoever believes in him will not be put to shame.' . . . The stone that the builders rejected has become the cornerstone". (1 Pet. 2:6-7; Isa. 28:16; Ps. 118:22) "You yourselves like living stones are being built up as a spiritual house, to be a holy priesthood, to offer spiritual sacrifices acceptable to God through Jesus Christ" (1 Pet. 2:5) "Do you not know that you [plural] are God's temple and that God's Spirit dwells in you? If anyone destroys God's temple, God will destroy him. For God's temple is holy, and you are that temple" (1 Cor. 3:16-17). "Or do you not know that your body is a temple of the Holy Spirit within you, whom you have from God? You are not your own" (1 Cor. 6:19). Mixed metaphor abounds, and the mix of resonance creates a scientific array of meanings. Freemasonry is replete with such references to the nature of the conversation between God and mankind. What could be clearer in meaning as to the nature of quantum redemption.

Throughout this chapter, the term sentience is employed as the best descriptor of the (absolute/spiritual) source of scaled degrees of awareness in a cosmic field that incorporates "intelligence" ranging from sub-atomic to superhuman scales. Within this essentially electromagnetic dreamlike reality of self-fields. "Self-fields do not have a life of their own—'You are not your own'—they are created by a source and are permanently attached to that source. Examples of self-fields are the gravitational field of the earth, the electrical field around an electron or proton, and the strong and weak fields around a nucleon." (Brooks, p. 132)

Perhaps the most accessible way to address scalar sentience within the Field is to employ the metaphor of the modern technological ***media-sphere***. The nature of the media-sphere may be understood in terms of the ***Metaverse*** which is defined as a collective virtual shared space, created by the convergence of virtually enhanced physical reality and physically persistent virtual space, including the sum of all virtual worlds, augmented reality, and the internet. In other words, everything is connected in the Field.

The technological *media-sphere* is become the human reality, but this virtual reality has many precedents in the mystical languages of theology. Like the seed metaphors of myth, the media metaphor has all the necessary components for understanding sentient media models—especially the concept called "belief" in scripture. Though the media metaphor is limited by current understanding, study of diverse "media models" within the reality of Logos allows for authentic research on the qualitative dynamics of sentience as defined in terms of frequency or bandwidth in the quantum field. If this hypothesis be approved, ancient science (astrology, astronomy, Pythagorean mathematics, alchemy) and modern science appear to communicate the same message.

In order to experiment with this theory and to generate a scientific hypothesis, the existence of an infinitely broad range of sentience may be assumed (the mathematics of which is based in quantum fields, *Hilbert space, and the Higgs Field*).[7],[8] Sentience may be defined as an interactive response to the environment, and such "interactive" response models may be understood as the language of "intelligent" behavior. Within this contextual framework, the autonomous nervous system (ANS), atomic and molecular electro-magnetic interaction, and whatever force-fields articulate "intuition" can be understood as sentient. The teachings of Maharishi Mahesh Yogi (and of Deepak Chopra, 1995) suggest that higher sentience is in the nature of self-fields--referred to as having a context of "self-reference" in which the True Self or Soul is a super-human conscious (coherent-sentient) state of being. Such a state seems to have characteristics that coincide with superconductivity[9] and the "Flow" experience"[10].

Notwithstanding the decades of research on the nature of consciousness and the recent corroboration that the source of consciousness resides in quantum neural microtubules that have the capacity for coherence (Hameroff, Stuart & Penrose, Sir Roger, 2014) or in cell membranes that function as brains (Lipton, Bruce, 2016, p. 86) sentient capacity is the potential to feel, perceive, or experience subjectively. In the modern context, sentience becomes a synergy of *perception or sensation* (Jungian perceiving functions) and *feeling and thought* (Jungian judging functions). Eighteenth-century philosophers used a similar concept to distinguish the ability to think (reason) from the ability to feel (sentience). In modern Western philosophy, sentience is the ability to experience sensations (known in philosophy of mind as "qualia").

In Eastern philosophy, sentience is a metaphysical quality shared by all "things" (Life) and Life requires respect and care. The concept is central to some important socio-political issues such as animal rights, abortion, and environmental science in general. A truth that most humans do not wish to understand is that the evolution of sentience is based on the ability to suffer (willingness to be sacrificed for the common good), and this ability confers certain rights. As a basis for discussion of this vast field of inquiry, the word sentience should replace the word "conscious". In the Quantum Field reality, it is time to re-introduce the concept of divine guidance. Atoms and molecules that exist in the various human functions (mind, emotion, sensation, and intuition) are well-known to affect the whole organism.

RECOMMENDATIONS

Redemption by Coherent Choice

In the modern context, the choices made by human individuals and collectives may be considered the key to redemption and the resolution of the current Global crisis. In a dual reality where each choice represents a consolidation of conscious and unconscious "sentient" impulses that coalesce in a moment of synergistic "agency"[11] an "initiation"[12] takes place--a transcendence from duality to unity—a baptism. Therefore, with every choice a human makes, the seeming limitation of duality becomes the dynamic for redemption.

Nevertheless, the evolutionary-educational value of the human condition—making choices within the human illusion and suffering the consequences of bad choices—is rarely appreciated, so humans fail to understand their own ***responsibility*** within the dynamic process of evolving sentience--redemption. Consequently, human limitation and erroneous belief in its own dominion has created a global-cosmic crisis for all integrated degrees of sentience including sub-human and super-human scales in exactly the same way that human choices have almost destroyed the natural environment of the planet. In today's transitional reality (transition tends to create an illusion of failure because success is defined by new and unfamiliar parameters), scriptural correspondences with the languages of science can no longer be ignored. Science can no longer be characterized as amoral, and it is scripture (from many sources) that provides the moral component which is presently missing from much science.

With some open-minded explication, mystic and scientific ontologies become very similar. The salient correspondences between the two worldviews are universal "unity" and their corollary "Hierarchical integrity"[13] in which multiple bandwidths of sentience conspire (breathe together) to achieve coherent redemption (evolution). As Jacoby points out (1973, p. 2):

Theology, psychology, history, physics, biology, and numerous other disciplines can all in equal degree serve as starting point for the investigation of being; they are interchangeable and to a certain degree they may even be transposed, each having its relevance according to the problem under discussion or the investigator's personal viewpoint. Jung…builds on his profound knowledge of psychic reality, so his edifice is no abstract theory born of speculative intellect, but a structure resting entirely upon the solid foundation of experience. Its two main pillars are: (1) The principle of psychic totality, and (2) The principle of psychic energism (Jacoby, p.2)

Just as many people ignore environmental science, they ignore the empirical value of experience. It may be stipulated that in Western culture, life-emphasis has been increasingly focused on so-called material dimensions rather than psychological dimensions. Observe how people read the sacred teachings. Observe which passages they avoid and try not to notice. In-particular, people close their eyes to everything referring to betrayal and psychic murder. They do not even wish to consider that they can harm at a distance by their thoughts and feelings. Thus, people avoid learning about that error of which they are most often guilty. One does not have to be a giant of thought in order to harm, and those who dismiss this language as ridiculous should review the meaning of prejudice. Even a mediocre thought poisoned by dis-harmonious frequencies may shred coherence.

An experienced seer—though seers are now considered quacks—will verify that poisoning by thoughts is even more harmful than narcotics, and infection by thought transmission is more contagious than bio-weapons—especially as transmissions are amplified with modern technological mediation. Notwithstanding the obvious impacts of advertising and propaganda; today, the impacts of such mediation—going viral—should be clear to everyone. In this emergent media age, it is the human mandate to research transmission and quality of thought-feeling and to *redeem* its effective resonances at a distance. Evidence of the importance of "positive thinking" (coherence) is longstanding, and the concept appears everywhere in the self-improvement literature as-well-as clinical research. (Galeon, 2016; McCraty, & Childre, 2010; Morris, 2010; Gillin et al., 2007; Tomasino, 2007; Chopra, 1995)

Cultural Problem-Solving With Scriptural Patterns of Belief

Often, the term "redemption" is misconstrued, and this has a negative influence on problem-solving. People tend to interpret scripture irresponsibly; i.e. they take the easy way out. The following biblical scripture demonstrates the point. "In Christ we are made wise, righteous, we are sanctified, and we are redeemed." (I Corinthians 1:30) Readers tend to like the idea that they are special, so they take the view of redemption in terms of salvation, absolution, or grace and try not to notice that the term may entail personal effort such as retrieval of something lost or recovery.

In other words, the required dedication to self-motivated responsibility in the face of trials is ignored. From the latter interpretation, redemption could involve rigorous and disciplined effort (selfless service to all life) on the part of the disciple. Such dedicated service tends to maintain the harmony of functional human agency, thereby achieving "acceptance" (attunement with Christ consciousness). "We have been accepted in Christ Jesus, for through his blood we are redeemed and forgiven our sins." (Ephesians 1:6,7) In this context, "acceptance" could be understood as a range (bandwidth) of benign resonance in thought-feeling and action patterns. Further, the objective of being accepted is to achieve a degree of harmony with the essence of Christ's vitality—its vital coherence. Eventually—through personal effort—the disciple achieves harmony or coherence with all sentient scales of "cosmic Life". This achievement manifests as "redemption" of all sentient bandwidths impacted by the self-field of the accepted disciple (learner). Though redemption may be based in "grace" (opportunity) it requires personal effort and responsibility.

This quantum field interpretation of the term redemption constitutes a reversal in the general understanding of the concept. Introducing the factor of personal "responsibility" in no way dilutes or demeans *Jesus* or the *Father* because it is this *Spirit Energy*—acting within un-redeemed reality—that guides life experience to a state of redemption. This is *Good News* that is exemplified in the life of Jesus—the epitome of the *"Savior"* principle—which demonstrated how humans should function in a quantum-sentient universe that is not limited to three dimensions.

Very truly I tell you, whoever believes in me will do the works I have been doing, and they will do even greater things than these, because I am going to the Father. [13] *And I will do whatever you ask in my name, so that the Father may be glorified in the Son.* [14] *You may ask me for anything in my name, and I will do it. (John 14:12-14 New International Version (NIV)*

The meaning of this message could not be more clear except by added reference to the quantum field reality. The *Savior's* mission demonstrated that—sometimes—in order to achieve "atonement with the father" it is necessary to experience the crucifixion of responsibility. "In (the coherent bandwidth of) Christ" (by witnessing the travail of Jesus) "we are made wise, sanctified, and redeemed". This was his gift—his good news—but the redemption may require crucifixion. The message seems very clear just as the semiotics in the story of Jesus are clear. Moreover, the symbolic significance does not actually end with the crucifixion. Ascension and at-one-ment with the father *("I and the Father are One.")* represent super-human states of sentience that may be understood as degrees of *redemption* that lead to rebirth or resurrection. The same process may be applied to cultural problem-solving and planetary renewal—in a very literal sense.

Alternative reinterpretations can be applied to scripture without demeaning the message or the messenger. Evolution changes the environment including the environment of language. Viewed from the language of modern media technology, "coming in the clouds" might be viewed in terms of tuning into a specific bandwidth. "We shall see the son of man coming in the clouds in all power and glory, so lift up your heads; for your redemption draws nigh." (Luke 21: 26-28) In this context, redemption may be understood in terms of the achievement of coherence in order to tune-into the bandwidth of the Christ. Indeed, such achievement is a gift, but it is also learned as an evolutionary accomplishment.

"John the Baptist ushered in the times of Jesus, saying 'Repent—the kingdom of God is at hand.'" (Matthew 3: 1,2,11) In current terms, the Kingdom of God is always at hand as the fundamental quantum archetype, and redemption may be understood as a process of maintaining coherence with the Kingdom—engaging in conversation with God in order to recover the coherence that has been lost. Metaphorically, John baptized with water, but Jesus would baptize with the Holy Ghost and with Fire. (Derivation of the term ghost is "gast" or guest, so the Holy Ghost may be understood as "intuition") This calls to memory the classic elements—earth, air, water, and fire—that symbolize the human functions of flesh, mind, emotion, and spirit (intuition).

St. Paul's conversion is a good example of how people tend to ignore unusual reality factors. If the existence of super-human bandwidths of sentience be accepted in practical terms, Paul's conversion due to super-conscious experience may be understood as literal. After his conversion, he was constantly guided, "You will be told what you must do." In this scriptural story about superhuman sentience, it is very clear that redemption and "active responsibility" are directly linked.

Much confusion exists relative to human understanding of scriptures because folks reject what should be obvious from common experience. For example, even myopic observers can surmise that their thoughts and emotions have even greater impact upon their surroundings than their physical actions. Everyone has experienced the impact of moods upon a gathering (mob mentality or euphoria), but if empirical observation is ignored, responsibility for such human agency may be temporarily avoided. However, avoidance may lead to considerable error and the pains of Karma; i.e., many humans once accepted the scientific idea that the earth is flat and the center of the universe. Misinterpretations are commonplace.

Scripture is replete with direct irrefutable guidance—a one-sided conversation that is often ignored relative to multi-dimensional human agency: The Golden Rule, "[Behave] unto others as you would have them [behave] unto you."—Christ; the Silver Rule "Don't do unto others that which you don't want done to you."—Confucius; the Ten Commandments (Don't lie, don't steal, don't covet, don't kill, don't make idols; instead, respect father and mother, keep the Sabbath, have no other God)--Moses; "Those who have much should help lift the burden of those who are less fortunate." (charity)—Mohammed. (Smith, Huston, 1958) All-of these theological-scriptural principles rest upon human life committed to what the one God desires of us. It is precisely this axiom that is presently missing in the scientific understanding of reality. However, all-of these scriptures can be easily understood from both metaphysical Monotheistic and quantum field ontologies and their related axiology and phenomenology.

It must be mentioned in passing that a secondary source of access to meaning and belief in scriptural teaching is the world's literature which constitutes an amplification of contextual opportunity for reaping the wealth hidden in scripture. The novel, Moby Dick, provides a wealth of scriptural language references and dramatizes the essential human misunderstanding of absolute (Divine) sentience. We see the following:

There is, one knows not what sweet mystery about this sea, whose gently awful stirrings seems to speak of some hidden soul beneath. (Melville, Herman & Moncur, Michael, 2003-2012, p. 111)

There's a sermon now, writ in high heaven, and the sun goes through it every year, and yet comes out of it all alive and hearty. (p. 99)

Consider all this; and then turn to this green, gentle, and most docile earth; consider them both, the sea and the land; and do you not find a strange analogy to something in yourself? (p. 58)

What is it, what nameless, inscrutable, unearthly thing is it; what cozening, hidden lord and master, and cruel, remorseless emperor commands me; that against all natural lovings and longings, I so keep pushing, and crowding, and jamming myself on all the time; recklessly making me ready to do what in my own proper, natural heart, I durst not so much as dare? Is Ahab, Ahab? Is it I, God, or who, that lifts this arm? But if the great sun move not of himself; but is as an errand-boy in heaven; nor one single star can revolve, but by some invisible power; how then can this one small heart beat; this one small brain think thoughts; unless God does that beating, does that thinking, does that living, and not I. (p. 132)

The above quotes were chosen because the speaker, Ahab, shares the cynicism and doubts of many Moderns and serves as an exemplar of perceptual error. That error is his misunderstanding of Divine sentience—and, therefore of Divine meaning and purpose. His lack of understanding is not a fault, it may be understood as a necessary learning process. Nevertheless, the dissonances in frequency resulting from a limited perspective explains why humans suffer. Suffering is self-inflicted (the process of learning from error and developing reason by experience of Karmic cause-and-effect)—lack of coherence between Divine/human scales of sentience. A temporary lack of understanding is acceptable, but blaming God amplifies error to the scale of blasphemy. Redemption is impossible without at-least some understanding and acceptance of the nature of the error. Both literally and figuratively, human error leads to crucifixion, but *crucifixion leads to redemption*—as expressed in the Christ-drama.

If this perspective is contemplated, severe discipline can be applied to the redemptive process-as-spiritual-responsibility. Shopping 'till you drop, profiteering at the expense of others, superficial dedication to integrity, lying about everything, contaminating cosmos with every breath and every movement, and blaming God for every discomfort will lead inevitably to total- destruction. However, total-destruction is not the end of the drama. We are told that our errors come back forever—until they are redeemed with manifest coherence which allows individual sentient beings to be part of the super-human bandwidth.

A vast scope of literary themes and references expand the opportunities for contextual redemption. The language of literature and scripture provides an infinite potential for understanding how to apply or not apply the beatitudes of spirit. The lessons of literature are infinite, and for this reason, the universal de-emphasis in humanities education in favor of technological education has severely restricted human opportunities to understand their responsibilities in a monotheistic-quantum reality. Writers might adopt the themes discussed in this chapter in-order to energize the process of collective redemption.

The scripture of the Hindu Bhagavad Gita--the Hindu Bible (Yogi, 1990)--is a sophisticated case-in-point. The Gita may be understood as both literature and inspired scripture. It is the story of the conversation between the contextual Indian Prince, Arjuna (the disciple), and Lord Krishna his charioteer (the Soul). The story is about how to behave in the immanent battle of Kurukshetra that symbolizes the final war between cosmic good and evil (Armageddon). The premise of the story is particularly relevant because the spiritually advanced prince—surveying the amassed armies—sees family members, heroes, revered teachers, and great men on both sides, so he is brought to a standstill—not wanting to war with family. Krishna's teaching informs Arjuna how to behave in the context of conflict within a quantum-unified reality, and—ultimately—this is how the scriptural story must be approached in order to plumb its meaning. Paraphrased, Krishna's advice to Arjuna is: "You should know better! You must act! But you must act with proper motivation because of its qualitative agency in a quantum field and because of its redemptive capacity."

Redemption According to Belief

Today, based on robust precedent and scientific research, the scalar integrity of Cosmos (A "monotheistic" spectrum of frequencies?) may be stipulated. Given the infinite complexity of cosmic integrity, the capacity to "believe in Scripture", instead of being considered mystical and ignored, should be recognized as a scientific virtue. Because "belief" is a kind of sentience which is open to that which is not yet conscious, it rises to a high-order and practical scale of sentience. Within the infinite complexity of quantum-spiritual cosmos, that which is unconscious is vast, so "belief" is not gullibility. Open mindedness is akin to belief, but belief is a far more powerful, focused, and proactive form of human agency than relatively passive open-mindedness. True belief is not fanaticism—which has an entirely different frequency and resonant quality.

Because the scope of human agency has expanded exponentially, so should the scope of the human sense of responsibility in a cosmic-sentient hierarchy. Acceptance of responsibility can only be accomplished in accordance with "spiritual" direction in the form of belief. Both agency and responsibility depend upon belief, so the practical significance of mysticism must be re-evaluated. In this integrated context, "Belief" in the existence of Divine scriptural guidance becomes an essential form of intelligence that may (unexpectedly) supersede even reason. From this new scientific perspective, scriptures can be viewed in terms of quantum dynamics that are measurable in psychological-linguistic-metaphorical terms. When this understanding becomes fact, humanity can begin to establish the virtue of accepting practical divine guidance relative to problem-solving behavior. Many theological-philosophical debates will be resolved from this perspective: the nature of "spiritual labor", the seeming paradox of free will vs. destiny", "faith vs. acts", "turning the other cheek" (remember Krishna's advice to Arjuna), "the paradox of existence of evil in a perfect cosmos", and a plethora of other theological dilemmas can be better understood from the perspective of the vectors of quantum sentience.

Re-evaluation of "mysticism" becomes the most reasonable affirmation of such Truths as "A nation's welfare begins in the heart of each person" (thinking with the heart). This concept is universally implied and disseminated fractally in the World's scriptures. From the perspective of unified field dynamics, the interactivity between the collective and its distributive components is a sociological-statistical fact that binds religion and science. A direct correlation exists between mind and heart—between the illusion of cosmic duality and the reality of cosmic unity.

The correlation has to do with a specific quality of the heart that is known as "coherence". The term coherence is variously defined in different fields of science-philosophy, but its basic meaning is "harmony", and harmony can exist in various "harmonic" scales and in various frequencies (bandwidths). Currently, heart coherence is being widely researched and this quality can now be measured in terms of heart-rate variability (McCraty, Rollin & Childre, Doc, 2010) This basically ineffable heart coherence has been correlated to such sentient states as Jungian "Compensation", mystic "transcendence", and the cognitive "Flow" state. (Schafer, 2018 & 2016b; Schafer & Yu, 2011) The dynamic of coherence may be witnessed in evolution, the Hero's Journey, the process of mystical-scientific Initiation, the metaphor of the Sun's journey through the heavens, and the ***qualitative nature of Belief***.

Yes, the concept of coherence may also be defined in the theological context of "belief"; i.e., a certain potential of harmony of understanding between the known and the unknown can be achieved if belief is a factor, and now researchers have come closer to being able to measure coherent cognitive states as harmonious frequencies. So, this form of coherent sentience becomes a way for defining and recognizing "belief" as a category of sentience. This is important because scientific methods that have been dominated by the need to quantify and to replicate experiments will take a new trajectory in which paranormal experiences which cannot be replicated such as prophesy, belief, telepathy, and action-at-a-distance may be researched as valid scientific fields of cosmic sentience.

At this stage of development, the existential validity of engaging in conversations with God will become indisputable science. In-order to support exploration of the emergent quantum-cognitive integrated reality, thousands of years of spiritual teaching and precedent can no longer be ridiculed. Such teaching exists in the world's spiritual texts. The difference between cultural sanity and psychosis is ultimately defined by a culture's "contextual"[14] perspective that begins "literally" in the heart of each person. (Alabdulgader et al., 2018; Anonymous, 2018 a, b, & c)

Coherent states of sentience may be understood in terms of the Eternal Verities—harmony of truth and beauty—of philosophy and scripture. To follow this logic, it is necessary to synthesize a number of research trends: Coherence which confirms the priority of the heart rather than the brain in human function; Jungian Individuation (Jacoby, Yolande, 1973), the Hero's Journey (Campbell, 1968); research on target groups for purposes of advertising (Anderson & Kanuka, 2003; Schafer, 2016c); Myers-Briggs personality types (Personality types explained, ND); and philosophical-scientific perspectives on the

"Eternal" according to quantum field theory, mathematics, and optics. (Reddy, 2018; McCraty & Childre, 2010; Hameroff & Penrose, Roger, 2014; Alexander, Boyer, Alexander, n.d.; Anonymous, 2018a, b, & c; Galeon, 2016)

FURTHER RESEARCH NEEDED

Contextual Media Biofeedback to Achieve Coherent Collective Psyche

In other words, the cognitive dialogue between Jungian *archetypes of the unconscious* and their *archetypal representations* that are perceived as reality can be correlated with software algorithms (unconscious) and the virtual realities (conscious) which they project as archetypal representations. The term context (Aiello, 2014) is critically important because it is this cognitive dynamic that ultimately contributes to a global government of conscience defined as autonomous diversity within unity. (Schafer, 2018a, b, & c; Schafer, 2016a, b, c, & d) We use the term "context" as it was defined by Carl G. Jung where context includes the totality of "functional" Psyche that exists according to "frequency bands" in corresponding dimensions (self-fields). This fourfold synergistic complex includes both unconscious and conscious energy patterns (archetypes & archetypal representations) that may be correlated with the energetics of sentient inspiration which takes place during the contextual reading of scripture.

This strategy for analyzing mediated images in order to understand unconscious psychological complexes can also be applied to analyzing scripture. Like insight as to the meaning of dreams, insight as to the meaning of scripture can heal the contextual "patient-learner" and can actually raise the harmonic quality of human energies that are generated from unconscious (quantum) dimensions and projected into conscious images that we call experience. One assumption underlying the dynamics of psychic conversation is that the Jungian Psyche is the key to functional structuring of the *Psych*ecology. Mapping the Psychecology can be accomplished using archetypal algorithms addressed to "personality profiles" (such as Myers-Briggs Typologies) of contextual collective target groups to project archetypal representational images of reality. The process is similar to stereotyping or target grouping, but because the motives differ, the consequences are not the same.

The meaning underlying infinite Cosmos is far beyond human understanding, but 20th century science has gifted us with the elements necessary for authentic research. Progress toward the objective of creating a culture of conscience guided by super-human sentience has been assisted by a legion of scientists in diverse fields. But among them have been giants who helped establish the operative foundation. Of course, everyone knows the name Einstein who changed everything, but his colleague, David Bohm (1985), is less well known even though he established the concept of the holographic universe (Giorbran, 2006). Among these giants are others like Karl Pribram who established the concept of holonomic mind (Pribram, 1971 & 1991) and Doc Childre who discovered the key to coherence with his research on heart rate variability. (Childre, 2006).

These giants have created the virtual worldview which can be converted to Godlike understanding of cosmic reality. Simulations in the Media-sphere can now be employed to a more coherent planetary life that accords with natural sentient life. One trajectory is research on alternative education and commerce using *Psych*ecology video games (PEG). (Schafer, 2016a; Schafer, 2016b; Schafer 2016c; Schafer & Yu, 2011a; Schafer, 2008; Schafer, 2007) Abundant research already confirms the foundational parameters of "contextual" reality, but it seems that serious threat of planetary extinction is the only incentive that will favor redemption.

Global problem resolution depends first on recognizing how powerful the human energies of cooperation and goodwill can be. All civilization is a consequence of these energies, but now we have the technological tools to research them more fully—and the keys required to actually manifest a global culture of conscience. Just as scripture has done in the past, the media-sphere could provide cognitive biofeedback in order to achieve coherent states in contextual collectives. Researched by the *Institute of Heartmath*, collective coherence can now be studied directly and adjusted in an ongoing process.

The media-sphere can be used to increase the degree of synchronization between different oscillating systems of sentience. (Anonymous, 2018a, b, & c) Where two or more oscillatory systems operate at the same basic frequency, they can become either phase or frequency-locked as occurs between the photons in a laser. (The emerging science of human-data interaction, 2015a) A quantum version of computer code can be written and synchronized. (Strogotz & Stewart, 1993) Such cross-coherence will foster coherent levels of frequency that will influence human choices and behavior in benign ways while preserving human dignity and individual autonomy of choice. Human energy—especially as amplified through the Media-sphere—has enormous impact on Field energy; however, technological developments offer the possibility for coordinating with the frequencies of super-human sentience and articulating these energies with a degree of predictability and with specific targeting of unconscious problems. Understanding the scriptural sentience discussed in this paper could enhance current cognitive research and the sophisticated technology that mirrors the Essene understanding of cosmic Law.

Many contemporary scientists believe that it is the underlying state of our physiological processes that determines the quality and stability of the feelings and emotions we experience. The feelings we label as "positive" actually reflect body states in which "the regulation of life processes become efficient, or even optimal, free-flowing and easy." (McCraty & Childre, 2010; Damasio, 2003)

Combining and focusing these coherent mediated energies in Celebrations of Humanity can be understood as direct adherence to the directives of God. Shouldering this responsibility could provide timely optimization of the conversation between benign spiritual powers and human thought-feeling. (Schafer, 2018) The annual focus or "Celebration" could amass collective energy for Good (reference peaceful global demonstrations for improvement of the status quo). Simultaneous with the Celebration, conferences consisting of the leaders of humanitarian organizations would be positioned for accurate reception of higher cosmic energies (expressed as ideas) for global problem resolution in the most efficient and practical ways. An ongoing Coherence Project generated from the celebration would provide a naturally growing organism for ongoing action (labor) to address critical planetary issues cooperatively and under the direct supervision of God.

This recommendation is neither politically naïve nor experientially untested. It is not utopian. Rather, because of its more coherent overview, it is much more practical and cost-effective than virtually all solutions presently being proffered. Salient features are (1) Spiritual reality as energy (2) Timeliness and scope which can be understood in terms of the "butterfly effect" of Chaos Theory (3) The powerful dynamic influence of the cognitive Media-sphere as a global biofeedback mechanism, and (4) The potential for focusing the preexisting network of tens of thousands of "humanitarian" organizations worldwide. All-of these features can be extensively rationalized and scientifically proved, but successful redemption will require enormous dedication and—at this generative moment—the commitment of a benign hierarchy of leaders with global stature. I use the term hierarchy because all cosmic Law—order and creativity—is based on hierarchy. Only chaos as classically defined "evil" is based on disorder and destruction.

CONCLUSION

The Media-Dream Analogy for Conversation With God

Scripture is a fundamental "sentient" media model; i.e., from the perspective of modern science, the claim that spiritual writing is "inspired by God" should become defensible to even those with little faith. The capacity to "inspire" is the exact opposite of being "literal". Like Jung's definition of "dead symbols", being literal equates to being "ossified", whereas the living quality of inspirational writing has to do with the living "contextual" ***psychological perceptions*** of those who are being inspired. Therefore, having some understanding of the psychology of perception is valuable for understanding the living sentience of "the word" and of the category of sentience called "belief". Today's science has provided us with the tools to understand the dynamics of the conversation with God. A collective, living psychological conversation with God manifests as the *Media-dream* which can be correlated with the dynamics of scriptural cognitive-conversation.

"Believers" already have a rudimentary understanding of sentient inspiration, but the cultural collective must begin to understand their authentic responsibility within the context of conversation with God. Belief in the conversation with God should now be considered mystical science. If the Media-sphere is structurally analogous with the quantum-cosmic Hierarchical sentience that Jung called Psyche, employing the global media-sphere as biofeedback that can foster coherent psychological redemption in contextual collectives may be understood as a conversation with God. (The purpose of the conversation would be a harmonious reframing of cognitive value structures according to principles of neural plasticity). The dynamics of the conversation may thereby be "scientifically" known with unprecedented authority, while the scriptural conversation can now be verified for even non-believers.

ACKNOWLEDGMENT

This chapter may not have been attempted except for the implacable belief in Biblical scripture that the author's brother has maintained throughout his lifetime. The author is well-aware of the historical disjunction between mysticism and science and of the longstanding scientific bias against sciences like astrology, alchemy, and the various mystery schools such as that of Pythagoras. Therefore, the author's academic publishing has soft-pedaled the scriptural aspects of belief in order to avoid the implacable prejudice of modernity against all things spiritual.

REFERENCES

Aiello, L. M. (2014). How Yahoo Research Labs studies culture as a formal computational concept. *MIT Technology Review*. Retrieved from http://www.technologyreview.com/view/529521/how-yahoo-research-labs-studies-culture-as-a-formal-computational-concept

Alabdulgader, A., McCraty, R., Atkinson, A., Dobyns, Y., Vainoras, A., Ragulskis, M., & Stolc, V. (2018). Long-Term study of heart rate variability responses to changes in the solar and geomagnetic environment. *Scientific Reports*. Retrieved from https://www.heartmath.org/assets/uploads/2018/03/long-term-study-of-hrv-responses-to-changes-in-solar-geomagnetic-enviroment.pdf

Anonymous. (2018a). The energetic heart. CA: *Institute of Heartmath*. Retrieved from https://www.heartmath.org/articles-of-the-heart/science-of-the-heart/the-energetic-heart-is-unfolding/

Anonymous. (2018b). Study shows that geomagnetic fields and solar activity affects human autonomic nervous system functions. *Heartmath Institute*. Retrieved from https://www.heartmath.org/articles-of-the-heart/study-shows-geomagnetic-fields-solar-activity-affect-human-autonomic-nervous-system-functions/

Anonymous. (2018c). Global coherence initiative. *Heartmath Institute*. Retrieved from https://www.heartmath.org/gci/

Baars, B. J. (1997). In the theatre of consciousness: Global workspace theory, a rigorous scientific theory of consciousness. *Journal of Consciousness Studies*, *4*(4), 292–309.

Bateson, G. (1972). *Steps to an ecology of mind: collected essays in anthropology, psychiatry, evolution, and epistemology*. University Of Chicago Press.

Bohm, D. (1987). Hidden variables and the implicate order. Quantum Implications.

Bradley, R. T. (2006). The psychophysiology of entrepreneurial intuition: A quantum-holographic theory. *Institute for Whole Sciences; Institute of Heartmath; e-motion Institute*. Retrieved from https://www.heartmath.org/research/research-library/intuition/psychophysiology-of-entrepreneurial-intuition-a-quantum-holographic-theory/

Brooks, R. A. (2010). *Fields of color*. Silver Springs, MD: Universal Printing, LLC.

Campbell, J. (1968). The hero with a thousand faces (2nd ed.). Princeton University Press.

Caughill, P. (2017). Boeing Just Took a Major Step Toward Autonomous Electric Flights. *Futurism*. Retrieved from https://futurism.com/boeing-just-took-a-major-step-toward-autonomous-electric-flights/

Childre, D. (2006). *De-stress kit for the changing times*. HeartMath LLC.

Chopra, D. (1995). *Body, mind, and soul. Turner Home Entertainment*. Atlanta, GA: PBS Home Video.

Damasio, A. (2003). *Looking for Spinoza, sorrow, and the feeling brain*. Orlando, FL: Harcourt.

Galeon, D. (2016). For the first time in history, we are seeing alien worlds (with spectroscopy). *Futurism*. Retrieved 7/11/2017 from https://futurism.com/for-the-first-time-in-history-we-are-seeing-alien-worlds/

Gillin, M., La Pira, F., McCraty, R., Bradley, R., Atkinson, M., Simpson, D., & Scicluna, P. (2007). Before cognition: The active contribution of the heart/ANS to intuitive decision making as measured on repeat entrepreneurs in the Cambridge Technopol. *Australian Graduate School of Entrepreneurship & Heartmath Institute.* Retrieved from https://www.heartmath.org/research/research-library/intuition/before-cognition-the-active-contribution-of-the-heart/

Giorbran, G. (2006). Everything forever: Learning to see timelessness. *Gevin., ISBN10*, 0979186102.

Glassner, A. (2004). *Interractive storytelling*. Natick, MA: A.K. Peters.

Gleick, J. (1987). *Chaos*. Viking Penguin Inc.

Goffman, I. (1974). *Frame analysis: An essay on the organization of experience*. New York: Harper and Row.

Hall, M. P. (2003). *The secret teachings of all ages*. New York: Tarcher/Penguin. Retrieved from http://www.sacred-texts.com/eso/sta/sta36.htm

Halpern, E. H., & Wolf, A. A. (1972). Speculations of Superconductivity in Biological and Organic Systems. In K. D. Timmerhaus (Ed.), *Advances in Cryogenic Engineering. Advances in Cryogenic Engineering* (Vol. 17). Boston, MA: Springer. doi:10.1007/978-1-4684-7826-6_11

Hameroff, S., & Penrose, S. R. (2014). *Discovery of quantum vibrations in "microtubules" corroborates theory of consciousness*. Retrieved 1/16/2014 http://phys.org/news/2014-01-discovery-quantum-vibrations-microtubules-corroborates.html)

Houran, J., Lange, R., & Hooper, D. (2018). Cross-examining the case for precognition: Comment on Mossbridge and Radin (2018). *Psychology of Consciousness*, *5*(1), 98109. doi:10.1037/cns0000126

Introduction and Research Review. (n.d.). *Maharishi International University Fairfield, Iowa, U.S.A.* Retrieved from https://www.mum.edu/pdf_msvs/v01/alexander.pdf

Jung, C. G. (1933). *Modern man in search of a soul*. Harcourt Brace & Company.

Lakoff, G. (2002). *Moral politics*. Chicago: The University of Chicago Press. doi:10.7208/chicago/9780226471006.001.0001

Lakoff, G. (2008). *The political mind*. Viking Penguin.

Lipton, B. (2016). Molecules of emotion. Santa Rosa, CA: Mountain of Love/Elite Books.

McCraty, R., & Childre, D. (2010). Coherence: Bridging personal, social, and global health. Alternative Therapies, 16(4).

Melville, H., & Moncur, M. (2003-2012). Moby Dick chapter 132. *The Literature Page*. Retrieved from http://www.literaturepage.com/read/mobydick-565.html

Micadei, K., Peterson, J. P., Souza, A. M., Sarthour, R. S., Oliveira, I. S., Landi, G. T., & Lutz, E. (2017). *Reversing the thermodynamic arrow of time using quantum correlations*. Ithaca, NY: Cornell University.

Morris, S. M. (2010). Achieving collective coherence: Group effects on heart rate variability coherence and heart rhythm synchronization. Alternative Therapies, 16(4).

Mossbridge, J. A., & Radin, D. (2018). Plausibility, statistical interpretations, physical mechanisms and a new outlook: Response to commentaries on a precognition review. *American Psychological Association*. Retrieved from https://www.academia.edu/38151856/2018_Response_to_Commentaries_cns_5_1_110.pdf?email_work_card=view-paper

Personality Types Explained. (n.d.). Retrieved from http://www.humanmetrics.com/personality/type

Pribram, K. (1971). *Languages of the brain; experimental paradoxes and principles in neuropsychology*. Englewood Cliffs, NJ: Prentice-Hall.

Pribram, K. (1991). *Brain and perception: holonomy and structure in figural processing*. Hillsdale, NJ: Lawrence Erlbaum Associates.

Reddy, J. (2018). Shashi Kiran (2018). The emergence of mind as a quantum field phenomenon. *NeuroQuantology: An Interdisciplinary Journal of Neuroscience and Quantum Physics*, *16*(11), 68–78. Retrieved from https://www.academia.edu/37922315/The_Emergence_of_Mind_as_a_Quantum_Field_Phenomenon?email_work_card=title

Reeves, B., & Nass, C. (1996). The media equation. Stanford, CA: Center For the study of Language and Information & Cambridge University Press.

Schafer, S. (2007). Premise: The Key to Interactive Storytelling. *Game Career Guide*. Retrieved April 3, 2007 from http://www.gamecareerguide.com/features/357/premise_the_key_to_interactive_.php?page=1

Schafer, S. (2008). *The Psychecology Game I-V; Using Story-based Video Games to Capture Veridical Data on Subliminal States of Being*. Toward a Science of Consciousness 2008, sponsored by the Consciousness Center & the University of Arizona. Retrieved from https://bandura.sbs.arizona.edu/login/consciousness/pubreport.aspx?aid=27

Schafer, S., & Yu, G. (2011). Meaningful video games: drama-based video games as transformational experience. In M. M. Cruz-Cunha, V. H. Valvalho, & P. Tavares (Eds.), *Business, Technological, and Social Dimensions of Computer Games: Multidisciplinary Developments* (pp. 312–329). Hershey, PA: IGI Global. doi:10.4018/978-1-60960-567-4.ch019

Schafer, S. B. (Ed.). (2016). *Exploring the collective unconscious in a digital age. In Exploring the Collective Unconscious in a Digital Age*. Hershey, PA: IGI Global. doi:10.4018/978-1-4666-9891-8

Schafer, S. B. (2016c). *The media-sphere as dream: Researching the contextual unconscious of collectives. In Exploring the Collective Unconscious in a Digital Age*. Hershey, PA: IGI Global. doi:10.4018/978-1-4666-9891-8

Schafer, S. B. (2016d). Fostering psychological coherence with ICTs. *International Journal of Reliable and Quality E-Healthcare*, *4*(4), 1-31. DOI doi:10.4018/IJRQEH.2015100101

Schafer, S. B. (2018a). *Generating a superconductive culture of conscience*. Dusseldorf, Germany: LAP Academic Publishers. Retrieved from: https://www.morebooks.de/store/gb/book/generating-a-superconductive-culture-of-conscience/isbn/978-620-2-07643-2

Schafer, S. B. (2018b). Synergy with the World Soul: Establishing the humanitarian organism. International Journal of Civic Engagement and Social Change, 4(3). Doi:10.4018/IJCESC.20170701

Schafer, S. B. (2018c). *Gaming cultural atonement*. Lambert Academic Publishers. Retrieved from https://www.morebooks.de/gb/search?utf8=%E2%9C%93&q=+978-613-9-89469-7+

Smith, H. (1958). *The world's religions*. New York: Harper Collins.

Strogatz, S., & Stewart, I. (1993). Coupled oscillators and biological synchronization. *Scientific American, 269*, 102–109. doi:10.1038cientificamerican1293-102 PMID:8266056

The emerging science of human-data interaction. (2015a). *MIT Technology Review*. Retrieved January 11, 2015 from http://www.technologyreview.com/view/533901/the-emerging-science-of-human-data-interaction

Tomasino, D. (2007). The psychophysiological basis of creativity and intuition: Accessing "the Zone" of entrepreneurship. *Institute of Heartmath & e-Motion Institute*. Retrieved from https://www.bing.com/search?q=Tomasino,+Dana+(2007).+The+psychophysiological+basis+of+creativity+and+intuition%3A+Accessing+%E2%80%9Cthe+Zone%E2%80%9D+of+entrepreneurship.+Institute+of+Heartmath+%26+e-Motion+Institute.&src=IE-SearchBox&FORM=IESR4A&pc=EUPP_

Trinity in the World's Religions. (2013). *SPEAKINGTREE.IN*. Retrieved from https://www.speakingtree.in/blog/trinity-in-world-religions

Yogi, M. M. (1990). *Maharishi Mahesh Yogi on the Bhagavad-Gita: A New Translation and Commentary*. Penguin Arkana.

ENDNOTES

1 Notwithstanding Nazi misinterpretations of the complex truths of mystical science, synthesizing science with mysticism is the current mandate. The mandate is not limited to the Nazi interpretation of racial superiority but to an evolutionary principle of racial synthesis.

2 Plato, Aristotle, Socrates, and Kant all contemplated the concept of *Lysis* (friendship) and its importance to civilization. (Oviedo, Michael, 2008) In the present context, the term is used as a salient factor in dramatic structure that serves as "the resolution" of the dramatic *exposition* (statement of the dramatic problem). Moreover, the term *Lysis* is applied to a much expanded concept of "sentient-civility" that touches upon the dynamic of cosmic sentience.

3 The prospective function is an anticipation of future conscious achievements like a plan roughed out in advance. (Jacoby, p 79) The more comprehensive archetypal prospective function must be distinguished from the compensatory function that is "purposive".

4 Though it may not be in vogue, the author does not see a conflict between the conclusions of Chomsky and Lakoff. These two geniuses are simply "believing" in the same truth from different perspectives.

5 Actually, domains of belief are very subtle--economics and engineering can also be considered moral domains.

6 The author does not pretend to be expert in the mathematics of physics. Rather, his expertise resides in a more linguistic approach. Nevertheless, a logical compatibility appears to exist within field calculations of metaphorical resonance.

7 Quantum fields are not described by simple numbers. They are described by vectors in what mathematicians call *Hillbert space* and their dynamics by *operators* that obey partial differential equations. Thus, while classical field intensity is described by a simple number, in QFT the simple number is replaced with the expectation value of field intensity.

8 The Higgs field quantum (called the Higgs boson) was detected in 2012. It interacts with other fields and, because it has a large vacuum expectation value, it generates the effective mass of those fields. It was the final ingredient that made electroweak unification possible.

9 **Superconductivity** is a phenomenon of exactly zero electrical resistance and expulsion of magnetic flux fields occurring in certain materials, called superconductors, when cooled below a characteristic critical temperature. However, since its discovery, superconductivity--a **microscopic** effect caused by a condensation of Cooper pairs into a boson-like state--has been linked to Higgs fields and discovered to exist at body temperatures. It is "speculated that superconductivity might exist in certain biological systems and specifically in the aromatic organic long-chain molecules which have conjugated double bonds containing π electrons (itinerant electrons)." (Halern, E. H. & Wolf, A. A., pp. 109-115)

10 Named by Mihály Csíkszentmihályi in 1975,***flow***, also known colloquially as being in **the zone**, is the mental state of operation in which a person performing an activity is fully immersed in a feeling of energized focus, full involvement, and enjoyment in the process of the activity. In essence, flow is characterized by complete absorption in what one does, and a resulting loss in one's sense of space and time.

11 ***Agency*** is the capacity of an actor to act in a given (contextual) environment. The capacity to act does not at first imply a specific moral dimension, so moral agency is a distinct concept. As debated in sociology, an agent is an individual engaging with the social structure. This debate concerns, at least partly, the level of reflexivity (interactivity) an agent may possess within a context. Agency may either be classified as unconscious, involuntary behavior, or purposeful, goal directed activity (intentional action).

12 **Initiation** is a rite of passage marking entrance or acceptance into a group or society—a formal admission to adulthood in a community. In an extended sense it can also signify a transformation in which the **initiate** is "reborn" into a new role.

13 The existence of hierarchical scales of evolution were addressed by Alfred North Whitehead (Foster, David, 1985, p. 24) and implied in Darwin's theory of evolution. Whitehead's hierarchical scales include electrons, atoms, molecules, organic molecules, cells, tissues, organs, and the whole body. Due to discoveries in modern science, correlated hierarchies can be extended to include super-human scales (as referenced in the world's theological hierarchies) throughout the frequency spectrum of quantum field theory (QFT).

14 Carl Jung defined contextuality as the entire psychic environment of a patient's unified energy field—which made the concept hard for contemporaries to understand. Jungian contextuality translates to unique individuality or "specificity" (Foster, David, 1985, pp. 60-85) To a degree, contextuality has been quantified with Myers-Briggs personality types.

Chapter 3
Language Patterns and Cognitive-Sentient Reality:
Certainty/Uncertainty in Cognitive-Sentient Exploration of Reality

Florin Gaiseanu
Independent Researcher, Romania

ABSTRACT

This chapter explores correlations between modern and ancient science as they apply to cognitive-sentient reality. The ancient wisdom and worldviews were based on a combination of reason and introspection—the frontier between certainty and uncertainty. This exploration was paramount in Egyptian, Greek, Chinese, Indian, and other cultures and philosophies. This contemplation is actually universal, and virtually all culture has contributed to the structuring of fundamental sentient patterns of cognitive reality. Taking into account some recent scientific discoveries at micro/macro levels, the concepts of certainty/uncertainty, and information science, the chapter argues the coincidence between the ancient model of chakras and a newer cognitive-informational model. Based on these correlations, a new model of cognitive-sentient exploration of reality (CSER) is discussed relative to its use as in analytical/synergistic model for research on cosmic sentience.

INTRODUCTION

The issue of human reason vs. universal mind is longstanding, yet unresolved. From time immemorial, philosophers have sought answers to questions about the nature and extent of human-interaction with its sentient environment and with the influence on his environment on the human condition. In Greece, philosophers like Socrates held that, "*The soul, endowed with all knowledge, existed before the body*". His disciple, Plato—inspired by Pythagoras and by the Egyptian hermetic philosophy—is known for his *Theory of Forms* in which "forms" are ideas. Ontologically, the *Theory of Forms* is, perhaps the most relevant to the modern reality of information and communication technologies (ICTs). Plato's

DOI: 10.4018/978-1-5225-9065-1.ch003

non-physical forms/ideas represent the most authentic reality, and ideas have the structure of "sacred geometry" which plays a primordial role in creating and affecting matter.

Anaxagoras (adept of the atomistic structure of matter) advanced the concepts of Democritus, but believed that a "creative divine mind"—organized and ordered matter—constitutes a thinking mind. Like Plato, Pythagoras believed that mathematics is the most sacred and exact of all the sciences, and demanded that all who came to him for study be familiar with arithmetic, music, astronomy, and geometry. He combined ancient principles of sentience in new ways to impose order on the universe's chaos through the patterns of mathematics and music (Hall, 2001).

In China, the 500s BC saw Lao Tzu and Confucius (initiating and developing the Taoist Yin/Yang philosophy), about the same time with Buddha (initiating the reincarnation and "channel-wheels" (chakras) philosophy) in India and a little before Socrates in Greece. Further west in Babylon, the Jewish prophets Ezra and Nehemiah were writing the Bible, and Zoroastrianism was having enormous influence. (Smith, 1958; Hall, 2001).

Scholars are familiar with the ontological postulates of cultures past including Egypt, Greece, and Mesopotamia, but the ontologies of these cultures are often obscured by lack of modern understanding. The ancients did not dissociate subjective and objective reality. Rather, they sought to discover their relationships. When ignorance of ancient language-symbolism—such as cuneiform and the hieroglyphics of the Incas, Aztecs, and Mayans—is added to modern misunderstanding, the result is chaotic. Misinterpretation of ancient ontology has led to the conclusion that ancient civilizations were scientifically primitive—a conclusion based on a false premise. Generally, ancient cultures had psychological subjective ontologies; i.e. they perceived reality as cognitive-sentient. Accordingly, ancient ontology is drastically different than modern ontology that has been based on material objectivity. However, the modern scientific ontology is changing rapidly due to a paradigm shift in scientific principles inaugurated with the advent of a quantum-based media reality.

Ironically, ancient ontology may prove to be more accurate than its modern offspring. The reason for this is that ancient worldviews were based on a combination of reason and introspection—the frontier between certainty and uncertainty--that may have been more accurate than modern dependence on technology, statistics, and a basically materialistic scientific method. In other words, during the so-called age of reason, modern scientists have abjured their most exquisite instrument for reasoning—their own multi-scalar sentience.

The greatest error that modern scientists make is their refusal to recognize how superhuman sentience influences human thought that constitutes the dynamic for evolution. As Norse mythology puts it, evolution happens "at the edges of extremes". As Carl Jung puts it (Jung, 1933; Schafer, 2016) the cognitive conscious/unconscious is engaged in a perpetual conversation, and that conversation is the frontier of evolution. In other words, the "value intensity" of the energetic interaction results in evolutionary ideas. If both sides of this cognitive conversation are not appreciated, no human progress can be expected. Generally-speaking, the ancients were less susceptible to scientific hubris than modern scientists who have long ignored what they consider "mystical" and, therefore, unscientific indicators of reality. If "introspection" is mystical and irrelevant, science becomes a façade (Schafer, 2016).

The modern scientific dependence on scientific method assumes that great cultures of the past were unscientific, but anyone who has delved into ancient methodology (alchemy, astrology, Pythagorean mathematics, etc.) knows that this view is a canard. This ephemeral perspective that scientific validity depends upon repeatability is puerile—too inconsistent with modern discoveries in the physics of quantum fields. Today's science is in process of disclosing a wealth of "hard-science" that is consistent with the mystical science of the ancients.

Re-discovery of correlations between ancient and modern science would enhance modern research on the universal mysteries. For example, study of the chakras in Ayurvedic science developed over thousands of years would explain how cognitive-sentient channels within the human organism serve to extend human participation with supernatural forces that guide evolutionary processes in the universe. Appreciation of the legitimacy of such cognitive models can already be understood in the context of scientific method. (the Ayurvedic science of Maharishi Mahesh Yogi (Schafer, 2016)

Focusing upon the frontier of cognitive certainty/uncertainty, this chapter explores the current trajectory of human sentience using the recently reported concepts on information and informational modeling of consciousness (Gaiseanu, 2016a,b; 2017b,c; 2018b,d). This proposed model of the consciousness structure based on information appears to be unique. Although various models of consciousness have been proposed—the quantum mechanics of mind computation mechanism (Hameroff, 1998; Hameroff, & Penrose, 1996, 2014), and the mechanism of quantum entanglement necessary for explaining remote communication (Radin, 2006), they are partial models. Other recent analytical studies suggest that informational modeling could consistently and coherently explain the properties of mind (Chalmers, 1995; van Lammel, 2006; Meijer, 2013a). Such necessary models have since become more concrete (Tononi, 2004; Gaiseanu, 2016a, 2017b,c, 2018b,d).

Among these models, only the Informational Model of Consciousness (IMC)--based on recent research concerning the matter/antimatter nature of the universe (Hajdukovic, 2013) and the disembodied microparticle in quantum physics (Denkmayr et al. 2013; Aharonov, et al., 2013)—deals with the properties of consciousness at different degrees of scale (Gaiseanu, 2016a, 2107b). Recently, IMC was extended to understand the mind connectivity with the body itself, contributing substantially to understand the main mechanisms within the mind-body problem (Gaiseanu, 2019b). According to this model (Gaiseanu, 2016a, 2018c), consciousness is composed of seven cognition centers (Gaiseanu, 2018d), which would explain the near-death experience (Gaiseanu, 2017b, 2018d) earlier described (Fracasso & Friedman, 2011), as well as the extra-cognitive power of mind recently called the "real magic" phenomenon (Radin, 2018). According to IMC, phenomena such as premonition, remote communication (telepathy), remote diagnosis (Myss, 1996), geographic landscape exploration (Dulcan, 2009; Meijer, 2013b), and even psychokinetic phenomena (Gaiseanu, 2016b; 2018d) can be also explained.

Such expansion of research premises has produced encouraging indicators to use this model, which was already formulated, described, published and known as the Informational Model of Consciousness (Gaiseanu, 2016a,b; 2017b,c; 2018a,b,c,d; 2019a,b), as a tool to understand the participative contribution of "sentience" within the scientific analytical research and language discourse, which is traditionally performed, or believed to be performed, especially by means of (limitative) mental logics or rational type algorithms and concepts. More explicitly, when we refer to this domain, we have to take into account also the personal subjective contribution of the researcher as an integrative sentience detector and creator, as it will be discussed in detail within this chapter. We will refer to such a "sentience" contribution as to a Cognitive-Sentient Exploration of Reality (CSER), closely correlated with the ancient "sentient" chakras model. The CSER is distinct from the Informational Model of Consciousness (IMC), which includes much larger meanings and concepts, as mentioned above. In other words, we will use some concepts of IMC correlated with ancient chakras model (Gaiseanu, 2018d), to understand CSER, which is actually only a particular application of IMC, besides the various other applications mentioned above. Basically, at the border between principles of certainty and uncertainty, the term "sentience" may provide for a greater degree of subjectivity within the domain of analytical research without diluting scientific integrity. The concept of "sentience", discussed in detail within this chapter, may reveal deep and multi-facet meanings. Essentially, the concept of sentience expands the concept of "intelligence at different scales" in the human experience of "reality" (Schafer, 2016).

BACKGROUND

It may be stipulated that language is an informational tool that humans use to communicate within a social community, but current experience emphasizes the difficulty of even human intercommunication. The complexity and the expressive power of the human language is spectacular. There are more than 6000 distinct languages on our planet, and it is cliché (England and America are two countries divided by a common language) that even within a single language like English an infinite variety of meaning and expressions exist. *The Cambridge Grammar of the English Language* weights in at over 1700 pages (Hurford, 2008). As a survival mechanism, from time immemorial the development of language has been essential. It may be stipulated that language is the mother of invention and civilization. Language is a fundamental tool for access of personal and collective meaning and for the expression of discovery that is essentially subjective and even subliminal. The sentient-processing of informational signals--external and internal—has been formalized as the *Informational Model of Consciousness (IMC).* (Gaiseanu, 2016a & b; 2018d).

As an informational tool, technological communication can be analyzed in terms of informational accuracy of content in messages. Information science has defined content accuracy using concepts of entropy and quantity of information within the transmission systems (Shannon, 1948, Gaiseanu, 2018d). The following elements do exist within a communicational process: (1) an issuing source and an encoding method that is "colored" by the personal skill of the author, (2) a channel or means of transmission--oral, written, fax, email, etc., (3) a decoding method that includes the skill of the receiver (McPheat, 2010). Shannon notes that in an informational (electronic) system (Shannon, 1948), where the noise of the transmission chain is expressed by entropy, the accuracy of the received message (expressed by a quantity of information) can be altered (shaded). Some specific distorting "noise"-inducing processes could be introduced by different cultures, different expectations, different experiences, different perspectives, or different communication styles. Therefore, even if the communication source (the issuing author) conveys a message as accurately as possible, a perception of the message will be "contextual".

The receiver will interpret (decode) the meaning of the message in a unique way that is often very different than the intended meaning. This subjectivity of meaning is complex because both the initiating source of information and the receiver of information are uniquely different. Not only are visual, auditory, and cognitive processing systems complex in themselves, the contextual experiences, attitudes, prejudices, and feelings of each agent in the communication process differ, and they must be synchronized as effectively as possible (McPheat, 2010; Pokorny, 2012; Lohrey, 2019).

Moreover, as communicated information rises to a higher mental order of complexity such as philosophy or scientific research in which the shared meaning of words becomes critical (Krishna, Vijay & Morgan, 2004; Lohrey, 2019), the process of maintaining sufficiently accurate comprehension of information is essential for the discovery of meaning. In order to avoid significant errors in the communication process and in the decisions resulting from the process, a degree of uncertainty must be eliminated from the process. But complexity is not limited to mental communication. Communication in even more subjective realms like emotion, sensation, and intuition requires a metaphorical language that is embedded with entangled meaning.

Therefore, a more comprehensive scientific language is necessary to decipher entangled meaning. This chapter addresses the challenge with an attempt to define entangled meaning with use of the term *SENTIENCE.* In spite of various attempts, there is no generally accepted definition of the term sentience (Clark, 2000). Maxwell refers to sentience as a collection of personally experienced, "*sensations of pain and pleasure, tactile, visual, olfactory or auditory sensations*" (Maxwell, Nicolas, 2001). Abram argues from a more ecological perspective where sentience may be applied to "language in a more than human world"—a philosophical return to a worldview based on ecology. This perspective is dynamically "participatory". It incorporates subtleties of perception, the mindful life of the body, the body's silent conversation with things, and the rediscovery of participation with the entire geosphere (Abram, 1997, pp 44-65). In other words, the world of sentience is larger than human language can express. The language of sentience sojourns beyond the limitations imposed by the current man-made communicational systems and incorporates the impulses to explore the mysteries of the unknown—of nature and the cosmos.

Every attempt to definitively say "what language is", is subject to a curious limitation. For the only medium with which we can define language is language itself. We are therefore unable to circumscribe the whole of language within our definition. It may be best, then, to leave language undefined, and to thus acknowledge its open-endedness, its mysteriousness. Nevertheless, by paying attention to this mystery we may develop a conscious familiarity with it, a sense of its texture, its habits, its sources of sustenance (Abram, 1997, pp. 73).

In-order to complete the panorama of meaning embedded in the concept of sentience at different scales of sentient *vitality,* recent research suggests that—at the scale of molecular organisms—single-celled organisms display behaviors which can be compared to cognitive actions of the neural system in higher organisms (Pereira, 2016). In other words, though such behavior is not conscious to humans, it appears to be "conscious" behavior from the perspective of single-celled organisms. In order to appreciate the meaning of such "sentience", it is foolhardy simply to ignore its significance to human "consciousness".

Recently, research by Stuart Hameroff and Sir Robert Penrose proposed ORCH-OR theory by suggesting the mechanisms of "consciousness" at quantum levels (Hameroff, 1998). The implications of this proposal of quantum-level consciousness and other research (Bradley, 2018) verify the need to develop a new language to express the sentience. Moreover, consciousness in animals is being reviewed. A large systematic review of the scientific literature concerning the sentience of animals shown that the subjective lives of animals are not only a vital part of human medical research, but are regularly measured and studied with scientific rigor (Proctor et al., 2013). Finally, the treatment of "alien" sentience in science fiction narratives has predisposed humans to the consideration and acceptance of sentient non-human life forms in both legal and ethical frameworks (Scerri & Grech, 2016).

Within the complex modern realm of technological language electromagnetically managed, perhaps the ancient and highly sophisticated percept of Chakras would prove useful. The chakra system originated in India between 1500 and 500 in the texts called the Vedas. Evidence of chakras, a spelled chakra, is also found in the Shri Jabala Darshana Upanishad, the Cudamini Upanishad, the Yoga-Shikka Upanishad, and the Shandilya Upanishad. The word chakra is defined as a spinning disk or wheel. On the human body, a system of seven primary chakras or energy wheels run along the spine, and the health of one's chakras is directly connected to the health of the physical body, the mind, and the emotional body.

In other words, the ancient system of chakras provides a blueprint of the multi-functional human being which connects the human microcosm with the macrocosm. Fractal geometry correlates not only the microcosm and macrocosm, but all the scales of universal sentience. This is why Westerners have misconstrued this ancient science as religion which is inferior to science. However, given the modern evolution of holographic quantum reality, it is reasonable not only to correlate this ancient science with advanced modern science, but to understand the chakra system as a challenging issue -- superior and more longstanding model for universal mediation (Schafer, 2016).

FUNDAMENTAL PATTERNS OF COGNITIVE REALITY ASSESMENT

One cannot explain the great mental achievements recorded by the ancient Egyptian and Greek civilizations without some reference to "super-human" help. Indeed, how did our ancestors develop the great civilizations of the past without guidance from super-human dimensions of "mind"? How did they develop superior civilizations without the help of quantum physics, astronomy, and of some of the high techniques and technologies which we have today? The answer is simple, in order to know and understand the mysteries of the universe and its laws, they trusted in subjective perceptual capabilities of the mind, correlated and applied them to the rational experiences of interaction with their environment.

The ancients employed their intuitive sentience in conjunction with their waking "consciousness" to interact with multiple scales of the natural environment necessary to survive dangers, to provide food and shelter, to avoid suffering, and to evolve. Although the science of subjective-objective cognitive functions has been attributed to the visionary philosophical investigation of the Greeks, this approach to science had been strongly established more than four thousand years ago, Egyptian civilization was the source of all later known advances (Dulcan, 2009). The Greek philosopher Pythagoras, who studied for 34 years in Babylon in Egypt (Hall, 2001, pp. 169-227), was the primary Greek contributor to the foundation of subsequent sciences. Pythagoras "intuited" his famous theorem in geometry and developed the theory of music and color (Hall, pp. 169-227), the symbolism of numbers (Hall, 2001, p. 191-205), the multiplication table, and the aphorisms of geometry (Hall, 2001, p. 195) which allowed for a geometrical model for distinguishing between good and evil in "Monadic" philosophy.

Pythagoras initiated the foundation for scientific experimentation with what today may be understood as cognitive-sentience. Plato, as a disciple of Socrates and a successor of Pythagoras, developed the philosophy of ideas—the power and the penetrating force of the thought as the primordial component in universal evolution. He was, therefore, the quintessential philosopher who discovered and emphasized what today can be called the cognitive-sentient force of the mind as a tool for understanding the vast scope of environmental reality and for maintaining a scalar coherence (frequency) with cosmic intelligence. His work provided not only a tool for expanding the knowledge of reality, but also for accessing the models of superior intelligence in the universe.

Nevertheless, the esoteric (secret), sacerdotal ideas underlying Pythagorean philosophy emanated from the Egyptian civilization. Egypt constituted the germinal point (the seed; school of the Mysteries) for all subsequent religious models (Dulcan, 2009). In fields of mathematics, geometry, physics, and the rational sciences, the contributions of the Greek philosophers like Pythagoras, Tales, Euclid, Archimedes, Democritus, Heraclitus, Anaxagoras, and Aristotle provide the groundwork for subsequent understanding of the empirically verified cognitive-sentient intelligence of Cosmos (Dulcan, 2009). The mysterious Alchemical axiom, "As it is above, so it is below" (Hall, 2001, pp.494-524) constituted the axiom with which ancient scientists were able to employ the *Law of Correspondence* to the universe of knowledge (intelligence) ranging from the sentient Monad to the sentient material life. In other words, they reasoned that if human-beings consist of earth (flesh/material), emotion (water), mind (air), intuition (fire); so too—by the law of correspondence—should the Monad (cosmos) consist of these multiple functions of sentience (Schafer, 2016).

The distinction between noumenal and phenomenal sentience is simply one of scale. As they are sorted out, seemingly divers ontologies merge. Platonic doctrine—based primarily on a model of idea and thought—and Aristotelian doctrine—based on a model of rational thinking—are both seeded in the concept of sentient patterns. In today's reality where quantum patterning of Plato's *Idea* exists, the ever-growing Western approach to the oriental thinking mode (Satsangi & Brugnoli, 2018) accepts the importance if cognitive perception/observation of the world by using the latent inner-potential of the mind. For the wise men of the Orient, the truth about us and the Universe lies inside us, in the depths of the spirit somewhere at the emotion-intellect interface (Dulcan, 2009).

Acceptance of the ability of the human mind to scrutinize the universe with what today can be understood as cognitive-sentience, is essentially a time-honored method employed in Taoism (Odoul, 2014). Before the age of reason, this cognitive model was universal--based on the understanding of man's unity with nature to the understanding of opposing and complementary energies (Yin/Yang) defined in Buddhism and Brahmanism (Ignatenco, 1994). The meditative model explored in the concept of chakras-as-spiritual-energy-centers in the human body is consistent with and superior to modern science. Today, the same ontology may be glimpsed in environmental science which has been extended to the chaos theories of physics and is beginning to be addressed in quantum field theory. The salient point is this: The concept of Cosmic sentience is not new. It is simply being ignored in much modern Western science.

Today, quantum physics shows us that the world of the elemental particles is different from human images to which we are accustomed due to our cognitive/perceptual filters. This divergence in perception requires that Modern science must play a dual role. It's research must incorporate both models of sentience--rational thinking and intuitive contemplation of *Nous*. If increasing uncertainty as to the nature of reality (such as the reality of the media-sphere) is presently confounding civilization, we are obliged to reinterpret and integrate ancient subjective ontology with modern subjective scientific ontology. We must search for ways to correlate them. Perhaps the best way to accomplish this scientifically is to employ models of scaled sentience instead of assuming that humans are the sole sentient beings inhabiting cosmos. Scientists already employ rational thinking to abstract theory about the nature of cosmic reality. It should not be surprising that the same reality is revealed differently not only from one species to another, but also from one person to another. Varying perceptions must be considered from the standpoint of how differing media models (frequencies) can create a coherent reality. In this context, a revelation model of sentience should be addressed (Dulcan, 2009).

INFORMATIONAL MODEL OF CONSCIOUSNESS AS A TOOL FOR UNDERSTANDING OF COGNITIVE-SENTIENT REALITY

As it was recently explained in a series of articles (Gaiseanu, 2016a, 2017 b, c, 2018d, 2019a), consciousness could be described as an Informational Model of Consciousness (IMC). This model would indicate that consciousness is based on a structure of seven informational centers that are directly correlated with the energetic wheels of the chakra system described above. For the convenience, to understand the concepts already defined earlier within IMC which we refer to, used now as instruments to analyze CSER, we will shortly present again the Informational Model of Consciousness as in the cited articles (Gaiseanu, 2018d; 2019a) and illustrated in Figure 1.

According to the IMC scheme, the human body is a bipolar living structure, composed of matter and information (Gaiseanu, 2018d; 2019a). The components of the informational system of the human body are illustrated in the figure below:

On the left side of the illustration:

1. Info-Genetic Generator (IGG), coming from the two parents (info-genetic input)
2. Genetic Informational System (GTS), for the genetic transmission to the next generation (info-genetic output)
3. Maintenance Informational System (MIS), sustaining the connection with matter (matter-connection pole) -- the nourishment of the body
4. Info-Emotional System (IES), managing the emotions

Figure 1. Schematic representation of the informational system of the human body (IM – Informed Matter)

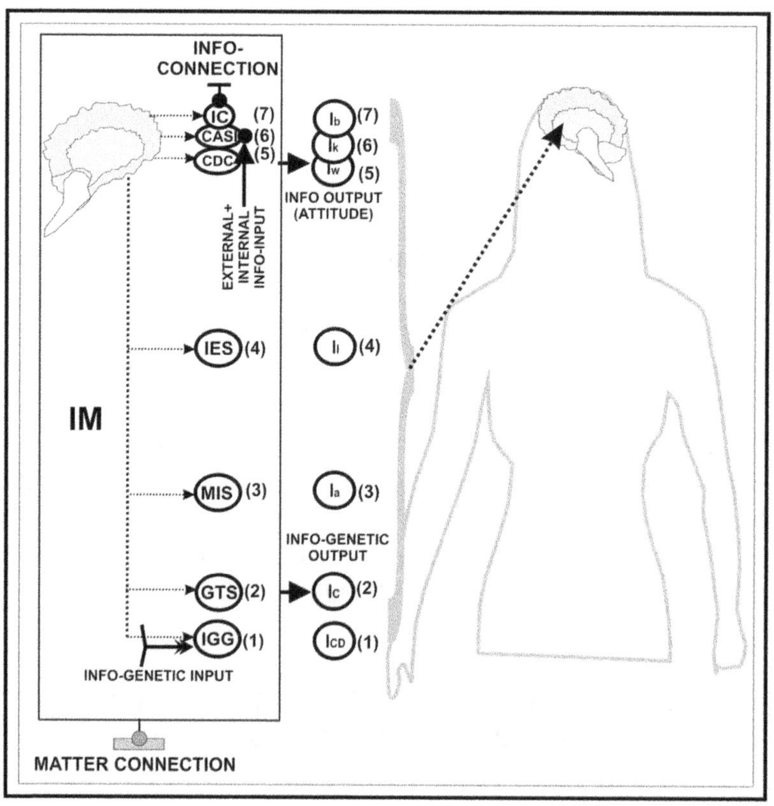

5. Center of Decision and Command (CDC), elaborating and transmitting the decisions (info output -- Attitude) to execution elements of IM (body)
6. Center of Acquisition and Storing of Information (CASI), the info-stored life experience
7. Info-Connection (IC), representing the connection with the informational field of matter/antimatter.

On the right side of the illustration are represented the corresponding cognitive centers associated to the informational systems described above, reflected in consciousness and suggestively called:

1. I-created (Icd)
2. I-create (Ic)
3. I-am (Ia)
4. I-love (IL)
5. I-want (Iw)
6. I-know (Ik)
7. I-believe (Ib).

The information received from the internal and external sensors is stored in the Center of Acquisition and Storing of Information (CASI), which becomes the personal data library, accumulating the entire life experience (Gaiseanu, 2019a). The data library can be projected as a component part of the cognitive center suggestively called I-know (memory). The personal "I" used to designate the cognitive centers highlights the subjective, personal, individual character of center activity. The existing information in CASI is ordered according to the necessities of adaptation to the environment and is converted into acquired decision-making patterns (Gaiseanu, 2018c).

This accumulated information can be recalled (retrieved) by means of the thought, which is defined within IMC as an informational operator on the informational data field, to become the input information for the so-called Center of Decision and Command (CDC). Its activity is reflected in consciousness by the cognitive center I-want—the decisional feature of output adaptation (attitude).

The Info-Emotional System (IES) deals with emotional activity and is reflected in consciousness by the center I-love. Within IMC, emotions have a reactive character, resulting as body-specific reactions to information (IM). This center plays an essential role in the sentient exploration of reality. It represents the personal body resonance ("frequency") with the perceived information. With participation of the center IL, received (virtual, theoretic) information becomes "living" information—a personal lived-sentient experience.

While the informational systems explained above are dedicated to a conscious connection with the inner and external environment, the informational systems presented below concern the automatic (autonomic or programmed) informational activity devoted to the body and species-survival. Although these systems are automatically programmed, as a result of the species acquirement, their activity is reflected also in consciousness by means of the cognitive centers I-am (Ia), I-create (Ic) and I-created (Icd), as represented in the right side of Fig. 1. These centers are associated respectively to the informational systems presented below (Gaiseanu, 2018d; 2019a).

Maintenance Informational System (MIS) relates to managing the connection with matter (food, water, air); MIS is reflected in consciousness by the cognitive center I-am (Ia), that confirms the self-own status.

The Genetic Transmission System (GTS) is responsible for the info-genetic transmissions corresponding to the cognitive center I-create (Ic) which concerns associative/social relations.

The Info-Genetic Generator (IGG) promotes genetic body-evolution. At conscious levels, this system is represented as the activity of the cognitive center I-created (Icd). It concerns predisposition and talents (Icd) (Gaiseanu, 2019a). During the first years of life, when the informational system is developed intensively, IGG is able to integrate a high quantity of information that contributes to personal characteristic behavior and the formation of mental pre-determinations of the future adult. It enriches Icd with deeply consolidated social, cultural and religious features (Gaiseanu, 2018c).

A special role should be attributed to the Info-Connection (IC) pole of the informational system which, according to IMC, assures the connection of IM with the informational field of matter and anti-matter (Gaiseanu, 2016a,b; 2017b,c). Within IMC, this pole is called I-believe (Ib) because it is sensitive to the very fine signals of the universe—like the anti-entropic force represented by the anti-gravitational field of anti-matter which is favorable to life-structuring (Gaiseanu, 2016a). This life-force could be associated with religious feelings and is detected in the cognitive center Ib as confidence and trust which protect *the living* against *the entropic* (destructuring) force of matter as determined by the second law of thermodynamics (Gaiseanu, 2018c). Moreover, this connection pole is a gateway for detection of paranormal (psy) phenomena, remote transmission by mind (Radin, 2006), remote diagnostics (Myss, 1996), and near-death associated phenomena like tunnel crossing, life-retrovision, extra-corporeal viewing, etc. (Gaiseanu, 2017b & c).

In informational jargon, the capture process for such very fine, low level signals that necessitate special conditions—like disconnection from internal and external environmental information sources, focused attention, and special training—could be treated as boundary processes for measuring certainty/uncertainty. Therefore, it should be considered a fine tuning and discrimination signal processor that can distinguish valid informational (certainty) signals from noise (uncertainty).

The human body can, therefore, be defined as an Informed Matter (IM) structure that integrates "matter" and information. The human body is like a media model with which humans communicate with multi-scalar sentient dimensions: the external environment by way of the neurobiological system of "sentient" neural transducers.

Modern arguments for supporting such a model are based on some recent discoveries in the field of physic at micro and macro (cosmic) scales. (Gaiseanu, 2016, 2017b,c).

- Dark matter effects could be attributed to anti-matter, generated in vacuum space as matter/anti-matter particle-pairs and polarized by cosmic bodies (Sun, Earth) (Hajdukovic, 2013).
- Anti-matter and matter show opposing properties such as reversal of the arrow of time and rejection of gravity.
- Information could be separated from matter itself (Aharonov et al., 2015; Denkmayr et al., 2013) which allows for considering information as a field irrespective of matter (Gaiseanu, 2016).

As may be seen in Fig.1, the architecture of the informational system of the human body described by IMC coincides with the empirical ancient model of chakras (Gaiseanu, 2018d) but supplemented and supported by modern *scientific* discoveries of physics at micro and macro scales of our universe. This is a very interesting point of coincidence between ancient "philosophy" and modern "science" that demonstrates the probability that philosophy and science are entangled in a sentient universal reality, and that their integration could point the way to discovery and understanding of our universe and human agency in that universe.

It can be seen that informational chakra centers are "sentient" in which information plays an effective role in all of its forms-- from "free" (virtual) types of information to matter-related (intrinsic, bounded) information (Gaiseanu, 2019b). The human body appears to be a bipolar info-matter (or informed matter) structure that is connected to both information (fields) and to "matter". Within such a structure, one the important cognitive center is the so-called I-believe center, which is the connecting pole for information, particularly with the anti-entropic field of anti-matter (supposed to be dark matter) (Hajdukovic, 2013).

According to IMC, the Info-Connection (IC) pole of matter and anti-matter fields is the assisting gateway for human agency--living, acting as an anti-entropic force against the entropic force of matter and thus assuring the body/mind equilibrium and the health of the microcosmic/macrocosmic organism (Gaiseanu, 2016b; Gaiseanu & Graur, 2018). This connection could explain the near-death experiences (NDEs) (Gaiseanu, 2017b) and other extra-normal properties of the mind (Gaiseanu, 2017c). I-believe could be associated with the feeling of trust and protection that is specific to the religious experience. I-believe is also connected with the hypnotic-based palliative therapies (Brunoli et al., 2017) and with the learning and psychic reprogramming processes of mind (Gaiseanu & Graur, 2018). This coherent informational model is well supported by the experimental evidences concerning NDEs and paranormal propertied of mind (Gaiseanu, 2018d).

The informational system of the human body-model has proved to be not only an active/reactive detector/transducer of its own body and ambient information, but also a fine and subtle detector of universal signals. Based on quantum mechanic and cosmic properties of matter and anti-matter at microscopic and macroscopic scales, IMC is an integrator model, able to explain by its own concepts, both the normal (current) and para-normal (psy) properties of the mind, including the characteristics of quantum-based and holistic vision of consciousness.

The Cognitive-Sentient Exploration of Reality (CSER)

From the perspective of modern science, the subtlety and discipline required to research cosmic sentience as energy is lacking. No matter how sophisticated modern technology may be, physicists and other scientists are rarely able to contemplate an ontology that is so alien to their own. Undoubtedly, it would come as a shock to them that such subtlety and discipline has been applied by ancient civilizations and the resulting science has been bequeathed to posterity over a period of thousands of years. Many modern "materialists" are still unaware of the scientific validity of meditation as it is applied to the model of the chakras.

CSER is an effort to begin a synchronization of ancient science with modern science by addressing similarities between the two. The visualizations (mental images) of meditative practices can be correlated with the imaging dynamics implied by treating *thought-as-vectors* in the informational field where the I-create function plays a decisive role. Stephen Brock Schafer has researched the correlations between the dream imaging dynamics of Carl Jung—the projection of pixel patterns as Jungian archetypal representations (Schafer, 2016). Also, the cognitive-sentient properties of a particular individual which distinguish him from others, could first be generated from the I-created center of sentient activity--a prototypical pattern of behavior genetically transmitted by previous experience of the parents and integrated with individual information. This would be consistent with Jungian psychology of the personal/collective unconscious and the evolutionary objective of Jungian "Individuation". Within this vast enterprise, CSER presents additional details that **are** closely related and can be explained and understood by correlating the activity of the chakras of sentience with the defined centers by IMC, so with the science of modern physics.

Animals have similar sentient properties. For example, the emotive reaction in dogs is very visible, and a great deal of research is ongoing relative to sentience in sub-human species. Sentient in these species was long considered unworthy of scientific research—just as the unconscious was considered to be non-existent. If some animal behavior approximated human "intelligence" it was attributed to instinct and ignored. Sub-special languages are being studied increasingly with amazing results which can no longer be ignored. Research on the chemical language of bees, the high-frequency language of dolphins and whales, and the electro-magnetic "language" of migrating birds provides a robust suggestion that animal languages should be extensively studied.

At the human scale, language abilities and concepts increase the cognitive-sentience properties of the mind enormously, but human language differentiation from animals no longer qualifies the human species as privileged. Sometime, human language skills have not contributed to evolution. We no longer know for sure if human language skills and educational systems result in the better understanding of abstract concepts such as for instance "presentiment". The human capacity to anticipate a problem or a solution may be trumped by animal sentience. At least these matters of sentience must be re-visited. They are matters of certainty/uncertainty balance. From the perspective of sentience in correlation with the terms of IMC, CSER may serve as a research model. The degree of certainty/uncertainty expressed by I-believe could play a decisive role during a decisional process.

CERTAINTY/UNCERTAINTY: INFORMATION AND THE ROLE OF CSER IN ANALYTICAL RESEARCH

The extra-conscious properties of the I-believe center potentiate the capabilities of the normal sensors (view, hearing, taste, smell, touch) in-order to perceive reality and to amplify the synergetic process. The lack of adequate language/concepts that apply to sentience and to subjective intelligence must be addressed and translated into mystical/technological language. In this regard, it would be wise to refer to those ancient systems that were successful in this effort. The chakra system is a case in point, and the languages of Egyptian and Oriental initiates would contribute to better understanding of sentient reality. These initiates have been able to project their wisdom-models into the future and to crystallize authentic concepts and symbology over time. Their scientific symbolism contains in its constituent elements an entire accumulation of useful data that was acquired after a careful, meticulous, and scrupulous effort. Of course, mathematics is a primary language for communicating authentic concepts.

Certainty/Uncertainty: Information

The behavior of microparticles in quantum mechanics provides guidance relative to the natural ways for disposal of multiple possibilities—the behavior of microparticles whose physics state can not be clearly detected from the all points of view. According to the indetermination (non-localization) principle of Heisenberg, locational probabilities of particles cannot be certain. Location can only be measured according to a probability distribution—the product of mass and velocity. Also, in any experimental measurement, there are supplementary uncertainties created by imperfect measurement processes that render deviations from normally expected behavior. The larger the standard deviation of this distribution, the more "uncertainty" it contains.

The human informational system operates with informational YES/NO units by means of CDC, reflected in consciousness by I-want (Gaiseanu, 2018c, 2019a). Some overlap exists between decisions related to physical movements and others to mental processing. The YES/NO binary alternatives are expressed as information units or "bits". Computers work with numbers in the binary system (0 and 1), not in the decimal system to which we are accustomed to use in everyday life. In computers, the binary system allows for the flow of energy through basic circuits, and YES/NO function as switches. The information unit (Bit) has been defined to characterize capturing, transmitting, and receiving information in an energy system. One Bit is an expression of probability (Shannon, 1948).

In the field of information science and technology, information is referred to as a measure of the certainty that exists among a finite number of alternatives in a binary system. The quantity of information acquired or delivered by a system can be evaluated as a difference between the entropy of the system associated with two distinct states. Television, telephony, the Internet, and computer systems are contributing to a prodigious penetration in our daily life of the notion of Bits that expresses the quantity of information which we receive and consume by way of such media. Including an amount of information into a physical system means the reduction of its entropy, and this determines the system's degree of disorder or uncertainty. In other words, the concept of information refers to the certainty that an event could take place, while that of entropy alludes to a degree of uncertainty. Information and entropy are inversely proportional. When entropy in an isolated physical system increases during a period of time, the amount of information decreases.

The Role of CSER in Analytical Research

In the informational operating system of mind, we choose between multiple possibilities/variants in both the internal and external environment. Choices grant credit—obtaining and generating information (through I-believe) only to some of them. This reduction, which in terms of quantum mechanics can be called a "collapsing" process, actually means the reduction of the entropy of the system (uncertainty) and the extraction of the amount of information (certainty) from it (Gaiseanu, 2018d).

Semiconductor materials, especially silicon, the basic material used today in manufacture of integrated circuits, sensors and microsystems, are a valid example of prepared structures for informational systems. Life itself is built on the skeleton of multiple combinations that carbon, with tetravalent atomic structure like silicon, can have especially with hydrogen, oxygen, nitrogen and sometimes with sulfur and phosphorus. This remarkable attribute of creative life is expressed by the capacity to allow for multiplication of the metastable states of carbon in various combination chains. Accordingly, living organisms are in a state of continuous transformation that show some sort of sentience. Life forms are able to compose, recompose, and to maintain some functionally stable structure for adaptation to their environment. So, life processes that are based on such structure are, themselves, capable of knowing and managing themselves. The reduction to a certain state (certainty) from the many potential (but uncertain) states, defines the informational system called life. In terms of physics and information science, life is expressed as a dynamic multi-scale difference in entropies. Therefore, although similar, information in living structures can be reduced, but not limited to electro-magnetic signals, like in microelectronics processors. Information from this point of view is much more than that, and refers to any kind of signal, independently of its nature (view, sound, smell, touch, taste…), quantum or not, at any micro and macro scale. Information in living structures is actually traduced in sentience—informational signals. This confers to IMC the universal and holistic character.

In the process of knowing, the mind investigates reality on the basis of certain existing references which provide the criteria for decision-making. These criteria are generated by the process, itself, at the border between certainty and uncertainty. Besides the information received by the internal and external common senses, information becomes perceptible as a form of revelation, of clairvoyance, or as an intuitive inspiration. According to IMC, this "transcendental" information can be characterized as depending on the activity of the informational center I-believe (connecting with the universe) and I-created (connecting with the species' acquired information).

In both science and art, virtuosity is manifested by inspiration which is obtained by perseverant work, exercises, and training. ***Virtuosity*** is a very personal/individualized process—what Carl Jung would call "contextual Individuation" (Jung, 1933; Schafer, 2016), so this creative process is determined by the individual's power of discovery. The strength and acuity of such a contextual virtuosity requires work on multiple levels of sentience akin to the ancient understanding of "evolution" in which a sentient unit progresses through all the stages of sentient development—from atom to galaxy. A more proximate illustration of the sentient evolution across different scales comes from the field of computational neural modeling. In this illustration, the digital, neural, and personality scales are integrated. George Lakoff describes the integration process from a cognitive standpoint for which the technical term is an 'executing schema,' or 'X-schema' for short (Lakoff, 2008, pp. 26-27). Lakoff explains how neural impulses move "upstream" and "downstream" in the evolutionary process for observing multiple scales of sentience and the virtuoso insights that qualify as intuitive discovery and the digital simulation of the process.

Some famous examples of human virtuosity Archimedes, the discoverer of the law concerning the repulsion force against a body submerged into a fluid; Newton, the founder of classical mechanics; and Mendeleev, discoverer of the ordination table of chemical elements. These discoveries were derived "spontaneously". All of them discovered their unique "evolutionary" laws on the basis of unconscious germination of accumulated data in a substrate. Who is to say that the evolutionary transformation of viruses is not the result of sentient virtuosity at that scale?

Silicon "Doping"

Metaphorically, who is to say that the random movement of atoms by association/dissociation in vacancies (positions without atoms) in the ordered lattice of semiconductors (in particular silicon structures) and mathematically expressed by the correlation factor necessary to define the coefficient of diffusion in the material, is not similar to the individual's movement in the social network? "Choices" at some sentient scale (mostly mathematical-physical-chemical) are based on similar stable, consolidated criteria that is essential to incorporate in order to attain something new (Gaiseanu, 2018c). In every case, the correlation factor for the evaluation of the diffusion coefficient (relating the "effectiveness" of penetration into the silicon lattice), expresses certainty (information) with respect to the uncertainty represented by the multiple virtual possibilities that an atom will follow a specific path through the migrating vacancy (calculated according to the possible trajectories on the order of 1, 2, 3 ...) in the vicinity of the associated atom.

Silicon doping (the behavior of silicon can be changed by "doping") is a distribution process that can be explained and analyzed in terms of a balance between certainty and uncertainty. Doping is based on the diffusion process whereby atoms of distinct species migrate throughout the ordered silicon lattice. This is done to obtain a coherent course for a particular diffusion process into the silicon bulk, describable by a certain specific distribution curve of the impurity concentration. The curve represents the concentration distribution of the migratory (diffusion) atoms within a given depth into a semiconductor.

By applying a mathematical equation called the *least squares fitting technique,* the degree of certainty of the experimental data can be increased. However, the technique, itself, may be understood as a sentient process. By the application of this technique, a physicist recognizes that "reality" at this scale must be described by approximations.

The best measure of the evaluation and control of sentient processes is statistical—based on certainty/uncertainty. The technological process of silicon doping—the distribution of the concentration of implanted atoms into the silicon depth by using a certain implantation energy and dose (quantity of atoms) can be described by a Gauss ("bell") distribution that applies to many other processes in nature. Therefore, "virtuosity" may be understood to exist at atomic-molecular scales—a consequence of the interaction of the implanted atoms with the atoms of the host lattice of a semiconductor.

The scientific language used to express the cognitive-sentient accumulation of data can be correlated with the language of meditative research on the chakras, but this language is so alien to modern scientists that its extensive use would obscure meaning. This correlation cam be done because according to IMC, all, or part of the cognition centers participate to the sentient exploration of reality, allowing new scientific discoveries, by a certain "successful" attitude with respect to a certain objective (Gaiseanu, 2018c). Therefore, the language and dynamics of physics are used with an understanding that the language of energy and force fields can be correlated with the ancient language of the chakras. An unbiased comparison of what is known about the chakras with what is known as discretized data accumulated with modern research and expressed in continuous (non-discretized) mathematical curves would yield Valuable knowledge regarding degrees of scaled sentience. If the value of the accumulation of thousands of years of scientific data be fully appreciated as the painstaking scientific discoveries of myriads of individual researchers, their work on the chakra model of sentience could become scientifically predictive.

The associative process for data acquisition that is accessible by CSER "I-know" is based on similarities among new with respect to the stoked data, so this mechanism plays a significant role in this memorizing process. For example, the intimation of the possibility to explicitly describe the phosphorus (donor-type) diffusion in silicon on the basis of a diffusion coefficient with three distinct forms on three diffusion regions may be understood as a sentient transition. CSER would treat this transition as a process from lower to higher and from simple to complex reality simulation—if it relates to the boron (acceptor impurity) in silicon, characterized by a diffusion coefficient with only two distinct forms of concentration dependence (Gaiseanu, 2013, 2017a).

In order to be triggered, the CSER process needs a target and a reference system where it can run, so training of the brain (meditation) within the framework of the referenced system—accumulated data. Thought—as an informational operator of the mind—becomes a vector in the informational field. Seen in this way, thought is a way to adequately manage this process (Gaiseanu, 2016b).

I-create also plays a decisive role in the management process. When studying the cognitive-sentient capacities of special children who "see/feel" the solution to a mathematical problem before all others, or who are the only ones who actually detect it, or who have a particular aptitude. Viewed from the perspective of education, these children need to be trained to achieve maximum performance—a system of management that has been almost entirely ignored or misapplied in current educational systems.

Developing better systems based on chakras and CSER becomes possible. Intensive data accumulation in the I-know center could provide suitable decision-making criteria and sufficient information to the operational/decisional (intelligence) center. I-want is therefore also a determinative-active informational center involved in such a process. Among other characteristics I-want determines the velocity and the degree of resolution required for the final solution. The affective center of consciousness, I-love could also contribute to the process because it is capable of mobilizing organic-sentient capabilities that are not presently being utilized in educational models. An effective adherence (I-love) to a particular ideal (a "dream" project) would energize the path to discovery by engaging optimal internal forces to achieve the goal. On the other hand, I-am frames the entire evolutionary power trajectory of the contextual organism which provides support for the performing process. Furthermore, the function of I-create could provide a generator for the transmission of acquired/integrated characteristics to posterity.

Language Meaning and Symbols in the Scientific Discourse

As it was remarked in the *Background* section, language as an informational tool is subjected to a certainty/uncertainty balance. Conversational discourse (at whatever scale) carries both explicit and implicit meaning, (Lohrey, 2019) so it can be analyzed in terms of information science. Explicit meaning reveals something while implicit meaning simultaneously conceals something (Lohney, 2019).

Within scientific discourse, a virtuoso author's identity is known by his/her discoveries and theories (explicit meaning), but also by his personal style of thinking (a mind model) with respect to the framework implicitly "embodied" in his research (Lohrey, 2019). Individuation (identity) is manifested by language on some sentient scale (atomic, molecular, chemical, linguistic). How a person thinks is a self-characterizing feature that is reflected in language, but many implicit degrees of sentience are embedded in the energy of language.

According to IMC, the mind operates on the basis of stereotypical thinking chains, acquired by association with other older similar information, and frequently repeated. This behavioral mode allows us to expend less energy, but it tends to obscure the more personal (contextual) aspects embedded in "behavioral" discourse. This tendency obstructs optimal learning and the consolidation of new information (evolution) as a critical aspect of life experience and the development of problem-solving criteria acquired through adaptation.

In other words, the stereotypical sentient mode impedes the evolutionary sentient mode which both exist on multiple levels of energy interaction. Adjusting the stereotypical trend to a more evolutionary trend could be accomplished within the CSER chakra model of sentience. The scope of such a scientific discourse would be more in attune with the complexities of a quantum-sentient worldview—if for no other reason than that it would generate a differentiated base for analysis—that requires diversity of perspective.

Within a complex sentient communicational process, subjective factors range far beyond an individual's subjectivity, but subjective variances are still ruled by metaphorical correspondences within the universal framework of "As it is above, so it is below". Within our CSER model, vast complexity can now be calculated optimally, with algorithms from many scientific fields, which would need to be harmonized. Moreover, certain recent discoveries relative to the sentient qualities of art would need to be incorporated.

Ironically, in this regard ancient ontology may prove to be more accurate than its modern offspring. The reason for this is that ancient worldviews were based on a combination of reason and introspection—the frontier between certainty and uncertainty--that may have been more accurate than modern dependence on technology, statistics, and a basically materialistic scientific method in its reality-determinations. In other words, during the so-called age of reason, modern scientists have abjured their most exquisite instrument for understanding—their own sentient introspection.

The greatest error that modern scientists make is their refusal to recognize how superhuman sentience influences human thought that constitutes the dynamic for evolution. As Norse mythology puts it, evolution happens "at the edges of extremes". As Carl Jung puts it, the cognitive conscious/unconscious is engaged in a perpetual conversation, and that conversation is the frontier of evolution. In other words, the "value intensity" of the energetic interaction results in evolutionary ideas. If both sides of this cognitive conversation are not appreciated, no human progress can be expected. Generally-speaking, the ancients were less susceptible to scientific hubris than modern scientists whose stereotypical prejudice has long ignored what they consider "mystical" and, therefore, unscientific indicators of reality. If "introspection" is mystical and irrelevant, science becomes a façade (Schafer, 2016). According to this argumentation (Schafer, 2016), recent research on Plato's sacred geometry (Pope, 2019) suggests that such science is actually "psychotic and carcinogenic".

Although in the 21st century the Science-Art Research Center successfully demonstrated that the life-force's use of Plato's sacred geometry recorded within the earth's seashell fossil record was evolving to infinity the prevailing scientific world ignored that information because it completely contradicted Einstein's worldview. This is now a complex medical problem. Emotionally, Einstein and his atomic science contemporaries had no option but to develop their symmetrical mathematical tribal science worldview. Since Neolithic times human tribes have been obliged to evolve intellectually by inventing weapons of conflict in obedience to the survival of the fittest, kill or be killed, paradigm observable in nature. However, a single species harming itself is obviously suffering from a neurological disease. (Pope, 2019, p. 4)

The modern scientific dependence on scientific method assumes that great cultures of the past were unscientific, but anyone who has delved into ancient methodology (alchemy, astrology, Pythagorean mathematics, chakras, etc.) knows that this view is a canard. This ephemeral perspective that scientific validity depends fundamentally upon repeatability is too inconsistent with modern discoveries in the physics of fluid dynamics and quantum fields. Today's science is in process of disclosing a wealth of "hard-science" that is consistent with the mystical science of the ancients.

Re-discovery of correlations between ancient and modern science would enhance modern research on the universal mysteries. For example, study of the chakras in Ayurvedic science developed over thousands of years would explain how cognitive-sentient channels within the human organism serve to extend human participation with supernatural forces that guide evolutionary processes in the universe. Appreciation of the legitimacy of such cognitive models can already be understood in the context of scientific method (the Ayurvedic science of Maharishi Mahesh Yogi https://www.amazon.com/Ayurvedic-Healing-Contemporary-Maharishi-Ayurveda/dp/1848190697).

The language of art has been orphaned in Western scientific method, but this scientific outcast now becomes the prodigal son. Artistic method, with its subjective, indirect, contradictory, and provocative hidden messages becomes obviously more consistent with authentic human cognition and cosmic sentience which is characterized by an evolutionary invitation to reflection and contemplation.

This is an implicit mode of exposition. By definition, the "hard problem" in science proposes a rational, explicit and clear style, even if it refers to models of uncertainty revealed in quantum mechanics. However, let us not forget that we do not actually live in the world of microparticles. We are their product. We are constituted from them, but actually we don't represent their arithmetic sum. Rather, we represent a coherent assembly of distinct macroscopic components with synergic functions directed towards a single primary purpose—to survive and flourish. A holistic science requires a holistic model of sentience that has only been recognized in environmental science.

Each of the body organs, made up of billions of atoms, subjected to other billions of chemical reactions and physical interactions, fulfills a common purpose in which each component part fulfills its own (non-quantum) function and role in time and place (Gaiseanu, 2016a, 2019b). These billions work together as a unity—imperceptible to humans—an aggregate with unitary, interdependent but coherent features that are related through information—informed through sentience. No matter how fascinated we may be with the spectacular discoveries of quantum physics, biology, and molecular genetics, we should not become pray to the temptation to stereotype human function as identical to microparticles.

Human agency is special and carries with it special responsibility. However, human responsibility is only possible with language, some understanding of cosmic sentience, dedication to eternal disciplined striving, and a revived appreciation of the concept, "As it is above, so it is below." The new scientific discoveries bring added value to the language of human sentience, making it more distinct, specific, and authentic. However, the concepts and models of science must now be expanded to sentience. To do so requires art.

Unlike science, in art the range of limitations on the exploration of reality are less rigid. The role and the "magic" of art may consist precisely in a balance between certainty and uncertainty—the characteristic balance of sentience. In art, the participant is invited to dream, to add his/her own contribution to the sentient product. Dedicated more to an implicit message, art has the role of provoking the participant's own sense, his/her own means of reception. The sentient participant is valued not only according to the explicit prism but by an implicit prism of receptive agency powered by imagination. In other words, sentience is infinitely expanded. The range of the sentient "game" is much larger than current science permits, so current science—in spite of its subtlety--must be attuned to a yet more subtle sentient science. Art has the role of initiating / promoting exploration of the unknown by way of the infinitely fractal subjective participation of the receptor and by canonizing personal experience, choices, and interpretations.

Instead, science, takes the role of expressing itself by means of certainty, thereby weighting the ratio toward certainty and away from uncertainty. From the perspective of philosophic view by means of ancient cultures and according to quantum mechanic perspective of the microparticles universe (Schafer, 2016), this role is what has been called illusory because within the emergent ontology there is no a real certainty, only a probability. Moreover, "certainty" tends to disallow personal "growth" in favor of entropy. A thorough discussion of Plato's sacred geometry as it applies to asymmetrical vs. symmetrical mathematics is not possible within the constraints of this paper. However, CSER and the chakra model appear to be consistent with the evolutionary tendency rather than the tendency to entropy as defined by the atomistic culture alone. As Einstein himself pointed out in an article in 1919 in the Times of London, a theory of principle is based on empirical observations without the need for a particular model whereas a constructive model (like the atomistic Boltzmann approach) will first make assumptions about a fundamental structure then will built a mathematical description of that structure that hopefully will give relationships between the empirically observed parameters (https://secondlawoflife.wordpress.com/2009/07/12/what-einstein-thought-about-thermodynamics/).

A new humanity may no longer be constrained with the algorithms of the past. They must be taken into account in order not to deform the future. However, the future is more implicit and dependent on what Jung would call personal "individuation". In order to achieve "transcendent" sentience—new models and methods for human agency—previous baggage accumulated during the years of preparation must not be allowed to misdirect science.

In a sense, in order to penetrate the hidden (implicit) meaning of art/science and to achieve one's own meaning through transcendent sentience, one must freely express, without prejudgments, individualized creativity and personality.

CONCLUSION

Cognitive Sentient Exploration of Reality was recorded during the Egyptian civilization where CSER was formulated in concepts, models, and symbols that demonstrated the relationship between humans and cosmos. Egyptian schools of the wisdom reserved this wisdom for initiates and priests of the Temple. In these schools of the Mysteries, the rational sciences of mind, ideas, and thought which were adopted by the Greeks, amplified by the Initiate Pythagoras, and incorporated in-part to later religions.

Today, based upon the principles of mathematics, empirical observations (sound, and color), the physics applied to the quantum mechanics of microparticles demonstrates that the true nature of the SCER reality can only be revealed with approximations that may be designated as a balance between certainty and uncertainty. Therefore, any understanding of the reality by means of CSER must be addressed both rationally and transcendentally. The Informational Model of Consciousness (IMC), provides and defines seven distinct, but interrelated cognitive centers—neurodynamic energy centers called chakras—capable of explaining Near Death Experiences and other extra-normal cognitive phenomena. As a concrete application, IMC can also serve as a basis to approach and explain the mechanisms of CSER processes--intuition, revelation, clairvoyance, and premonition--hovering at the border between certainty and uncertainty. The personal "flair" (the capacity to see or smell the future) is being applied in many scientific and design fields, design technology, control and evaluation processes in industry and education, intelligence services, business management, medicine and private life.

The capacity to "predict" is an important objective of almost all fields of human knowledge, so that which has been considered anomalous in the past has become a primary factor in the "modern" reality of CSER. Predictability can be explained by the synergistic contribution of cognitive center activities defined and described within IMC. Summarizing, the centers I-created (inherited abilities), I-want (definition of a target and a system plus relational/decisional intelligence) and I-believe (connecting to the informational phenomena like premonition, clairvoyance, intuition, revelation and the pole reducing uncertainty to certainty) determine the predictability process. Predictability dynamics can be applied to education/training, i.e. the development of the data acquisition system (Iknow), which improves the access database for the sensory-operative system and the acquisition and analysis/processing of information. The center I-love, resonating with information in a sensitive way, determines the affective adherence to the proposed objective and mobilizes the specific internal resources of the body to achieve the proposed goal.

Semiconductor materials, particularly silicon, are atomic/electronic systems which can be analyzed in terms of certainty/uncertainty balance and information, like the living structures. The atomic diffusion and implantation in silicon is such a process. A specific illustration of the process is the distribution of concentration resulting from the association/dissociation mechanism of point defects such as vacancies and interstitial atoms with the host atoms during the atomic transport into the semiconductor bulk. The uncertainty level of the experimental data could be reduced by mathematical techniques like the best square fitting, but the subjective CSER of the researcher could be deeply implied in order to discover the optimal way to research, from lower to superior (upstream) and from simple to more complex levels. The principle may be applied in the case of the analytic simulation problem of phosphorus diffusion with respect to that of the boron diffusion in silicon, as an already applied example.

The communication process by language can be analyzed in terms of information, as defined in the science of information, where the implied factors are: the transmitting source (the issuing author), a codifier (his/her personal skill), a transmitting line (the transmission means) and the receiver and his decoding system (personal skill). As the language itself is a result of the certainty/uncertainty balance, coming from the definition the words and from the particular expressing/meaning style of the issuing author, the scientific discourse could be clarified by use of the symbols. The using of symbology, applicable especially in mathematics, physics but also in other sciences, is a powerful method because may eliminate or reduce the implicit uncertainty. The receiver of the scientific discourse should follow therefore rigorously the meaning of the symbols, by running the implicit previous algorithms concurring to the symbol definition. The receiver's preparation plays a determinant role for the suitable capture of the transmitted message. On contrary, the implicit meaning in arts should be discovered by the receiver, accordingly with his/her sentience potential, this becoming a co-participant of the sentient-exploration of the reality transmitted by the art message.

By its powerful potential of reality exploration, IMC opens a wide gate for further investigations and research related to the correlation between the CSER model and the CHAKRAS, with large perspectives of application in medicine, neuroscience, life sciences, psychology, philosophy and even in physics, allowing to clarify the possible inter-dependence between mind and the subtle field/particle support of information at micro and macro scale, within the so called mind-body problem.

ACKNOWLEDGMENT

The author addresses his thanks to Stephen Schafer for the incentive, guidance and contributing discussion on this topic, especially on the philosophical discourse concerning the sentience.

For their close and warm support, he thanks his son Adrian Gaiseanu and his daughter Ana Maria Gaiseanu.

He is highly grateful to his eminent and exemplar parents, Professors Emanoil Gaiseanu and Florica Gaiseanu.

REFERENCES

Abram, D. (1997). *The spell of sensuous perception and language in a more than human world.* New York: Vintage Books, A Division of Random House Book, Inc.

Aharonov, Y., Popescu, S., Rohrlich, D., & Skrzypczyk, P. (2015). Quantum Cheshire cat. *New Journal of Volume Physics, 15*, 1-5. Retrieved from http_iopscience.iop.org_1367_2630_15_11_113015_art.pdf

Anonymous. (2009). *What Einstein Thought About Thermodynamics, The Second Law of Life*. Retrieved from https://secondlawoflife.wordpress.com/2009/07/12/what-einstein-thought-about-thermodynamics/

Bartholomew, Y. B. (2018). The evolution of consciousness. *Environment and Ecology Research, 6*(4), 259-269, http://www.hrpub.org

Brugnoli, M. P., Pesce, G., Pasin, E., Basile, M. F., Tamburin, S., & Polati, E. (2015). The role of clinical hypnosis and self-hypnosis to relief pain and anxiety in severe chronic diseases in palliative care: a 2-year long-term follow-up of treatment in a nonrandomized clinical trial. *Ann. Palliat. Med.* doi:10.21037/apm.2017.10.03

Chalmers, D. (1995). Facing up to the problem of consciousness. *Journal of Consciousness Studies, 2*(3), 200–219.

Clark, A. (2000). *The theory of sentience*. Oxford Press. doi:10.1093/acprof:oso/9780198238515.001.0001

Denkmayr, T., Geppert, H., Sponar, S., Lemmel, H., Matzkin, A., Tollaksen, J., & Hasegawa, Y. (2013). *Observation of a quantum Cheshire Cat in a matter wave interferometer experiment*. Retrieved from http://www.arxiv.org/pdf/1312.3775

Dulcan, C. D. (2009). *Inteligenta materiei [Intelligence of matter]* (3rd ed.). Cluj, Romania: EIKON.

Floyd, K. (2011). *Interpersonal communication*. McGraw-Hill Education.

Fracasso, C., & Friedman, H. (2011). Near-death experiences and the possibility of disembodied consciousness: Challenges to prevailing neurobiological and psychosocial theories. *NeuroQuantology: An Interdisciplinary Journal of Neuroscience and Quantum Physics, 9*(1), 41–53. doi:10.14704/nq.2011.9.1.389

Gaiseanu, F. (2013). Contributions to the modelling and simulation of the atomic transport processes in silicon and polysilicon and applications. *Proceedings of the Romanian Academy, Series A, 4*(4), 376–384. Retrieved from http://www.acad.ro/sectii2002/proceedings/doc2013-4/15-Gaiseanu.pdf

Gaiseanu, F. (2016a). Consciousness as informational system of the human body: Consciousness and life physics. *Cosmology and Astrophysics Journal, 16*(1), 14–25. Retrieved from http://physics.socionic.info/index.php/physics/article/view/227/182

Gaiseanu, F. (2016b). Informational subsystems of the consciousness. *The Science of Consciousness, Abstract Book, TSC 2016 TUCSON, 9*, 29. Retrieved from http://www.consciousness.arizona.edu/documents/TSC2016_BOOK_of_Abstracts_for_emailWeb.pdf

Gaiseanu, F. (2017a). Modelling and simulation of the impurity diffusion and related phenomena in silicon and polysilicon systems in microfabrication and micromachining technologies. *Annals of the Academy of Romanian Scientists, 10*(1), 41–78. Retrieved from http://aos.ro/wp-content/anale/IFVol10Nr1Art.4.pdf

Gaiseanu, F. (2017b). Quantum-assisted process of disembody under near-death conditions: an informational-field support model. *NeuroQuantology, 15*(1), 4-9. Retrieved fromhttp://www.neuroquantology.com/index.php/journal/article/view/971

Gaiseanu, F. (2017c). *An information based model of consciousness fully explaining the mind normal/paranormal properties*. NeuroQuantology 15(2),132-140. Retrieved from https://www.neuroquantology.com/index.php/journal/article/view/1040

Gaiseanu, F., & Graur, A. (2018). Cognitive centers related attitude: Application for an iterative evaluation method in music-based therapy process. *Abstract Book of the Conference on Science of Consciousness.* Tucson (Arizona) April 2-7. http://www.consciousness.arizona.edu/documents/TSC2018AbstractBook-final3.pdf

Gaiseanu, F. (2018a). Near-Death experiences and immortality from the Perspective of an informational modeling of consciousness. Gerontology & Geriatrics Studies, 2(3). 1-4. Retrieved from https://crimsonpublishers.com/ggs/pdf/GGS.000538.pdf

Gaiseanu, F. (2018b). An informational modeling of consciousness and cognitive centers. *Proc. of the Human Project Brain (HPB) International Conference: Understanding Consciousness, A scientific quest for the 21-st century.* Barcelona, June 21-22. Retrieved from http://hbp-ic.com/scientific-information/programme/

Gaiseanu, F. (2018c). Destiny or free will decision? A life overview from the perspective of an informational modeling of consciousness Part II: Attitude and decision criteria, free will and destiny. *Gerontology & Geriatric Studies*, *4*(1), 1–7. doi:10.31031/GGS.2018.04.000576

Gaiseanu, F. (2018d). Information: From philosophic to physics concepts for an informational modeling of consciousness. *Philosophy Study J.*, *8*(8), 68–82. doi: 10.17265/2159-5313/2018.08.004.Retrieved from http://www.davidpublisher.org/Public/uploads/Contribute/5c06323653cd2.pdf

Gaiseanu, F. (2019a). Destiny or free will decision? A life overview from the perspective of an informational modeling of consciousness Part I: Information, consciousness and life cycle. *Gerontol & Geriatric Stud.*, *4*(3), 1–6. Retrieved from https://crimsonpublishers.com/ggs/pdf/GGS.000586.pdf

Gaiseanu, F. (2019b). The informational model of consciousness: Mechanisms of embodiment/disembodiment of information. *NeuroQuantology: An Interdisciplinary Journal of Neuroscience and Quantum Physics.* (in press) 17(4)

Hajdukovic, D. (2013). Can observations inside the solar system reveal the gravitational properties of the quantum vacuum? *Astrophysics and Space Science*, *343*(2), 505–509. doi:10.100710509-012-1303-3

Hall, M. (2001). *The secret teaching of all ages: An encyclopedia outline of Masonic, Hermitic, Qabalistic and Rosicrucian symbolical philosophy.* Retrieved from http://www.sacred-texts.com/eso/sta/index.htm (Original publication 1928)

Hameroff, S. (1998). Quantum computation in brain microtubules? The Pennrose - Hameroff". Orch OR" model of consciousness. *Philos. Trans. Royal Society London, Ser.A. Math. Phys. Sci.*, *356*, 1869–1898. doi:10.1098/rsta.1998.0254

Hameroff, S., & Penrose, R. (1996). Orchestred reduction of quantum coherence in brain microtubules: A model for consciousness. *Mathematics and Computers in Simulation*, *40*(3-4), 453–480. doi:10.1016/0378-4754(96)80476-9

Hameroff, S., & Penrose, R. (2014). Consciousness in the universe: A review of the 'Orch. OR' Theory. *Physics of Life Reviews*, *1*(1), 39–79. doi:10.1016/j.plrev.2013.08.002 PMID:24070914

Hodges, D. B., Aylet, K., & Reeves, S. (2008). Analytic research - Discourse Analysis. *BMJ (Clinical Research Ed.)*, *337*, a879. doi:10.1136/bmj.a879 PMID:18687729

Hurford, R. J. (2008). The evolution of human communication and language. *In Cap 14*. In P. d'Ettorre & D. P. Hughes (Eds.), *Sociobiology of Communication, An Interdisciplinary Perspective* (pp. 249–264). Oxford University Press. doi:10.1093/acprof:oso/9780199216840.003.0014

Ignatenko, A. (1994). Cum sa devii un fenomen [How to become a phenomenon]. Bucharest: Ed. Biodova.

Jung, C. (1933). *The collected works of C. G. Jung: The real and the surreal*. Retrieved from http://www.ihp.sinica.edu.tw/~medicine/book/Jung.html

Krishna, V., & John, M. (2004). The art of conversation: Eliciting information from experts through multi-stage communication. *Journal of Economic Theory*, *117*(2), 147–179. doi:10.1016/j.jet.2003.09.008

Lakoff, G. (2008). *The political mind*. Viking/Penguin.

Lohrey, A. (2019). *The principles of nonlocal realism*. Retrieved from https://www.academia.edu/38239068/Nonlocal_Realism.docx

Marchetti, G., Giulio, M., & Alharbi, A. (Eds.). (2015). *Attention and meaning – The attentional basis of meaning*. New York: Nova Science Publishers, Inc.

Mariella, S., & Grech, V. (2016). Sentience in science fiction 101. *SFRA Review*, *315*, 14–18.

Maxwell, N. (2001). Evolution of sentience, consciousness and language viewed from a Darwinian and purposive perspective. In The Human World in the Physical Universe: Consciousness, Free Will and Evolution. Rowman and Littlefield.

McPheat, S. (2010). *Effective communication skills, training & venues*. Google/Scholar.

Meijer, D. (2013a). Information: What do you mean? On the formative element of our universe. *Syntropy*, *3*, 1–49.

Meijer, D. (2013b). Immortality: Myth or becoming reality? On the conservation of information. *Syntropy*, *3*, 168–207.

Myss, C. (1996). Anatomy of the spirit: The seven stages of power and healing. Crown Publishers, Inc.

Odoul, M. (2014). *Ghidul complet al durerilor fizice - Terapii psihoenergetice, Prefață Dr. Medynski, Thierry* [The Complete Physical Pain Guide - Psycho-Therapeutic Therapies, Preface by Dr. Medynski, Thierry]. Bucuresti, Romania: Ed. Paralela 45. Retrieved from https://caleaechilibruluii.files.wordpress.com/2014/04/michel-odoul-ghidul-complet-al-durerilor-fizice.pdf

Pereira, C. (2016). Is it quantum sentience or quantum consciousness? A review of social behaviours observed in primitive and present-day microorganisms. *NeuroQuantology: An Interdisciplinary Journal of Neuroscience and Quantum Physics*, *1*, 16–27.

Pokorny, M. (2012). Sentience, awareness consciousness. *Organon F*, *19*(Suppl.), 51–63. Retrieved from https://www.academia.edu/28728892/Sentience_Awareness_Consciousness

Pope, R. (2019). Plato's map of the future omni artifact. *Science-Art Research Center of Australia.* Retrieved from https://connect.xfinity.com/appsuite/#!!&app=io.ox/mail&folder=default0/INBOX

Proctor, H., Carder, G., & Cornish, A. (2013). Searching for animal sentience: A systematic review of the scientific literature. *Animals, 3*, 882-906; doi:10.3390/ani3030882

Radin, D. (2006). *Entangled minds, extrasensory experiences in a quantum theory.* Paraview Pocket Books, a division of Simon Shuster, Inc.

Radin, D. (2018). *Real magic: Ancient wisdom, modern science, and a guide to the secret power of the universe.* New York: Ed. Harmony Books, Crown Publishing Group, Penguin Random House LLC.

Satsangi, A. K., & Brugnoli, M. P. (2018). Anxiety and psychosomatic symptoms in palliative care: From neuro-psychobiological response to stress, to symptoms' management with clinical hypnosis and meditative states. *Annals of Palliative Medicine, 7*(1), 75–111. doi:10.21037/apm.2017.07.01 PMID:28866901

Schafer, B. S. (2016). *The Media-Sphere as Dream: Researching the Contextual Unconscious of Collectives. In Exploring the Collective Unconscious in the Age of Digital Media.* IGI Global.

Sen, S., Mahendra, A., & Priyadarshi, P. (2015). *Communication and language skills.* Cambridge University Press.

Shannon, C. (1948). A mathematical heory of communication. *The Bell System Technical Journal, 27*(3), 379–423. doi:10.1002/j.1538-7305.1948.tb01338.x

Smith, H. (1958). *The world's religions.* New York: Harper Collins.

Tononi, G. (2004). An information integration theory of consciousness. *BMC Neuroscience, 5*(1), 42. doi:10.1186/1471-2202-5-42 PMID:15522121

Van Lammel, P. (2006). Near-death experience, consciousness, and the brain: A new concept about the continuity of our consciousness based on recent scientific research on near-death experience in survivors of cardiac arrest. *World Futures, 62*(1-2), 134–151. doi:10.1080/02604020500412808

Chapter 4
Archetypes of Corporativity and Hierarchy in the Internet

Denis Lvov
Siberian Federal University, Russia

ABSTRACT

C.G. Jung defined the concept of a collective unconsciousness (CU). Fundamentally, Jungian archetypes of the unconscious may be understood as quantum electrodynamic (QED) patterns that exist at deep levels of human individual and collective psyche. Unlike the concepts of Sigmund Freud, comparative research demonstrates that Jungian dream therapy—which is based on archetypal dynamics—accords with the most recent research in the field of neurolinguistics. Archetypal dynamics of the collective unconsciousness have direct applicability to research on sentience in a quantum electrodynamic field according to quantum field theory (QFT). Increasingly in today's science that which has been considered unconscious must be understood as "sentient." Before Freud and Jung, science did not contemplate the unconscious because a "nullity" cannot be measured.

INTRODUCTION

Carl Jung provided a unique way to deal with archetypal dynamics. By doing so, he created an indispensable tool for significant research on nullity, but such subjects are difficult for people to understand. This paper addresses the many different connotations that exist relative to the use of the term "archetype.

The concept of "archetype" is used throughout the disciplines of modern humanities and social sciences, but it is rare for different authors to have a clear understanding of these archetypal "patterns of the unconscious. Indeed, it is rare for researchers to distinguish archetypes which are entirely unconscious from "archetypal representations" that are more-or-less conscious projections of archetypal patterns.

Understanding this dynamic relationship between *archetypes of the unconscious* and *"archetypal representations"* is of critical importance to scientific research during the paradigm shift into a mediated age. The dynamic is critical because during an age of illusion, access to "contextual meaning"—as scales of sentience—is the key to evolution. Archetypes are like the software program patterns that are projected as the pixel patterns that appear as images. Currently, understanding of Jung's concept includes

DOI: 10.4018/978-1-5225-9065-1.ch004

a wide variety of nuances, and these also vary depending upon the context (contextuality) of usage in different fields. Naturally, this causes a significant variation in the definition of the term and leads to confusion—especially as it relates to the subtle nuances involved in correlating QED with the concept of "sentience". In this regard, it is appropriate to undertake an analysis of the use of the term "archetype" in the titles of research projects supported by the Russian Science Citation Index.

BACKGROUND

The units of analysis used in this research were selected publications that appear in the website search engine of the scientific electronic library http://elibrary.ru/. This resource has been selected because it supports the system of the *Russian Science Citation Index*. In addition, access to the resource is free.

The following parameters were set to search for the word "archetype" in the title during the 2015-16 years of publication. Because the term "archetype" is a marker for content. Indication of the term in the title underlines its importance for the text and suggests the relevance of the concept in the publication. Of course, a more thorough of content would entail the use of the term throughout the text, but the elibrary.ru system is not always open access. Often, the full texts of the publications are available only for a fee, and in many cases only the output data are available. In this context, and in view of the already mentioned limited resources, it was decided to focus on the title search, thus leveling the search procedure and equalizing all publications to a certain extent.

With these constraints, the analysis of publications that appeared in the elibrary.ru search engine was undertaken. On the 31 of January in 2017 in response to a request by the word "archetype" in the title of the publication over the years 2015-16, 320 positions were displayed. Further study revealed 5 redundancies. For some reason the 5 publications were included - for example, the article "The archetype of fish in Russian culture" (Oshchepkova). Thus, 315 unique pieces were found.

The subjects of these publications are diverse and extensive. In order to achieve a rough systematization, the whole array was divided into conditional sections. Publications themes were categorized on the basis of key concepts involved in the titles. Again because of the already indicated unavailability of full texts for evaluating a significant part of the material, additional analysis of the output data release was conducted only in the case of thematic ambiguity (as for example: Tikhonov). Finally, the following statistics were obtained:

Sections are arranged in descending order of the number of publications. Publications defined in the "literature and folklore" section are united as they focus on the study of texts. They are specifically separated from the "art studies" section to show their dominance. Publications in the section of the "art studies" incorporated all kinds of art from ancient sculpture to cinema. Architecture is listed as a different category because of its more practical orientation and its function that relates to the organization of physical human space.

THE CONCEPT OF THE ARCHETYPE IN RUSSIAN-LANGUAGE PUBLICATIONS DURING THE YEARS 2015-2016

As can be seen from the table, the term "archetype" is used almost exclusively in the humanities (in the broadest meaning). The only publication in biology can be regarded as negligible deviation. With

Table 1. Distribution of thematic categories where the word "archetype" is found in the title. (elibrary.ru database for 2015-16)

Theme section	Amount	Percentage
Literature and folklore studies	119	37.77
Culture and ethnic studies	55	17.46
Art studies	35	11.11
Philosophy	30	9.52
Psychology and pedagogics	28	8.88
Advertising, public relation and marketing	21	6.66
Legal and law studies	9	2.85
Religion studies	7	2.22
Architecture	5	1.58
Management	5	1.58
Biology	1	0.3
Total	315	100

an impressive margin "literature and folklore", the term archetype appears in more than a third of all publications. Thus, we can conclude that the concept of "archetype" is extremely convenient for the studies of literary texts.

Apparently, this concept is considerably convenient for the description of certain phenomena in all areas related to culture and art. This is evident from the fact that these humanistic categories occupy the top three places in the list. It can be deduced that the concept of the archetype reflects-phenomena that are important for understanding the vague concept of "second nature"- a characteristic or habit in someone that appears to be instinctive. Already, the concept of "archetype" for Russian-speaking researchers seems to reflect an important aspect of artificiality that characterizes human nature. In other words, this vague and distorted usage of the term archetype suggests that the term is specific to all humans rather than a difference related only to such things as differences in language; i.e. simple tracing of etymology from Greek into Russian language, which is traditionally translated as "first form". The first form for copying or imitation may well act as a natural phenomenon of "sentience", absolutely independent of the human being. However, the connotations of the usage of the term "archetype" in research titles-- at least for the Russian-language authors—seems specific to the human species. Therefore, the analysis of this concept can be specifically useful for the philosophical understanding of human existence.

It is worth noting that a clear definition of the concept of "archetype" is rarely given in the open texts. The term is treated as if it is "known" and does not require definition, but this is a significant error that leads to vast confusion. At best, in-order to understand the meaning of the term, it is necessary to refer to the ambiguous context in which it is used.

Literature and Folklore

Usually, within the context of the category of "Literature and Folklore", the term archetype refers to a dramatic pattern in literature. Carl Jung is often given credit for the term, but the narrative-mythic significance of the term may be credited to Joseph Campbell who gave it the name of the Hero's Journey (1968) as a universal pattern in the world's mythology. Actually, the source of this Journey was defined by Jung as "Individuation"— the primary archetype of human evolution. Therefore, if we pay attention primarily to the publications from section "Literature and Folklore", then the word "archetype" very often refers to some general scheme (the Hero's Journey), which occurs in a variety of artistic texts.

This is a good example of how Jung's scientific specificity is diluted and misunderstood. Just as in the publications of the other sections here it was often found that references were made to C.G. Jung, whose legacy greatly influenced the reception of the archetypal pattern of the Journey. "However, in modern times, a mention of the term "archetype" has been reduced to a simple designation of common mythological motifs and structures, used in all the artistic works of literature without reference to Jungian philosophy" (Baturin D.A., 2016, pp. 177-180). Thus, in most cases, even if there is a direct reference to the Swiss psychoanalyst, understanding of any relationship with the deep structures of the collective unconscious (which is central to Jungian research) is lost. In this sense, as revealed in the relevant publications in this section as a whole, the use of the concept "archetype" is usually understood in terms of a literary pattern rather than as a universal cognitive pattern. For most authors this cognitive basis for universal sentience is not important. Superficial literary exegesis is perceived as being more important than any understanding of cognitive sentience. Realization of universal-sentient commonality goes far beyond its commonplace use as a pattern in literature.

Culture and Ethnic Studies

The same emasculation of the concept "archetype" is often observed in publications from the "culture and ethnic studies" category. Instead of being a fundamental cognitive dynamic, the term archtype becomes merely a label to suggest some vague authenticity in an author's hypothesis. A reference to the collective importance of the archetypal phenomenon lends an essential credibility to an author's hypothesis relative to the dynamics of social collectives.

Even though the cognitive dynamic may be superficially understood, these dynamics become an important aspect of their research. The term "cultural archetype" is often used in publications from this section. Use of the archetypal concept not only emphasizes the relevance of the subject, but also loads the concept with additional support for an assumption of mutual experience – some kind of supra-individual nature that is essential for research on any social collctivity.

Here again, mentions of C.G. Jung occur without understanding the deep cognitive meaning of the archetypes as they are defined by the Swiss scientist. The following example is characteristic:

Under the cultural archetype the author understands the phenomenon, which classical interpretation was formulated by the founder of analytical psychology Karl Jung. It was he, who revealed the connection between the human unconscious and cultural processes. In this context, cultural archetype is a basic element of culture, which is forming a constant model of spiritual life" (Privezentseva, 2016, pp. 198-202)

The importance of realizing that there exists a connection between individual and collective cognitive dynamics remains unexplained, but their importance to understanding culture is observed. Already in the next sentence it is said: "There are distinguished universal cultural archetypes:-discovery of fire, chaos, creation, the union of male and female, change of generations, "Golden Age ", etc." (ibid). The consequences of connection are listed, but no effort is given in order to trace the ties between unconscious archetypes and conscious patterns.

In another publication the archetypal examining concept is misused in a way directly contrary to-Jung's understanding of archetypes as the structures of the collective unconscious: "National consciousness is laid at the genetic level, in the form of symbols and signs, archetypes of consciousness, constraining and organizing the ontological world (the material world) and the world of consciousness (ideal world)". (Pshenichnaya, 2016, pp. 307-312) The reference to "Archetypes of consciousness" would be an oxymoron for C.G. Jung.

Art Studies

In the publications devoted mostly to graphic art. Jungian symbolism is often used as direct references to classical Jungian archetypes. For example, there are references to "Anima" in the article by T.I. Erokhina on gender aspects in the representation of the Russian Silver Age in mass culture. However, this text focuses not so much on Anima itself, as on its more specific example – Persephone. Further analysis divulges the common practice of deferring to literary examples of archetypes rather than archetypes employed in the writings of Jung. In other words, the archetypes of art are constructed from archetypes in literature. "We believe that this archetype of the Poet can therefore be correlated with the literature archetypes" (Erokhina, 2016, p. 333). Artistic appeal to the "literature archetype" is quite characteristic - a concept, which has some commonality with the already analyzed above "cultural archetype". Also, in the publications dedicated to the fine arts you can find the term "visual archetype" (Nikolenko, 2015) characterizing the invariance of certain forms of images. Jung would have considered such invariance of symbolic images as "dead symbols" due to their lack of vital interpretation.

Thus, in the largest categories under consideration, in most cases the concept of the "archetype" serves merely to emphasize similarity of concept perception rather than meaningful discovery which is the entire purpose of archetypes. Archetypes become just a convenient "common space". Jung would address such a cultural space as a "collective persona" because "A people, in the doctor's eyes, presents only a somewhat more complex picture of psychic life than the individual." (Jung, 1933, pp. 209-210) Jung's realization that, "Disruption in the spiritual life of an age shows the same pattern as radical change in an individual." (p. 202) allows for the deep cognitive study of cultures Perhaps quite aptly, the potentials of this correlation between individuals and cultures is generalized by T.S. Zlotnikova (2015) In cultural studies, archetypal patterns serve research in several capacities:

- As a focus for study of psychological dynamics and their correlated representations
- As contextual parameters for designing cognitive models (matrix/mechanism/algorithm) for other representations such as myth, cognition, consciousness, drama, language, and even ontology.
    - As a fundamental ontological dynamic for clarification and expansion of research in the problematic field of cognition and sentience (Zlotnikova, p. 34)

In the publications that were reviewed, the concept of "archetype" is often interpreted in terms of Zlotnikova's second category where usage of the word "archetype" usually refers to the concept of "archetypal image". However, this usage is in comlete cotnradiction to Jung's principle. Jung specifically differentiated the terrms archetype and archetypal representation. Representations are projections into consciousness from unconscious dimensions of the archetypes that are only accessible from interpolation of their images. The term "archetype" is often misinterpreted as some very definite mythological image or motif. But motifs are nothing more than questionable representations; "It would be absurd to assert that such variable images could be inherited". (Jung, 1991, p. 62) Elsewhere, Jung stressed that — because they are unconscious — the archetypes cannot be meaningfully defined. Understanding of archetypes is limited to formal and conditional "relatively conscious representations". In other words meaningfully archetypes can only be inferred from conscious experience that is filled with conditional influences (Jung, 1998, p. 127).

In order to make full use of the potentials that the concept of archetypes accrues, the distinction between unconscious archetypes and conscious archetypal representations must be appreciated. This distinction allows the principle to be used in research on quantum electrodynamic (QED) "sentience". A paraphrase of this Jungian principle is that an archetype is embedded in the collective unconscious "program" that an individual perceives and acts upon. In Jungian terms, archetypal actualization is "contextual". In other words, an individual's response to archetypal patterns is entirely unique. Individual responses to a given situation depend upon unique interpretations by an individual (or a collective) of semiotic resonances originating in unconscious dimensions. Behavioral responses are best understood in terms of the structure of images. Such cognitive structure is known to be dramatic in nature. This dramatic structure of dreams has been verified and correlated with the dramatic structure of the cognitive unconscious (Schafer, Stephen Brock, 2016a; fer, Stephen, 2012; Lakoff, George, 2008 & 2002).

To illustrate this cognitive dynamic, in some typical situation one or more archetypal images occur, which symbolically describe the situation for the individual. An archetypal image is symbolically represented in the individual consciousness as a "contextual" image that shares the archetypal dramatic structure and resonates with contextual experience. The conscious experience of the person fills the archetypal pattern with conscious details that impel the individual to a specific sequence of actions that are determined by the laws of dramatic architecture. The concept of "cultural archetypes" follows the same dramatic pattern. George Lakoff said, "We live our narratives." that are a combination of both personal and cultural stories from which we create our own self-image.

Philosophy

Within the philosophy category, the concept of archetypes is detailed, analyzed, and defined according to the premises of various authors. It can even be said that it is typical for the majority of authors of the publications in this section to actually do exactly the same as in this article except that they often employ the concept of "archetype" not as a subject for study, but as a working generic term, used to refer to the archaic nature of the genesis of other phenomena that are of interest to respective authors, e.g., social physicality (Gutov, 2016) exploring certain archetypes as structure-forming promptings for the various phenomena of social life (Syutkina, 2016), or the concept of all-unity (Davydova, 2015) that relates to QED sentience.

Psychology and Pedagogy

The publications from the "Psychology and Pedagogy" category are similar to the Philosophy category. Usage of the "archetype" concept is applied as a generic pattern, and primary importance is given to the author's central concept. Publications that refer to the concept of "archetype" stay within the traditional parameters of Jungian psychology. It is worth noting that attention is also given to other Jungian authors, for example, to the C. Pearson's "12 archetypes", as well as to the works of J. Hollis.

Advertising, Public Relations, and Marketing

As a rule, authors of publications placed in the category "Advertising, public relation and marketing" are not inclined to a detailed analysis of the "archetype". Instead, the concept is used within a limited—mass communication—environment in an effort to identify target groups and to design advertisements that appeal to target audiences. Such designs are associated with various meanings, quite easily recognizable by target audiences. Advertisements, brands and images are frequently analyzed from the point of personality types in accordance with stereotypical archetypes. Authors refer to existing typologies (Gabrieljan, 2016), or propose their own (Chzhan, 2016). References to Jungian archetypes can also be found in the analysis of messages for specific products; e.g., the jewelry collection "Russian passion" (Turan, 2015).

Legal and Law Studies

In publications referred to the legal category, the concept of "archetype" often occurs in connection with ethno-cultural phenomena with the purpose of rationalizing vague or ill-defined concept underlying the Russian legal system. For example, "Domestic researchers believe that the most important aspect in the analysis of the basic foundations of legal mentality in modern Russia is the example of national archetypes" (Emel'kina, 2016) The emphasis of certain archetypes underlying the Russian legal system accentuate and justify the relevant status quo, so it would be more accurate to call them "representations". Such as archetypal representations as "Moscow - the Third Rome", "national calling", "baptism" (Mamychev, 2016) or "original values" (Kochetkov, 2015, p. 155) are employed to justify the relevant status quo.

Religion and Architecture

In the field of religion and architecture studies, it seems that the term archetype is often confused with the word stereotype or what Jung calls "dead symbols". Characteristic phrases are: "Russian cultural archetype" (Gribunin, 2016) or "Orthodox archetype". (Gribunin & Kanarskiy) Invariably, in this category, the psychological dynamics in the attribution of the term archetype are ignored. Such examples as "the archetype of the Orthodox wooden churches" clearly refers—at best—to a representation without much connection to a true archetype. (Permilovskaya, 2016).

Management

In publications referred to in the "Management" category, the concept of the archetype is used both to describe the most common organizational management models and the characteristics of individual role templates. In other words, in this category the term refers only to operational models and is almost completely divorced from any psychological implications. For example, "archetype of the system" refers to patterns of systemic behavior rather than to any analysis of the psychology behind the behavior. (Glazunova, 2016). However, N.F. Sirina and T.I. Bushueva give the following more psychological definition: "Archetypes are a particular model of human behavior that can be congenital, hereditary and universal". (Sirina, 2016, p. 94)

Summary

Based on the present analysis, we can conclude that the concept of "archetype" in the modern Russian scientific space is used primarily in the humanities and social studies with the following summary regarding usage of the archetypal concept:

1. Usage of the concept is more-or-less consistent with academic usage and is supplemented with quotations from Jung's work. However, traditional academic usage often does not grasp the deeper meaning and significance of Jungian theory. For example, though the term, "the archetype of the collective unconscious", is used, simultaneously references are made to conscious phenomena that cannot—according to Jung—be considered archetypes.
2. The important distinction between archetypes of the unconscious and archetypal representations of the conscious is rarely made. For example usage of the "archetype" concept to name empirical. Whether through ignorance or cursory examination, the distinction seems unimportant.
3. Usage of the term "archetype" for certain schemes, which are expressed in verbal or visual symbols and, according to the authors, are cultural markers are actually "archetypal images" or archetypal representations. For many practical purposes, the distinction is relatively unimportant, but in order to gain a deeper understanding of cultural archetypes and to learn about culturally unconscious dynamics, the distinction must be clear.
4. Usage of the concept of the "archetype" does not often rise above generic understanding. This serves to impart a rather vague archaic relevance to the subject being researched, but it provides little new information relative to the potential significance of the cognitive concept.

Tentative Conclusions

Regardless of the subject area, in any discussion of archetypes it is common to see a direct or implied allusion to ethnic and cultural specificity. In other words, it is the outer form — the representation — that is useful to writers. The problem is that when there is no understanding of how the collective unconscious influences the conscious reality, the entire edifice must necessarily be considered irrational and its validity questioned. At the same time, the archetypes are often irrationally associated with some symbols accepted as a universally valid. The consequence of this approach is not to throw light on the cognitive phenomenon, but instead to condemn sentience to something unknowable that exists on the border between the conscious and the unconscious.

The purpose of this analysis is to determine what needs to be done in order to correct or improve the general understanding of the archetypal cognitive dynamic and to suggest its relevance to the subject of universal sentience. Archetypes should be understood as a common experience preserved in the collective unconsciousness. The collective unconscious exists as an infinite, open-ended compendium of human experience, which can reflect a wealth of alternative ways of responding to typical situations occurring in human lives. This is how Jung understood the Psychic concept. Of course we can recall the most famous Jungian archetypes – Self, Anima, Animus and Shadow. But this list is not closed. In the terms used by C.G. Jung himself, the collective unconscious is ruled by the amplification and its resulting "Compensation" which is entirely different than the reductive Freudian version of Psyche. Jung's amplification method of using archetypes to heal not only patients but collective sentience is described in all his works, but "Man and his symbols" is a good place to begin. Jung says:

There are as many archetypes as there are typical situations in life. Endless repetition has engraved these experiences into our psychic constitution, not in the form of images filled with content, but at first only as forms without content, representing merely the possibility of a certain type of perception and action. When a situation ocurs which corresponds to a given archetype, that archetype becomes activated and a compulsiveness appears, which, like an instinctual drive, gains its way against all reason and will, or else produces a conflict of pathological dimensions, that is to say, a neurosis" (Jung, 1981, p. 480)

In one sense, archetypes may be expected to generate each typical human situation. And what is more typical for any human being than to communicate and interact with other human beings? Mankind's existential property is sociability. Or, as Aristotle wrote almost two and a half thousand years ago, "man is by nature a political animal". (Aristotle, p. 37) Therefore, every man or woman by their nature is deployed in social relations and must associate with others. This group will support its members to deal with the world outside. Sociability induces the group identity. Individuals begin to think as a common or shared entity. Yet every member must define his or her own place and function as a unique aspect inside this group. Sociability-may be described as a sentient community characterized by unity and diversity, but how is this seeming dichotomy to be resolved? Lvov and Gritskov have defined two archetypes: corporativity and hierarchy (Gritskov Yu.V., Lvov D.V.)

ARCHETYPE OF CORPORATIVITY

Constituted by the Self archetype of an existential image, or experience of alienation from the outside world ("the Self in the world") puts a person in a situation of existential loneliness, a tragic situation described by E. Fromm: "while a man used to be an integral part of the world, he/she was not aware of any possibilities or effects of individual actions, and he/she did not have to be afraid of the world. However, becoming an individual, he/she is left alone with the world, which is so stunning and frightening" (Fromm, E., 2007, p. 15).

Realizing the personal detachment from the world, a person cannot return to the blissful state of unconscious unity with it. The situation is aggravated by the fact that the world has to be confronted single-handed. But this very situation leads to the desire of the Self to break out beyond itself, to connect with another, more powerful Self, to confront a hostile world and "alien" selves with this Self. Hence, the archetype of the Self produces the need for existential communion with others — an evolutionary "amplification". It is vital for a person to identify themselves with any community.

Therefore the archetypal program of the Self is present in any social community, e.g. a family, a religious community or a commercial organization. Under this program, every community is perceived by its members (I-Selves) as allocated from the world, or an integral and unique entity of We-Self.

To this extent the existential interaction of the individual and the community can be understood as an interaction of I-Self and the We-Self, which may be linked to each other either in harmony or in discord. Obviously the appearance of the individual in the existential space of the collective Self is associated with the transformation of his/her own Self. An example of such a transformation of I-Self is noted by E. Durkheim's religious sanctification of man's relationship to society. Another example is that feeling a part of socio-cultural community, people unconsciously seeks to act as its representatives.

As it was already mentioned, one can fully realize himself or herself only in conjunction with the members of "his/her" social community (We-Self). In connection with this there is a task of distinguishing "us" from "them". However, as noted by A.I. Fet, "social instinct" is common to all higher animals, and this "social instinct"--which allows distinguishing between "us" and "them" — has evolved to a "global" scale of realization among people. This means that the person identifies "his/her own people" not in line with instinctual (genetically inherited) program settings, but rather in accordance with an amplified empathy that includes the inherited culture and traditions but is not limited to them. The instinctive program of recognizing other people as "our own" is still evolving to include larger communities and to incorporate them as part of the We-self.

An archetypal program constituting a situation of "we and others" can be called the archetype of corporativity. The archetype of corporativity opens the possibility of individual perception of themselves as a part of an expanded sociocultural community that provides them with a prism of belonging that allows them to expand the I-self archetype to an ever-larger We-self.

Relationships with individuals who are identified as "our own" are created by the standards of "most favored option", and the relationship with the "others" are made the contrary way. This is why realization of the existence of cosmic sentience is important. Sentience is the common denominator among living organisms.

ARCHETYPE OF HIERARCHY

Establishing position or "place" within the larger We-self is important. As shown above, the archetype of the Self governs the formation of the basic ideas about the world and building a relationship with it. However, Within the community, a person is faced with the task of determining her own place and the possible limits of their actions. If you recall the concept of the Self by G.H. Mead, that part of the human personality, which he refers to as "Me", allows a person to create a certain understanding of himself/herself as a member of the community. "Me" is an image of himself/herself in the manifold relationships with other members of the community. This image is based on an understanding of the individual's own position in the community and the expectations of the other members of the community, corresponding to their positions.

According to V. Chalidze, "The hierarchical structure appears in any group of interacting people" (Chalidze, 2005, p. 58). Moreover, the definition of an individual's position in the hierarchy happens almost unconsciously, "automatically". "This assessment is provided automatically, which allows each individual to define a hierarchy of wills in the community and their place in the hierarchy through successive comparisons of their wills to the will of other individuals (i.e., to set a number of hierarchical relations)" (Chalidze, p. 29).

The archetype of hierarchy is a program of the collective unconscious, which creates the image of any social community as a "pyramid" of unequal levels. Social interactions within this hierarchical community are ambivalent, at the same time they incorporate the seeming dichotomy, The desire to rule and the willingness to obey the rule. Aggression, and willingness to cooperate are components of the archetype (Gritskov, 2006, p. 87)

An important thing to remember about the archetype of hierarchy is that it can manifest as tyranny or beneficence. The tyrannical modality has to do with top-down authority. Whereas the beneficent modality values individual creativity. This is an important distinction that goes back to the biblical modality difference between good and evil. In other words, individuality is valued in the beneficent mode of hierarchy. Individual interpretation provides interactivity among hierarchical levels. Hierarchical dynamics like patronage and protection can be understood as positive. Most important, hierarchy provides structural order, and any sort of order is better than the chaos of evil. Hierarchy allows participants to optimize their creative participation in the evolutionary process, so — from the populist perspective — hierarchy is a blessing to both the I-self and the We-self. Finally, universal sentience is the key to evolution in a cosmic interactive hierarchy. Humans should re-evaluate their power to reach out to super-human realms of hierarchy for problem-solving assistance.

CONTINUITY BETWEEN ARCHETYPES OF CORPORATIVITY AND HIERARCHY

A superficial view of the archetypal programs of corporativity and hierarchy may seem to contradict one another, and this seeming contradiction must be understood in terms of the significance of the part within the whole. For example, within the hierarchical framework sentient humans are made in the image of God, but the sentient bacteria in the human digestive system perform a valuable function for the survival of the entire hierarchy. All sentient levels of the cosmic hierarchy can work for the integrity and stability of the community. An understanding of the dynamics of universal sentience becomes the first stage for respecting We-community values. Corporate groupings differentiate levels of the hierarchy, and — in ideal situations — each level interacts with other levels. In other words, in a t collective totality the archetype of corporativity will cultivate a sense of community and equal belonging to "their own" social group, while the archetype of hierarchy realizes the ordering of the positions and interactions that maintain group integrity. Otherwise, an existential conflict produced by the archetype of corporativity would confound the desire for equality and unity which is characteristic of the archetype of hierarchy — a desire for orderly harmony within the organized unity.

In addition, the contradictions between the archetypal programs of corporativity and hierarchy may arise due to free will. Hierarchical values may contradict corporate value structures. The overview maintained by the hierarchy archetype supersedes the more limited perspectives of corporate levels of the hierarchy. For example corporate free will may cause damage to other levels of the hierarchy just as the human level causes damage to the natural hierarchy of life systems (ecosystems). In this situation, limited "agency" loses track between the one who rules and the one who obeys. In such a relationship there is no existential sense of unity and authentic rulers tend to be viewed as unfair by obeyers. This jeopardizes the integrity of the group.

Many hierarchical organizations, such as modern corporations, tend to emphasize the tyrannical model. Corporations devalue the I-self in favor of the We-self. Ironically, corporate officers tend to give their own I-self priority in every situation. This is called selfishness. In this tyrannical setting, the value of individuality is denied, and a system of coercion takes its place. Instead of I-Self, representing the collective We-Self, an individual's social role is constrained within a certain set of expected responses. One of the most important aspects of human relations is compliance with specified standards of behavior set by their social position. However, if — as is often the case — an insular perspective takes precedence over maintenance of a harmonic unity in the collective whole of unique individuals who recognize the right of each other to be unique, hierarchical integrity is jeopardized.

The archetype of corporativity requires either overt or covert adjustment of individual values and meanings to some more generalized scheme of authentic harmony. Also, it can be argued that conformism — as a negative socio-cultural phenomenon – emerges due to the archetypal program of corporativity. In this negative manifestation a person takes the point of view of the group only in order to be included in the group legitimately, especially if his/her fundamental interests are not affected.

A negative consequence of such conformity to limited corporativity is due to fear that disagreement with a decision imposed by a group upon him/her being ostracized. A more positive aspect of corporativity could lead not only to a voluntary unity of free men who share a common vision, but also to autonomous restrictions of their I-Selves. To make We-Self viable, the existential comfort from belonging to it should significantly exceed the conformist costs of its members (individual Selves). Otherwise, there is an increased risk of weakening hierarchical unity and separation of dissident I-Selves who demonstrate dissident values and behaviors that are contrary to those that are generally accepted in a given corporate community.

These dissident Selves ("strangers among us") are potentially a source of acute internal corporate conflicts, as relative We-Selves tend to view them with more belligerence than "strangers". After all, while it is natural when one does not trust the actions of outsiders (this distrust is "legitimized" by their status of "alien Selves"), it is as natural to trust your "own" members, and this trust should be justified by "our own members" by loyalty not only to the world view and the core values within the group, but also to the existing hierarchy in the group. Dissidence destroys the group's relative We-Self from inside thus destroying the self-image established as part of a group with shared values. Such-failure of "loyalty expectations" creates preconditions for the perception of dissident members of the group as a "traitors".

However, it may be the case that dissidents are only trying to restore harmony between the archetypal programs of corporativity and hierarchy. But the renegade is already deprived of any internal privileges and is ostracized; now it is forbidden for him/her to publicly identify with the group and this person is divested of all previously assigned to him/her social roles (For example, in Christianity the excommunication from the community and existential comfort that allows staying in it is considered one of the most serious penalties). In fact, such a violent interruption of communication between the community and the individual is an execution that causes the death of a social I-Self by the relative We-Self — such as the case of Christ's crucifixion by the will of a Jewish elite.

Demonstrated contradictions generated by the archetypes of corporativity and hierarchy can become the basis both for the destruction and for the development of the unity of the integrated group and of its individual representatives. A clear dissonance between the archetypal programs can lead to a complete imbalance of intragroup relations with far reaching consequences — splits into separate independent groups, group disintegration, or perhaps the constructive resolution of such a deep-seated conflict and a qualitative adaptive change in the group's structure and its core values.

ARCHETYPES AND INTERNET

Due to their belonging to the collective unconscious, archetypes are very conservative and rigid. At the same time, they are only hollow schemes, which must be filled in from the conscious experience. This dialectic makes them quite adaptive to constantly changeable social environment. That is why even very new phenomena can actualize these ancient deepened psychic structures. In turn, archetypes will determine human's perception of the situation and therefore behavior in it.

It may be argued that the Internet has the same cognitive structure as Jung's Psyche or George Lakoff's "embodied cognition" that may be tracked in the nervous system. Both Psyche and embodied cognition have conscious and unconscious elements. For example, in the case of the nervous system, there is an autonomic nervous system (ANS) and a seat of complex consciousness that appears to be centered in the brain. Depending upon the type of receiver, the Internet can be tuned to certain frequencies that — at any time — can be understood as Internet consciousness. Simultaneously, the pervasive bandwidths that are not being received can be considered as the unconscious Internet. Moreover, all Internet content can be shown to have narrative (dramatic) structure. (Lakoff, George, 2008 & 2002; Jacoby, Yolande, 1973, pp. 82-83) that can be correlated with QED and with video game structure. (Schafer, Stephen, 2011; Schafer, Stephen & Palamides, Thomas, 2011; Schafer, Stephenm 2008 & 2007).

Therefore – though the Internet is a very new means for social communication – social media sites may also be understood as massively multiplayer online role-playing game platforms where cognitive dynamics are at work. Essentially, the Internet equates to Jungian Psyche in all its apparent structural dimensions, and unconscious elements of the Internet equate with Jung's collective unconscious. The Internet may be studied as just another place to research intersubjective and collective dynamics which are as old as humankind itself. Users of the Internet remain themselves — creative social I-Selves engaged in a process of evolution that is augmented by the Internet. Therefore, the Internet can become a very positive evolutionary dynamic — especially as they relate to evolution of both the I-Self and the We-Self. Social networks naturally form themselves into group identities and structures. In this context, archetypes of Self, corporativity, and hierarchy exist in vivo.

Given close scrutiny, new digital environments divulge new realities that before had been invisible. Scrutiny discloses new anomalies peculiar to the manifestation of archetypes. These anomalies must be explored from the perspective of a We-Self that far exceeds current perspectives on universal sentience. In the greater context, Jungian Psyche and neurobiological cognitive structures and dynamics appear to be accurate, but the new information provided by so-called anomalies is difficult to integrate into any orthodox worldview. The paradigm shift in worldviews must now encompass the basic assumption of the existence of a primal universal life-energy which can be called "sentient".

The energetic dynamics of the Internet and other digital media forms have begun to familiarize people with the dynamics of virtual and simulated realities. Even after all the digitalization and virtualization of our everyday reality, we still can see how auto-organizing online communities tend to form their inner hierarchies and identification codes that consist of specific values and norms. Such codes organize group entities and give them means to distinguish "us" from the "alien them". Within virtual worlds, human beings are still social by nature and; therefore, they tend to unite in social partnerships.

New systems of social relationship allow individuals and groups to re-invent themselves. The dynamics of virtual worlds allow humans to expand and extend their "agency"; i,e., humans gain powers and opportunities that have not been available to them in normal circumstances. One of the first things people like to do on the Internet is become anonymous — to conceal or alter their identity. Nothing is impossible. An individual can alter persona relative to appearance, origins, sex, age, race, voice, and so on. In other words, people can now project who they are with their own thoughts and desires. Perhaps this has always been true, but now it is obvious, and the dynamics for doing so are known.

From the context psychic interaction between archetypes and archetypal representations, we must ask ourselves what this new dynamic means. The details of Jungian "Individuation" change, but the principle remains the same. For example, digital avatars and profiles become easily interchangeable. Moreover, the counteragents of online communications can easily be concealed or confused. Intentional concealment is becoming a global problem; i.e., political hacking and the use of lies to confuse. Even such sophisticated and competent agents as commercial providers may become unable to identify their clients. For example, underage gamers may just use their parents' credit cards. Only a complex and expensive analysis will suffice to reveal clients according to the traits of their behavioral patterns as demonstrated by Ahn, Jeon and Yoo (2014).

The Internet opens new dimensions of freedom in the process of building the Self. Virtual reality may form collective entities that are not based on habitual hierarchy and identity scales. In his analysis of Second life islands, Keith McIntosh says: "These [virtual] spaces are, instead, given order by the cultural perspectives of their [virtual] residents, embedding virtual environments with a benign quality and cultural specificity that could never be achieved in real-world constructions. These virtual constructions provide freedom, paradoxically, from reproducing cultural conventions in the form of real-world symbols. These symbols are then displaced from their position in a capitalist order and allowed to take on new meanings and significance in the virtual world" (McIntosh, 2008, p. 210).

So, perceptions of the meaning of symbols changes, but — apparently — the psychic dynamics do not change. This suggests that cultural ties can be consciously amended in order to endorse strong repeated and benign communicative acts grown on the soil of common values and interests. Virtual communities can be designed in terms of digital mechanisms, but real in analog terms of personal involvement and shared feelings. Such design is not really difficult. Even such a basic media platform as forum may establish a new We-Self. As observed by Simone Gabbriellini: "A sense of belonging is often associated with the concept of online community. First people gather online to share interests, but developing a sense of belonging requires repetition — repeated interactions and contacts, and the evolution of a core of stable interacting users might be the first step in building such a shared feeling" (Gabbriellini, 2014, p. 819).

Creating online groups may be not just allowed but coerced and imposed by providers of the media. This imposition is usually motivated by profit--especially if grouping activities like massively multiplayer online role-playing games are considered to be a vehicle for increasing profits. Advertising dynamics of the "focus group" may be used to define a "target group" of gamers in order to design a game according to tastes embraced by specific target groups. In the process of defining how a group can fulfill its game expectations, we can see the birth of new realities based on subjective dynamics of virtual communities called MMORPG guilds (massive multiplayer online role playing games). However, in spite of the external stimulus to establish these guilds, instantiated in-group dynamics extend well beyond the narrow context of game-only interactions. Though a game format allows for efficient coordination of psychological target group relationships within the social structure of the game, many guilds have active forums or mailing lists with which group players "can speak about the game but also about general real life issues" (Rossi, 2008, p. 2).

This inter-mixture of realities can be used for political purposes to confuse because, psychologically, the hybrid reality is "real" and leads to real "offline" consequences exemplified by the following quote:

These virtual worlds seem to be...new tools of communication for the political parties, as [illustrated in] the French presidential election of 2007. The arrival of the Front National in January, 2007, caused a wave of demonstrations structured around two groups: AntiFNSL and Second Life Left Link (SLLU). If the first group were created for the occasion, the second was preexistent with the event, being instituted as a group of "all those which are close to [the] left wing" [and] carrying out of the actions denouncing liberal policies with images of the anti-G8 demonstrations (Mauco, 2009, pp. 16-17).

Thus, the archetype of corporativity applies to and influences the processes of online group formation. The same may be said about the archetype of hierarchy. Every person needs to define her own place in a community. In the virtual reality, which gives to a person much more liberty than the real world, a variety of restrictions is imposed by designers of video games. These design features have a variety of effects on the status of gamers in the group. Of course, status is important within the context of a gaming group, so there must be some guarantees that after reasonable contribution the individual may expect some recognition. This characteristic maintains for either online or offline communities. Nevertheless, this variety of design may be considered within the contextual frame of archetypal representations and analyzed for meaning as such. In other words, variations in game design represent changes in the unconscious archetypal patterns that generate the representations.

With regard to effects on the archetype of hierarchy within MMORPGs, if a low-status player decides to drop out of the game, the next higher-status player is demoted. This illustrates the effect on an individual player, but there is also an effect on the entire group. One dropout can create a chain reaction of dropouts and a diminishing of social interaction due to smaller network size (Kim & Lee, 2013, p. 4).

A highly suggestive and valuable characteristic of virtual design is that it can enhance reality in unexpected ways. Brigitte Jordan highlights the phenomenon of lurkers: "An interesting problem in online participant observation, often simply ignored, is the issue of the invisible presence of lurkers. Lurkers constitute a category of participant in cyberspace for which there is no parallel in real-world anthropological ethnography. Lurkers constitute an audience that at any moment may become active. Lying in wait out of sight, they work behind the scenes and wield invisible power to which producers of online content and advertisers orient. All participants know that lurkers are present and their presence can be confirmed by records of access to the Internet site, but, like ghosts, they leave no observable traces for the researcher" (Jordan, 2009, p. 187). Though the Lurker phenomenon cannot be explained with orthodox materialistic worldviews — because they are true representations (projections) of archetypal reality – these lurkers may be considered "real".

In fact, the presence of such invisible "ghosts" is not obvious from an orthodox-reality perspective, the possibility of their presence influences a user's perception of the game situation as well as the player's behavior in it. The influence of ghosts — disembodied realities — has a tangible effect on gamers and gameplay. How these ghosts perceive us manifests as the tendency to create "cool avatars" on social media sites and to post positive photos and happy selfies online. All of these tendencies fall within the influence of the archetype of hierarchy. All this is due to the aspiration to climb higher and higher on the social hierarchy. The discovery of meaning that is projected in digital attitudes toward social wellbeing fall within the prevue of the hierarchical archetype.

"The political problem is that it is easier to perceive Facebook as less alienating than a nuclear power plant that pollutes a river or lake. You can more easily use and enjoy Facebook, even when it exploits and monitors you, but you cannot swim in a nuclear contaminated lake or drink its water without seriously threatening your health. The damages caused by digital alienation are more indirect, mediated, long-term, and invisible. This does however not imply that they are unproblematic, but rather that they are more difficult to challenge and contest in social struggles. Users who argue that Facebook is great because it is social and that they therefore do not feel or do not mind digital exploitation, surveillance and exclusion express a partial truth, namely that social media can pose social advantages. At the same time their consciousness is reified because they cannot see beyond immediacy, only think about immediate individual advantages and not about the disadvantages that some users may unjustly suffer

for example from being excluded from obtaining a mortgage, losing their jobs or being suspected of terrorism because of data mined from social media and analyzed with predictive algorithms by credit scoring companies, employers or state institutions" (Fuchs, 2016, pp. 172-173).

CONCLUSION

This phenomenon of virtual online influence on our offline social reality may easily evade our perceptions and slip through our fingers. However, archetypal dynamics remain valid, and an archetypal approach to social issues – because the archetypal approach reveals unconscious dynamics of collective human behavior – may help to solve problems and to complete the whole puzzle.

Thus, as we see from the above analyses, the existence of new digital realms introduce alternative archetypal representations and innovations in group structure and inter-personal interactions. However, the dynamics of archetypes and their projections remain the same for patterns of collective human behavior. Even while each person acts in unique ways, not thinking of or perceiving the common interests, the entire social jigsaw is based upon the archetypal dynamics of group identities and hierarchies. Biological metaphors can be used to describe these "sentient" phenomena.

"No one is assigned to groups by a central authority. There are no rules, other than party size, for how groups must be structured. Instead, groups emerge in an entirely decentralised and self-organised way, through a process of negotiation between players, based on some loose norms and even looser relationships. This emergence follows the classic rules of emergence in biological systems, like an organism that 'spends much of its life as thousands of distinct single celled units, each moving separately from its other comrades, but then under the right conditions, those myriad cells will coalesce into a single, larger organism' and 'it becomes a they'. In these game universes, players, like other living organisms [are sentient], 'think locally and act locally, but their collective action produces global behavior'" (Galarneau, 2005, p. 5).

Because the dynamic archetypal correlations among dreams, video games, social structures and universal sentience have been tentatively established (Schafer, Stephen Brock, 2016 a & b; Schafer, Stephen Brock, 2012; Schafer, Stephen, 2011; Schafer, Stephen & Palamedes, Thomas, 2011; Schafer, Stephen 2008 & 29=007), it seems quite practical to follow the Jungian tradition of the archetypal approach to explore irrational patterns of sentient collectives in order to understand the expanded awareness that characterizes the paradigm shift.

REFERENCES

Ahn, D., Jeon, S., & Yoo, B. (2014). *What shows your age? An empirical analysis of identity fraud in online game classification systems.* GCCR Working Paper Series 14-8. Retrieved from https://ssrn.com/abstract=2442683

Aristotle. (2010). *The Politics: translated and with the introduction, notes and glossary by Carnes Lord.* The University of Chicago Press.

Baturin, D. A. (2016). Fentezijnyj neomif v kontekste teorii arhetipov K.G. Junga [Fantasy & neomyth in the context of C.G. Jung's theory of archetypes]. Novaja nauka: opyt, tradicii, innovacii, 2016(2), 177-180.

Campbell, J. (1968). The hero with a thousand faces (2nd ed.). Princeton University Press.

Chalidze, V. (1991). *Ierarkhicheskii chelovek [The hierarchical man].* Moscow: Terra.

Chzhan, Ts., & Tolkachev, S. P. (2016). Issledovanie arkhetipov v kitayskoy telereklame [The study of the archetypes in the Chinese TV commercials]. Obrazovanie i nauka v sovremennykh usloviyakh, 2(2), 29-31.

Davydova, T.R., Kasimov, R.Kh. (2015). Kategoriya vseedinstva kak arkhetip russkoy filosofii i sotsiologii [Category of unity as the archetype of Russian philosophy and sociology]. *Elektronniy nauchno-prakticheskiy zhurnal kul'tura i obrazovanie, 1.*

Emel'kina, I.V. (2016). Spetsifika zakona i prava v rossiyskom (russkom) arkhetipe [The specifics of the law and the rights in the Russian archetype]. *Mir nauki i obrazovaniya, 1.*

Erokhina, T. I. (2016). Gendernye arkhetipy «serebryanogo veka» v massovoy kul'ture [Gender archetypes of "Silver Age" in mass culture]. Yaroslavskiy pedagogicheskiy vestnik, 5, 329-334.

Fet, A. I. (2005). *Instinkt i sotsial'noe povedenie [Instinct and social behavior].* Novosibirsk: Sova.

Fromm, E. (2007). *Begstvo ot svobody [Escape from freedom].* Moscow: Academic Project.

Fuchs, C. (2016). *Critical Theory of Communication.* London: University of Westminster Press.

Gabbriellini, S. (2014). The Evolution of Online Forums as Communication Networks: An Agent-Based Model. *Revue Francaise de Sociologie, 55*(4), 805–825. doi:10.3917/rfs.554.0805

Gabrieljan, T. O. (2016). Rol' arhetipa v vizual'no-graficheskoj tipologii brenda [The role of the archetype in the visual-graphic brand typology]. Vestnik kemerovskogo gosudarstvennogo universiteta kul'tury i iskusstv, 1, 88-94.

Galarneau, L. L. (2005). *Spontaneous Communities of Learning: Learning Ecosystems in Massively Multiplayer Online Gaming Environments.* Retrieved from https://ssrn.com/abstract=810064

Glazunova, V. V. (2016a). Primenenie arkhetipov sistem v formirovanii innovatsionnogo potentsiala organizatsii [The use of systems archetypes in shaping the innovative potential of organization]. Naukovedenie, 8(2).

Gribunin, V. V. (2016b). Arkhetip stradaniya v russkoy kul'ture: religioznyy aspekt [The archetype of suffering in Russian culture: religious aspect]. Osnovnye tendentsii gosudarstvennogo i obshchestvennogo razvitiya rossii: istoriya i sovremennost', 1, 31-38.

Gribunin, V. V., & Kanarskiy, D. I. (2016). Transformatsiya russkogo pravoslavnogo arkhetipa v sovetskiy period [Transformation of Russian Orthodox archetype in the Soviet period]. Osnovnye tendentsii gosudarstvennogo i obshchestvennogo razvitiya rossii: istoriya i sovremennost', 1, 91-95.

Gritskov, Yu. V. (2004). *Obrazy stradaniia v stradaiushchem soznanii [The images of suffering in suffering consciousness].* Krasnoyarsk State University.

Gritskov, Yu. V. (2006). *Fenomen stradaniia i sposoby ego osvoeniia v kul'ture* [The phenomenon of suffering and ways of its learning in culture] (Unpublished doctoral dissertation). Omsk State Pedagogical University, Omsk, Russia.

Gritskov, Yu. V., & Lvov, D. V. (2015). *Interaction of In-Group Archetypes and Hierarchy while Forming in-Group Community. Journal of Siberian Federal University. Humanities & Social Sciences, 11,* 2325–2330.

Gutov, E. V. (2016). Transformatsiya arkhetipov sotsial'noy telesnosti [Transformation of social archetypes of corporeality]. Kul'tura, nauka, obrazovanie: problemy i perspektivy, 329-332.

Jordan, B. (2009). Blurring Boundaries: The "Real and the "Virtual" in Hybrid Spaces. Introduction to Special Section on Knowledge Flow in Online and Offline Spaces. *Human Organization, 68*(2), 181–193. doi:10.17730/humo.68.2.7x4406g270801284

Jung, C. G. (1981). *The Archetypes and the Collective Unconscious* (2nd ed.). Princeton University Press.

Jung C.G. (1991). *Arkhetip i simvol* [Archetype and symbol]. Moscow: Academic Press.

Jung C.G. (1998) *Psikhologicheskie aspekty arkhetipa materi* [Psychological aspects of the mother archetype]. Moscow: Academic Press.

Jung, C. G., von Franz, M.-L., Henderson, J., Jacobi, J., & Jaffe, A. (1964). Man and His Symbols, London: Aldus.

Kim, M., & Lee, B. (2013). *Are There Too Many Superheroes? Analysis of the Social Distance in Massive Multiplayer Online Role Playing Game.* KAIST College of Business Working Paper Series No. 2013-019. Retrieved from https://ssrn.com/abstract=2330090

Kochetkov, V.V. (2015). Konstitutsionalizm i arkhetipy russkoy vlasti (casus 1906 goda) [Constitutionalism and the archetypes of Russian power (casus 1906)]. *Rossiyskiy zhurnal pravovykh issledovaniy, 1,* 155-163.

Lakoff, G. (2002). *Moral politics.* Chicago: The University of Chicago Press. doi:10.7208/chicago/9780226471006.001.0001

Lakoff, G. (2008). *The political mind.* Viking Penguin.

Mamychev, A. Yu. (2016). Arkhetipy gosudarstvennoy vlasti v pravosoznanii sovremennykh grazhdan RF [Archetypes of the government in the legal consciousness of modern citizens of the Russian Federation]. Aktual'nye problemy sovremennogo prava, 325-329.

Mauco, O. (2008). Politicization of the On-Line Worlds as a Problematic of a Private Space Publicized: Comparative Study of MMORPG and Second Life. In *PCA ACA National conference.* Retrieved from https://ssrn.com/abstract=1329345

McCraty, R., & Schafer, S. (2016). *Exploring dimensions of the media dream: Functional context in collective personae.* Hershey, PA: IGI Global. doi:10.4018/978-1-4666-9891-8.ch001

McIntosh, K. (2008, Fall). The Social Construction of Virtual Space. *Michigan Sociological Review,* 196–214.

Nikolenko, N.N. (2015). Traditsionnyy ornament nomadov kak vizual'nyy arkhetip [Traditional ornament of nomads as a visual archetype]. *Universum: filologiya i iskusstvovedenie, 9-10.*

Oshchepkova, A. A. (2016). Arkhetip ryby v russkoy kul'ture [The archetype of the fish in Russian culture]. In VIII kirillo-mefodievskie chteniya "slavyanskoe slovo v kontekste vremeni", 101-106.

Permilovskaya, A. B. (2016). Derevyannyy pravoslavnyy khram kak voploshchenie etnokul'turnogo arkhetipa russkoy traditsionnoy kul'tury [Wooden orthodox temple as the embodiment of the ethno-cultural archetype of Russian traditional culture]. Yaroslavskiy pedagogicheskiy vestnik, 2, 245-250.

Privezentseva, V. A. (2016). Kul'turnye arkhetipy v angliyskom gumanizme vtoroy poloviny XVI - nachala XVII v [Cultural archetypes in English humanism in second half of XVI – beginning of XVII century]. Yaroslavskiy pedagogicheskiy vestnik, 2, 198-202.

Pshenichnaya, V. V., & Gatiatullina, E. R. (2016). Problema proyavleniya arkhetipov v natsional'nom soznanii [The problem of the archetypes manifestation in the national consciousness]. Sovremennye problemy upravleniya prirodnymi resursami i razvitiem sotsial'no-ekonomicheskikh system, 307-312.

Rossi, L. (2008). *MMORPG Guilds as Online Communities - Power, Space and Time: From Fun to Engagement in Virtual Worlds.* Retrieved from https://ssrn.com/abstract=2137594

Schafer, S. (2007). Premise: The Key to Interactive Storytelling. *Game Career Guide*: Retrieved April 3, 2007 from http://www.gamecareerguide.com/features/357/premise_the_key_to_interactive_.php?page=1

Schafer, S. (2008). *The Psychecology Game I-V; Using Story-based Video Games to Capture Veridical Data on Subliminal States of Being*. Consciousness Center & the University of Arizona. Retrieved from https://bandura.sbs.arizona.edu/login/consciousness/pubreport.aspx?aid=27

Schafer, S. (2011). Articulating the paradigm shift: Serious games for psychological healing of the collective persona. In *Business Social Networking: Organizational, Managerial, and Technological Dimensions.* IGI Global. Retrieved from http://www.igi-global.com/book/handbook-research-serious-games-educational/5

Schafer, S. (2012). Optimizing Cognitive Coherence, Learning, and Psychological Healing with Drama-based Games. In *Video Game Play and Consciousness*. Nova Science Publishers. http://www.facebook.com/insights/?sk=po_261980910540561#!/pages/Video-Game-Play-and-Consciousness/261980910540561?sk=wall

Schafer, S., & Palamides, T. (2011). Designing an Ethical structure for Social Influence Marketing. In *Business Social Networking: Organizational, Managerial, and Technological Dimensions*. IGI Global. Retrieved from www.igi-global.com

Schafer, S. B. (2016a). *The dream analog: Psychecology video games. In Exploring the Collective Unconscious in a Digital Age*. Hershey, PA: IGI Global. doi:10.4018/978-1-4666-9891-8.ch002

Schafer, S. B. (2016b). *The media-sphere as dream: Researching the contextual unconscious of collectives. In Exploring the Collective Unconscious in a Digital Age*. Hershey, PA: IGI Global. doi:10.4018/978-1-4666-9891-8

Sirina, N. F., & Bushueva, T. I. (2016). Sistemnye arkhetipy v upravlenii predpriyatiem zheleznodorozhnogo transporta [System archetypes in the management of railway transport organization]. Sovremennye nauchnye issledovaniya i innovatsii, 5, 94-99.

Syutkina, E. V., & Solodukhina, O. S. (2016). Arkhetip ottsa v sisteme formirovaniya doveriya v obshchestve [The archetype of the father in the formation of the system of trust in society]. Vektory psikhologicheskoy bezopasnosti sovremennogo obshchestva: doverie i sub"ektivnoe blagopoluchie, 71-74.

Tikhonov, A. V., & Duberman, Yu. E. (2015). Arkhetip [Archetype]. Sotsiologiya upravleniya: teoretiko-prikladnoy tolkovyy slovar', 28-29.

Turan, N. K. (2015). K voprosu ob izuchenii arkhetipov i mifov v reklame (na primere deyatel'nosti yuvelirnogo doma «freywille») [On the question of the study of archetypes and myths in advertising (for example, the activities of the jewelry house «freywille»)]. Molodoy uchenyy, 9, 1269-1274.

Zlotnikova, T. S. (2015). Arkhetip v filosofsko-psikhologicheskoy traditsii [Archetype in the philosophical and psychological tradition]. Kody massovoy kul'tury: rossiyskiy diskurs, 34-48.

Chapter 5
Jung's Collective Unconscious, Integrative (Mind-Body-Spirit) Yoga, and Self-Realization

Manoj Sharma
Jackson State University, USA

ABSTRACT

The collective unconscious is a construct presented by Jung to epitomize a depiction comprising of memories and impulses about which one is not aware and that is common to the entire humankind. An ancient system of mind-body-spirit practice, yoga, also implies the yoking of human consciousness to super-consciousness, which is an expanded form of the collective unconscious used by Jung. Super-consciousness is not only linked with the unconscious of the humankind but also to the entire nature or Universe all the way to the static primordial state in which there is no vibration and yet is the source of all creation. Yoga helps decipher this primordial state which is also called by some as self-realization. This chapter explicates the concept of the collective unconscious, the system of ancient yoga, a modern practical paradigm of kundalini energy yoga (KEY), and steps for self-realization to decipher and conclude this characterization.

INTRODUCTION

Among the early contributors to the discipline of psychology, as it is known in the West in present times, is Sigmund Freud (1856-1939). Freud was an Austrian neurologist and father of the technique of psychoanalysis. One of his students was Carl Jung (1875-1961), the Swiss psychiatrist whom he often referred to as his most brilliant student (Shor, 1944). Jung has written 105 German and 35 English titles and is well-known for his work in analytical psychology on introvert-extrovert characterization, word association tests, synchronicity, the "collective unconscious" and rebel psychoanalysis (Carvalho, 2014; Jacobi, 1943; Meredith-Owen, 2011). The beauty of Jung's work is that he does not come across as a competitor to Freud. His work complements Freud's work through building on the concept of superego and establishing proactive linkages with everyday realities.

DOI: 10.4018/978-1-5225-9065-1.ch005

Jung (1968) introduced the enigmatic construct known as "collective unconscious" which he used to describe the inner deeper layers of the unconscious mind that he contended was common to all humankind. Jung differentiated between "personal" and "impersonal" unconscious (Jung, 1968; Wechsler, 1971). According to Jung (1968), the "personal unconscious" is comprised of things that an individual encounters in one's lifetime such as the forgotten memories of childhood and the repressed desires or impulses as an adult. These aspects are often encountered in dreams. Freud (1923) also talked about this type of unconscious and the aim of his psychanalysis technique was to make this unconscious conscious to enhance adjustment in behaviors. On the other hand, the "impersonal unconscious" is not derived from one's experiential base but is inherent to all humans and is derived from the collective experience of the humankind. This is also called as the collective unconscious by Jung (1968). His conceptualization of collective unconscious did not receive much credence in his lifetime because it was difficult to empirically test it and was even ridiculed as being fanciful. The concept borders on mysticism which traditional psychologists despise (Freeman, 2016). However, recently greater interest in the exploration of mutual linkages of this concept of Jung's collective unconscious with contemporary psychology, anthropology, sociology, and other disciplines has resurfaced (Hunt, 2012).

Another system that has gained currency in recent times is the ancient system of yoga which is gaining in its popularity in its various forms and styles all over the world. Yoga often classified under the term mind-body intervention or mind-body-spirit intervention is a traditional practice but now an established part of integrative health also known as complementary medicine or alternative medicine. Some perceive yoga as a means to attain flexibility; others use it to heal bodily ailments; while some contend it to be a way of life mainly to be used to decipher the fundamental purpose of life or Self-realization. It is in this backdrop that this chapter explicates the concept of the collective unconscious as described by Jung, the system of ancient yoga, a modern practical paradigm of *Kundalini Energy Yoga* (KEY) and steps for Self-realization to decipher and conclude the characterization of both collective unconscious and Self-realization for oneself.

BACKGROUND

Jung's Collective Unconscious

Jung's collective unconscious has been referred by some as "the soul of the world" (Cashford, 2018). Jung (1968) talks about a dynamic process of "individuation" which is defined as a lifelong psychological pursuit of attempting to differentiate the conscious self from the unconscious self. Another way of looking at individuation is to look at it as the liberation of the self to contribute more freely in the "collective unconscious" at the same time managing everyday activities (Shor, 1944). Jung's work derives largely from the "transcendent, higher plane" of "psyche integrated into the Self" (Jacobi, 1943, p. 127). Jung draws critical archetypal matches with the Oriental symbolic representations (*Mandalas*), Yogic techniques, Jesuit practices, Buddhist rituals, philosophical underpinnings of Medieval Alchemy; and the psychotherapeutic linkages to the fundamentals of Faith (Shor, 1944). So the conceptualization of collective unconscious borders on science and philosophy with more leaning toward the ethereal esoteric that has prevented it from being widely accepted into the mainstream.

In order to build the empirical base for Jung's collective unconscious some recent studies have been done. One such study from linguistics is by Sotirova-Kohli and colleagues (2011) that examined some Chinese characters called *kanji* as being archetypal images in Japanese as an influence of the collective unconscious. The salient finding of this study was that the Chinese *kanji* characters when matched with their accurate meaning were recalled statistically significantly as compared to those that were unmatched. The study concluded the role of the unconscious collective knowledge of the meaning of the Chinese characters that helped regarding correct matching.

A different angle to collective unconscious has been presented from the field of microbiology. The human digestive system has a composite ecosystem of microbes called the *microbiome*. By some estimates, the weight of microbes consists of about one kilogram in the human body or about the size of the human brain. According to microbiologists, this forms part of the unconscious system shaping the functioning of the nervous system and behavior (Dinan, Stilling, Stanton, & Cryan, 2015). Studies in this regard are in their infancy but may have a relationship to the Jungian concept of collective unconscious which may extend this concept beyond just humankind to all forms of creation in the Universe and their impact on an individual's life.

A criticism of Jungian analysis that has been pointed out mainly in the minority literature pertains to its ineffectiveness in adequately capturing the trepidations of the indigent, suppressed and disenfranchised subgroups (Adams, 2002). This is certainly a valid point that alludes to the fact that basic subsistence needs need to be fulfilled before one can embark upon the onward journey of exploring collective unconscious within one's psyche. It is imperative that food, clothing, shelter and other basic needs are met before one can contemplate on higher-order functions in life. This is also in consonance with Maslow's (1943) hierarchy of needs theory that points to the fulfillment of lower order physiological and safety needs before embarking upon higher needs.

Collective Unconscious and Collective Intelligence

David Wechsler (1971) draws a parallel with Jung's collective unconscious with the concept of collective intelligence. Wechsler (1971, p. 904) presents the word collective as "pertaining to a quality or trait that is shared in any way by two or more persons or things." He differentiates between" collective behavior" in which two or more people may act in a certain way in a situation to accomplish any given task with the notion of "collective intelligence" in which two or more people share ideas that lead to greater insights among all involved. Collective intelligence owes its origin from sociology but has influenced psychology. Wechsler traces the concept of collective intelligence to the earlier seminal work by Gustave LeBon (1841-1931), a French polymath, who wrote *The Crowd: A Study of the Popular Mind* (Le Bon, 2002) and gave the concept of crowd psychology whereby while being part of a crowd, an individual automatically acts at the will of a leader probably driven by the unconscious drives and motives. The source of this automatic behavior has not been empirically traced; however, it is a widely known phenomenon.

Collective intelligence is defined in contemporary literature as "shared intelligence of a collective or group of people" (Brennan, Binney, Parker, Aleti, & Ngguyen, 2014, p.193). The concept of collective intelligence is ancient and dates back to the traditional cultures in which the wisdom of elderly guided decision making. In modern times, with advances in information technology and the Internet, this concept is being widely applied in computer sciences (Smith, 1994) where it is currently being used in developing assets for Virtual Reality (VR), Augmented Reality (AR), Mixed Reality (MR), and many other such applications. In health behavior research, several multi-theory models based on collective intelligence have evolved based on collective intelligence such as the integrative model of behavioral prediction (Fishbein, 2009); the theory of triadic influence (Flay & Schure, 2012), multi-theory model (MTM) of health behavior change (Sharma, 2015; 2017a). According to Brennan and colleagues (2014) a pragmatic contemporary collective intelligence framework consists of eight elements: (1) *diversity* of populations and opinions contributing to the outcome; (2) both formal and informal, both virtual and physical *structures* for cooperation; (3) presence of tangible and intangible (such as ideas) *exchangeable elements* between all involved; (4) *space* to participate openly and fully such as through blogs, websites, social media etc.; (5) *communicable elements* that comprise of specifics of what needs to be done, when, how, where, and by whom; (6) *shared language* comprising of signs, symbols, and other cues; (7) having a provision of *rewards and incentives* for all participating members; and (8) a firm *belief* in the power of collective and power of people – the participatory approach.

Collective Intelligence and collective unconscious are different. Though Wechsler (1971, p. 905) contends that collective intelligence is "impersonal" and is defined by Jung in his conceptualization of collective unconscious as a separate psychic system "consisting of pre-existing forms of non-personal character" that does not develop individually but is inherited. However, there is a salient and fundamental difference between collective intelligence and the collective unconscious. The collective intelligence is a product of participatory conscious minds while the collective unconscious is inherent unobvious, intuitive, deep-seated and insentient in its origin. In unraveling the collective unconscious, deep contemplation or introspective meditation is vital (Sharma, 2018). The concept of introspective meditations will be described later in this chapter under the Kundalini Energy Yoga (KEY) section.

Super-Consciousness

Henri Antoine Jules-Bois (1868-1943), a French writer, based on the work of French psychologists wrote about the concept of super-consciousness for the first time and differentiated it from Freudian concept of sub-conscious or unconscious or the personal unconscious described by Jung as elaborated earlier. As cited in *Autobiography of a Yogi*, Jules Bois wrote in 1928 about super-consciousness as "is the exact opposite of the subconscious mind as conceived of by Freud; and which comprises the faculties that make a man really a man and not just a super-animal." (Yogananda, 1998, p. 70). He further differentiated it from hypnotism that is mainly used to go into the sub-conscious mind. The superconscious mind is the storehouse of all that has happened from time immemorial and can be construed as the accumulation of all information in nature. It is also construed as the cause of all effects and has the predictive potential for the future. It is also responsible for gigantic breakthroughs and other extraordinary tasks of unselfish nature (Yogananda, 1998, p. 144). Vivekananda (1973, p. 83) talks about this super-consciousness as *Samadhi* (about which we will talk later) and describes it as "no feeling of I and yet the mind works."

Frederic W. H. Myers (1841-1901) founded the *Society for Psychical Research* in London in 1882 and authored the book *Human Personality and Its Survival of Bodily Death* (Myers, 1903). He talked about the 'Subliminal Self' that helped in functions such as automatic writing, precognitive dreams, the trance states, and other anomalous mental experiences. Yogananda (1998) also cites his work in providing credence to the notion of super-consciousness. He quotes Myers as calling it the "treasure house" hidden in the deep consciousness.

Vethathiri Maharishi (1986, p.iii), an Indian philosopher states that human beings are "potentially omniscient" and can develop their consciousness to reach the super-consciousness. The state of achieving super-consciousness is within the reach of every human being but we get distracted in our everyday mundane pursuits that this goal becomes distant and unreachable. Yogananda (1998, p. 562) explains vividly the difference between the various stages of consciousness as they relate to the body and breathing. In the conscious state of mind, one is aware of the breath as well as the body. In the subconscious state of mind, such as in sleep, one experiences a temporary separation between the body and the breath. In the superconscious state of mind, one is freed or transcends from the delusion that "existence" depends on the body and the breath. The breathless state as described in yoga is also the superconscious state. Super-consciousness is not a topic of intellectual debate or a theoretical proposition or a speculation matter or a belief system. Simply stated it is the Truth that can only be experienced. As the sixth century BC Chinese philosopher, Lao Tze (or Laozi), the founder of Taoism, has said. "He who knows, tells it not; he who tells, knows it not." We will discuss some practical aspects of attaining this state through yoga in a subsequent section in this chapter.

SYSTEM OF ANCIENT INTEGRATIVE (MIND-BODY-SPIRIT) YOGA

Defining Yoga

The word yoga has its origin in the ancient language Sanskrit. The actual pronunciation of the word is Yög meaning, 'union.' The word is commonly used in math where it is used to denote the sum of any two numerals or for the mathematical function of addition. The Merriam-Webster's Dictionary (2018) defines yoga as "a Hindu theistic philosophy teaching the suppression of all activity of body, mind, and will in order that the self may realize its distinction from them and attain liberation." To the uninitiated mind the Merriam-Webster dictionary definition of yoga serves little purpose and, on the contrary, provides confusion to the real meaning and purpose of yoga. Let us examine this definition a little closely. The first assertion in the definition is that it is something to do with "Hindu." The word Hindu itself is a misfortunate term in the literature, as the term was used by the European colonialists to refer to people living around the river Indus in South Asia, and in contemporary times it has become associated with a so-called religion. However, the practice of yoga has nothing to do with any religion, except that perhaps the origins of the generic techniques used in yoga were first used by the human settlers around the river Indus (Singh, 1983). So the reader must first be clear that yoga is not a religion or anything to do with any country. It is based on the collective wisdom of the ancient seers of humankind.

The second assertion in the definition is that yoga is a "theistic philosophy." Theism (Merriam-Webster's Collegiate Dictionary, 2018) means belief in the existence of god or gods, specifically the existence of one God viewed as the creative source of man and the world that transcends yet is immanent in the world. Again such a belief is not necessary or mandatory for the practice of yoga (Mahan, 1961). Therefore, the reader who chooses not to espouse such a belief should not feel excluded or detracted from the practice of yoga. One has to realize that yoga is beyond simply a matter of such beliefs. Furthermore, the word philosophy also connotes exclusion from practicality. The Merriam-Webster's Collegiate Dictionary (2018) goes on to describe the word philosophy as all learning exclusive of technical precepts and practical arts. Again, if one were to believe that yoga is merely a philosophy, one would start wondering on its practical utility. Matter of fact, yoga is one of the most practical tools available to humankind to reflect, solve and find rational solutions to everyday problems (Maharishi, 1998).

The definition in the Webster's dictionary goes on to describe yoga as a "teaching of the suppression of all activity in mind, body, and will." Such a description is clearly a misinterpretation. A rational mind questions if yoga is indeed a suppression of all activity in mind, body, and will then how is it different than clinical death or persistent vegetative states. Again, the purpose of yoga is to activate the mind, the body, and the will so that they function optimally and ultimately attain full potential. Yoga helps in enhancing the functioning of the mind and body and in no way aims at suppression (Mahan, 1981; Maharishi, 1989; Vivekananda, 1973).

Finally, the dictionary definition explicates the purpose of yoga as, "in order that the self may realize its distinction from them (body, mind, and will) and attain liberation." Unfortunately, to the newcomers, as well as even advanced practitioners, again this will prove to be confusing and subject to misinterpretation. The purpose of yoga is not to emphasize the distinction of the self from body and mind but to foster "union."

Let us ponder closely on what this 'union' means. If one was to believe in human existence as only existing at the body level, such a view would be supported in the physiological paradigm. Physiology is the study of the biological functioning of the various parts (organ systems, organs, tissues, and cells) in the body. In such a pristine physiological paradigm, yoga can be appreciated as the union of various organ systems and their components (organs, tissues, and cells) in the body. This is usually referred to as the *homeostasis*. All living beings aim at maintaining or preserving homeostasis. Severe deviation or lack of this homeostasis is described as death. Therefore, if one was to believe that human existence is nothing beyond the body than for such an individual, yoga becomes a means to achieve optimal functioning of the various organ systems in the body or union within the various organ systems or *homeostasis*. There is nothing right or wrong in adhering to such a belief and practicing yoga for this purpose.

If a person believes that human existence also includes the functioning of the mind, such a view would be upheld in the psychological paradigm. Psychology is the science of mind and behavior. In this paradigm, various models of the mind have been proposed. In the west, the earliest models included descriptions of the mind by Sigmund Freud (1894, 1923). According to this viewpoint, the psychoanalytic theory, the mind comprises of id, ego, and superego. The id is a completely unconscious part of the mind and is the source of instinctual drives and needs. The ego is the conscious part of the mind that acts as the mediator between the person and the reality. The superego is only partly conscious, the part of the mind that represents internalization of parental conscience and rules of the society that provides the judgment of actions as moral or immoral. Freud hypothesized the constant struggle between the three in the performance of any behavior. If one was to subscribe to this viewpoint, yoga becomes a means to establish a union in these three layers of the mind. More contemporary conceptualizations of the behavior are exemplified by theories, such as the social cognitive theory, which hypothesizes the reciprocal determinism between personal attributes, social environment and behavior (Bandura, 1986) or multi-theory model (MTM) of health behavior change (Sharma, 2017a) in which cognitive, conative and environmental factors are linked with volitional behaviors. In this context, yoga can be construed as a means to establish the union between the mind and the performance of behaviors in harmony with the environment. If one subscribes to the viewpoints in parapsychology and acknowledges the superconscious state of mind then yoga helps in achieving a union of the conscious mind, with the subconscious mind, with the unconscious mind, and with the superconscious mind. The simple word balance can describe such a state of the mind. Again there is nothing right or wrong in adhering to such a belief and practicing yoga for this purpose.

Both physiology and psychology are established sciences. So, together with the views espoused by both these disciplines can be categorized as the modern scientific paradigm. In such a paradigm the connection between the body and mind can be easily appreciated without getting stuck in the arbitrary demarcations of the disciplines. In such a paradigm, yoga can be interpreted as a means to achieve psychosomatic equilibrium or the union of the body and mind.

In most modern-day scientific disciplines, the highest degree that is awarded to one who has mastered the art and science of the field is known as the Doctor of Philosophy (Ph.D.). Based on this simple fact, one can admire that science and philosophy are indeed closely intertwined. No science is complete without philosophy. If one was to acquiesce to the philosophical paradigm and explore what is the union that yoga refers to, one would have to delve deeper and explore the meaning of human existence little more carefully. Is the human existence only that of the body and mind? If that was true then, how is that when the body and mind are at complete rest, like when we are sleeping yet we still exist? The physiologists may posit an argument that indeed the body is functioning at the basal metabolic rate and is not completely at rest. So what is this basal metabolism that keeps the body running even at rest? The science will call that as the energy. This energy is needed to perform all the actions from morning till evening and then even while we are sleeping. This energy is always functioning. What is the source of this energy? Scientists will say the origin of the energy is from the food that we eat, the air that we breathe, and the storage that we maintain in the body. So this energy assumes greater importance than the body and mind because it is this energy that is needed to run the body and the mind. Acknowledgment of this inner energy is what the philosophical paradigm bases its argument on. Philosophy when hued with religion refers to this inner energy as the "soul" or the "spirit." One can even call this "Factor X" as some scientists have labeled it (Bisht, 1985). If one can appreciate this argument, then yoga becomes a means to establish a union between the body, mind, and inner energy. This union results in what can be called as equanimity.

Let us take this argument a little further and explore the world around us. Let us start with other human beings. The body of different human beings differs outwardly and so does the conditioning of the mind, but is there any difference in this inner energy? Contemplating this issue we will come to the conclusion that this inner energy remains the same for all. Let us go a little further and examine other mammals. We will again find that the body and the mind may differ but in essence, inner energy that drives the organism remains the same. Further analyses of other living beings in the animal kingdom will, again, help us reflect on the similarity of this inner energy that makes the organism perform its functions. Reflection on the plant kingdom again we can appreciate the similarity of this inner energy. Even when we reflect on inanimate objects comprising of a different conglomeration of molecules which in turn are comprised of atoms, we find that the same inner energy is responsible for the movement of electrons within the atoms. So we find that this inner energy is universal to all. Everything and anything that we perceive in the whole universe is indeed a manifestation of this energy. Such an argument can be appreciated in the universal paradigm. In this paradigm, yoga becomes a union of the human body, mind, and inner energy with the all-pervading cosmic energy. It is in this context that Maharishi (1989, p. 1; 1992, p.4) describes yoga as, "the art of realizing the soul's relationship with the body, the society, the world, and the universe, of maintaining its harmony with, and finally getting it merged with, the universal soul." Similarly, Sivananda (1974, p. 498) describes yoga as, union with all. In the same vein, Vivekananda (1973) describes yoga as a fine state of perception in which one can perceive all the different mental states. Similarly, Mahan (1961) describes yoga as the complete merger or union of self with the primordial state. This is the essence of interactive human agency related to superconductive scales of "quantum sentience" (denominated the *Psychecology*) described in this book.

So whatever our belief system may be, we can appreciate the beauty of yoga. If we have a religious bent of mind, irrespective of what religion we subscribe to, yoga becomes a means of attaining union with Supreme Consciousness which we can call as Almighty, Lord, God, Ishwar, Shiva, Allah, Jehovah, Holy Spirit, Primordial State, Static State, Universal Soul, Super-Consciousness or any other name. If we do not like the association with religious beliefs, then yoga becomes a means of attaining union with total existence or nature. We can look at all the magnificent creation that is perceivable by the senses and yoga as a way to establish harmony or complete union with this existence. Sean Carroll (2016) reflects on the diminutive status of human life when compared to the magnanimity of the universe but it is the link of realization of one's consciousness with the totality through yoga that transforms a human in deciphering the true purpose of life. Death is inevitable for all and life is way too short in the universe but yoga makes this short life meaningful.

The utility of yoga has been tested in the scientific literature. This has happened mainly in the context of the prevention and treatment of ailments. Yoga has been found to be advantageous in the treatment of depression, anxiety, low back pain, and asthma and a promising approach that needs more testing for hypertension, diabetes mellitus type 2, and arthritis (Sharma, 2014). It is also helpful in stress management especially for cancer patients and other people suffering from acute and chronic stressors.

Asthanagayoga

Historically, the system of yoga owes its origin to the practices perfected by the settlers around the Indus (*Saraswati*) valley in the era 5000 years BC. Initially, the practices were perfected by few and the knowledge kept in close confinements. The practices were only taught from the master teacher (*Guru*) to selected disciples and relied solely on memory (*Smriti*) for transference. The earliest documented written versions relating to these practices are found in the aphorisms (*Yogasutras*) of Patanjali (200 BC), a saint who is considered as the first authority on the subject. Patanjali described the eight steps of yoga or *Asthangayoga*. The eight steps refer to one's societal conduct, physical conditioning, and psychic conditioning. Today all schools of yoga adhere to these eight steps, either completely or partially in some form or the other. The eight steps of *Asthangayoga* are (1) *Yama*, (2) *Niyama*, (3) *Asana*, (4) *Pranayama*, (5) *Pratyahara*, (6) *Dharana*, (7) *Dhyana*, and (8) *Samadhi*. The first four steps are often called the *bahirmukhi yoga* (or outward yoga) and the last four *antarmukhi yoga* (or inward yoga).

The Sanskrit word "*Yama*" means discipline. The practices described in *Yama* are the practices that help one restrain one's social conduct for living in harmony within the society. There are five *Yama* in yoga. First such practice is the principle of *ahimsa* (or non-injury). Avoiding deliberate injury to self or to others in the immediate present or with a potential to cause such injury at a future time is the practice of non-injury. The second *Yama* is *Aparigraha* (or receiving exactly what is right). This entails setting limits to one's needs and drawing the limit to one's wants. The third *Yama* is *Asteya* (or not desiring what another has or non-stealing). The fourth *Yama* is *Brahmacharya* (or sensual control). This entails not overindulging in sensory pursuits and transcending the consciousness of mere senses. The final *Yama* is *Satya* (or Truth) that entails being honest as far as possible in one's thoughts, words, and actions. These principles are fundamental and common to all religious beliefs and all logical reasoning. Most law and order in all countries around the world are developed to uphold and preserve these fundamental principles. Judaism, Christianity Buddhism, Jainism, Islam, *Sanaatan Dharma* (Hinduism), Bahai Faith, Confucianism, and other religious doctrines all advocate these same principles of *Yama*.

The Sanskrit word, "*Niyama*" means observances or rules for self. The *Yama* and *Niyama* go hand in hand. The first *Niyama* is *Shaucham* (or cleanliness) that entails cleaning both inner and outer body and mind. The second *Niyama* is *Santosham* (or contentment). This entails being completely contended in the present without regretting the past or worrying about the future. The third *Niyama* is *Tapas* (or practice) that entails self-discipline of regular practice to purify oneself. The fourth *Niyama* is *Swadhyaya* (or self-study). This entails examining the true nature of "Self." The final *Niyama* is *Ishwar Pranidhana* or giving ownership of outcomes of one's actions to Higher Power.

The third step in *Asthangayoga* consists of the practice of *Asanas*. *Asanas* are low impact physical activity postures that serve several purposes, of which the primary purpose is optimal maintenance of the body. Some schools of yoga, such as *Hatha yoga*, predominantly focus on this aspect (Vivekananda, 1973). There are close to 840,000 different postures that the human body can assume. In order to perfect all of these postures, even a full lifespan of 120 years would perhaps be deficient. One has to appreciate that there is absolutely no need to learn all the postures but select a few depending on the specific needs that one might have. Furthermore, not all of us can perform all the postures, so one has to judiciously select what one can do depending on age, gender, body flexibility, and other such personal characteristics and the ultimate purpose we wish to pursue.

The fourth step in *Asthangayoga* consists of the practice of *Pranayama*. *Prana* means the life force or the inner energy, *Yama* means discipline. It is the regulation and increasing the stock of the life force or inner energy that this practice aims at establishing. One of the sources of the inner energy is the air that we breathe. Intake of the air is accomplished by the respiratory system. As adults, we do not do a very good job of breathing and tend to breathe superficially using only the upper part of the lung. Such breathing is known as thoracic breathing. Infants and children breathe applying abdomino-thoracic breathing that expands the lungs completely thereby enhancing the air uptake and increasing the stock of *prana* or life force or inner energy. The aim of *Pranayama* is to simulate this expanded form of breathing to increase the stock of life force.

The fifth step in *Asthangayoga* consists of *Pratyahara* (or gathering inward). This has been defined by Sivananda (1974, p. 491) as "abstraction or withdrawal of the senses from their objects." An effort to control the mind through others or using hypnosis is not recommended in this endeavor (Vivekananda, 1973). Self-practice of detaching the mind at will away from senses strengthens this practice which must be done routinely.

The sixth step in *Asthangayoga* is the practice of *Dharana* (or holding the mind at certain points). When the mind is limited to one place it is called *Dharana*. When we write or do another creative activity we are in essence doing *Dharana*. In transcendental meditation, a popular technique in the West, this is done through the use of focusing on a *mantra* (Romas & Sharma, 2017). In Chakra meditation, the *Dharana* is done on various points corresponding to the endocrine glands in the body.

The seventh step in *Asthangayoga* consists of the practice of *Dhyana* (meditation). The starting point of meditation according to Sivananda (1974) is *manana* (or introspection or reflection) to purify the mind. Then meditation is bringing the electroencephalogram (EEG) waves of the mind to subtler levels. When mind and senses are consciously working the EEG waves are vibrant as beta waves at 14-40 cycles per second. As one starts to calm the mind through meditation (which is wakeful sleep) the EEG waves start to come to alpha waves (8-13 cycles per second) and then gradually come to theta waves (4-7 cycles per second).

The final or eighth step in *Asthangayoga* consists of the practice of *Samadhi*. The word Samadhi is derived from Sanskrit "*Sam*" meaning equal or together and "*Adhi*" meaning the primordial state. This has been referred to earlier as also super-consciousness (Vivekananda, 1973). The first state of *Samadhi* for a meditator (*yogi)* is called "*Savikalpa Samadhi,*" the second one as "*Nirvikalpa Samadhi*" and the final stage as "*Sahajavastha*" or "*Jivanmukta*" (Sivananda, 1974, p. 373). The *Savikilpa Samadhi* is exemplified by delta waves on EEG (<1 cycle per second) attained by few in the wakeful state but everyone in sleep (Romas & Sharma, 2017; Vivekananda, 1973). In this state, the meditator (*Yogi*), meditation (*Dhyana*) and the object of meditation (*Dhyeya*) become one and the same which provides bliss (Sivananda, 1974). Yogi must not stop here and reach the *Nirvikalpa Samadhi* where the mind completely melts into the totality. This is also known as the omega on EEG (or 0 cycles per second) or complete silence or stillness and attained by only very few in a wakeful state. As a result of attaining this state the "residue of actions is burnt" and the practitioner gets liberation (*Sahajavastha* or *Jivanmukti*) (Sivananda, 1974, p. 373). Clearly, since few have attained this state this concept is hard to establish in the mainstream but at the same time, one has to remember that exceptions prove the rule.

SOLUTIONS AND RECOMMENDATIONS

Any theory or philosophy unless it is practical is of no value. There is a need to make yoga practical and applicable in everyday life for everyone so that the experience of collective unconscious or super-consciousness can be achieved by all and not just a selected few. To accomplish this goal various schools of yoga have evolved over the years (Romas & Sharma, 2017). Some of these that start with meditation are *Kriya Yoga* developed by Self Realization Fellowship and its sister organizations; *Raja Yoga* as developed by Ramakrishna Paramhansa and Swami Vivekananda and their associated organizations; *Transcendental Meditation* developed by Mahesh Yogi and its associated sister organizations; *Kundalini Yoga* (Mahan, 1961; 1981) and its variant *Simplified Kundalini Yoga* (SKY) (Maharishi, 1989); *Mindfulness Meditation* developed by Jon Kabat-Zinn (Khoury, Sharma, Rush, & Fournier, 2015) and many other such schools.

The Indian philosopher Jiddu Krishnamurti (n.d.) however contended that "Truth is a pathless land" -- an idea that also has merit. Oftentimes, in adhering to the practices of one school of yoga, the practitioner gets influenced by the charisma of the leader or teacher and forgets the main purpose which is the attainment of the super-consciousness. The practitioner also gets sidetracked or distracted in the practices and rituals to the extent that the main goal becomes elusive. In order to overcome such obstacles, the practitioners of the system of kundalini yoga and Simplified Kundalini Yoga (SKY) have evolved a self-help approach in which there are just generic techniques from these schools and there is no teacher on a pedestal. Sometimes in the spiritual arena, some unscrupulous teachers act as "middlemen" and exploit the followers, a practice that has been shunned by all self-realized teachers. Hence, everyone is an "experimenting practitioner" self-discovering and evolving the generic techniques to enhance their own journey of consciousness. This paradigm is called *Kundalini Energy Yoga (KEY)*. Such a paradigm accepts all religions and practices known to different groups of human beings and only complements them. This paradigm does not advocate the formation of any "group" and is beyond such narrow conceptualizations. One can use the techniques of this approach that are derived from yoga and specifically kundalini yoga with any existing approach that they practice and modify these in any way to achieve the ultimate goal. However, it has to be understood that ultimately even the techniques have to be relinquished in reaching super-consciousness which itself guides the seeker more in lines with what Krishnamurti calls as the "pathless land."

Modern Practical Paradigm of Kundalini Energy Yoga (KEY)

Kundalini Energy Yoga (KEY) techniques have been tested empirically for some mundane outcomes with healthy participants (Sharma, 2001), osteoarthritis patients (Sharma, 2005), anxiety patients (Mehta & Sharma, 2012-13), and individuals needing smoking cessation (Sharma & Corbin, 2006). The results have been quite encouraging in these ordinary pursuits. More research is underway especially with regard to its potential to higher order uses. In addition, the personal experience of the author with the techniques and anecdotal experiences of co-practitioners lends credence to these techniques in unraveling the ultimate goal.

In this context, it needs to be appreciated that all of us come on this planet earth as uninstructed, unconditioned newborns. Everything that we learn we learn here. This is called conditioning of the mind that depends on the kind of parents we have, the family that we live in, the teachers that we learn from, the prevalent mores, norms and cultures in the society and other such factors. To claim that one conditioning is any better than the other is an argument in futility. However, we must make the best use of this conditioning to enhance our journey of unraveling the super-consciousness. In KEY paradigm, the practices of *Yama* can generally be interpreted to include respecting and adhering to the existing laws and edicts prevalent in the place that one resides. These practices change according to the country/place (*desh*), time (*kaal*), and circumstances (*paristhithi*). If one is looking for more guidance in order to self-introspect on some of the attributes of *Yama* and *Niyama* a 25 chapter reflective introspective mediation guide is available based on the ruminations of the author (Sharma, 2018). However, the ultimate self-discipline to pursue the goal of super-consciousness must come and be guided from within. Such introspective meditation forms the base or foundation on which all practices need to be built. Purity and selflessness go a long way in shortening the journey of consciousness.

The ideologies in *Niyama* pertain to the principles that help in maintaining the body and mind to foster better health and wellness. We often take our body and mind for granted. Is it not our duty to maintain the body and mind? Especially when one is engaged in the pursuit of yoga, sometimes one falsely starts believing that the body and mind are not important and yoga is some form of mere esoteric pursuit. It is for guarding against this misperception and performing the important duty of maintaining the body and mind that principles of *Niyama* are established. In KEY, these principles can be viewed as developing sound dietary habits, regular physical activity patterns, taking measures for prevention of diseases, avoiding the use of intoxicants, adopting safety measures to minimize the risk of possible harm from other living beings, climatic changes, accidents, and natural disasters. With regard to diet, a vegetarian diet consumed in moderation consisting of sun foods (such as fruits, vegetables, nuts, seeds, grains etc.) and ground grown produce (such as peanuts, potatoes etc.) can be experimented. *Asanas* and breathing techniques or *Pranayama* to enhance the life stock of energy may also be practiced and several techniques are available which can be expanded based on one's preference (Health for All, 2001).

In addition, one must practice a regular balance between work, sleep, and relaxation. For relaxation techniques such as *Shava Asana*, progressive muscle relaxation (PMR), humor, creative activities, autogenic training, self-hypnosis, biofeedback, visual imagery, and other such methods can be chosen (Romas & Sharma, 2017). It is very important to have sleep at night for at least seven to eight hours for adults. Likewise, it is very important for most able adults to work for seven to eight hours every day based on their chosen career. Some time from the remaining time may be devoted toward the pursuit of yoga goal and selfless service which are also very important.

For *antarmukhi yoga (Pratyahara, Dharana, Dhyana, Samadhi)*, meditation techniques can be utilized. Some of these (ideally developed and mastered sequentially) are (1) watching the breathing; (2) using a *mantra* (such as *Aum, Amen, Shanti, Santosham*, etc.); (3) focusing on the five elements (*Panchbhootas*) of earth, water, air, fire. and sky; (4) focusing on the six senses (*Chat Indriya*) of touch, taste, smell, hearing, sight and kinesthetic; (5) focusing on *Chakras* or endocrine glands of the body such as sex glands (*Mooladhara Chakra*), adrenal/suprarenal glands (*Swadishthana Chakra*), pancreas (*Manipuraka Chakra*), thymus (*Anahata Chakra*), thyroid/parathyroid glands (*Vissudhi Chakra*), pituitary gland (*Agna Chakra*), and hypothalamus (*Sahasrara Chakra*); (7) human-nature (*Jeeva-Brahma Samadhi*) meditation where expanding of conscious mind to super-consciousness is done; (8) Absolute meditation (*Thuriyateeth Samadhi* or *Nirvikalpa Samadhi*) where all techniques culminate and the static primordial state itself takes over (Romas & Sharma, 2017). Of course, it cannot be overstressed that all practices and meditation require concerted regular implementation and patience. Often several decades of regular practice are required before results start accruing. However, concerted practice, purity in thoughts, words and actions and selfless service drastically shorten this pursuit.

Self-Realization

Jung describes the model of the "self" in several of his works such as *Symbols of Transformation*, *The Archetypes of the Collective Unconscious*, and *Aion*, as a "numinous image of psychic wholeness that becomes manifest in any authentic mystical or religious experience." (Gullatz, 2010, p. 700). Clearly, Jung was influenced by religious and philosophical influences prevalent in his times. According to Jung, the ultimate goal of "Self-realization" is accomplished when "the meaning of life no longer lies in the ego" (Jacobi, 1943). In his philosophical positioning of Self-realization Jung does not consider the developmental stages of personality that consist of infancy, childhood, adolescence, early adulthood, middle adulthood and older adulthood (Shor, 1944). Instead, he talks about the innate nature of this disposition which is in consonance with Vedic philosophy of yoga according to which all human beings are Divine and potentiated with the urge of Self-realization.

In essence, Self-realization and linkage with super-consciousness is the same thing. Sivananda (1974) describes this state of Self-realization as being *Jivanmukta* or a person who has destroyed attachment to the outcomes of all desires and is perfectly balanced under all circumstances (unaffected by dualities) and practices *sama bhava* or *sama drishti* (oneness in everything). Vethahiri Maharishi (1986), Vedic philosophy, Buddhist *malas,* and others describe the process of reaching the goal of self-realization through purification of the three impurities. The first one is the impurity of ego or *anava* in Sanskrit. In this impurity, one associates itself with the body and mind or the *anu* (atom) or the "I" and "mine" separate from the totality. This impurity can be purified through meditation and linking with the Absolute. The second impurity is *karma* in Sanskrit or the imprints of previous actions. This impurity can be cleansed by superimposition of good and uplifting thoughts, words, and deeds. The third impurity is that of *maya* in Sanskrit or the illusion. This can be cleansed by the above two efforts whereby wisdom automatically ensues. It is the belief of the author that, while the seeker makes all efforts to achieve the state of self-realization, this is a state that is not achieved by only the efforts of the body and mind which merely pave the way but ultimately it is endowed by the superconscious itself to the right instrument. While Self-realization (or linkage with super-consciousness) is indeed a worthy goal to pursue but a more salient question is what to do after Self-realization? Is that the end or is it a means to work for the betterment of humankind? Clearly, if more people become Self-realized our planet will become a better place. There will be fewer inequities, fewer injustices, fewer disparities and greater harmony, contentment, peace, and satisfaction among humankind. So the techniques for Self-realization must become universally available through a myriad of schools and channels of yoga.

FUTURE RESEARCH DIRECTIONS

The research with unconscious was not possible in the times of Freud and Jung because of pragmatic difficulties. But in modern times science and technology have advanced and we have many more tools available for engaging in multidisciplinary research in this area. Such a multidisciplinary team needs to consist of yogis, psychologists, sociologists, health behavior experts, physiologists, neuroscientists, biostatisticians, physicians, medical researchers, and public health and health education experts. The yogis need to refine the techniques. KEY offers a finely refined scientific paradigm that can be easily implemented. The psychologists, sociologists, and health behavior experts need to design psychometric and other measurement tools to gauge the transition in constructs being reified. These would include knowledge about yoga, attitudes about yoga, theoretical constructs that may be antecedents to yoga behaviors and actual yoga-related behaviors. Physiologists, neuroscientists, and biostatisticians need to come up with objective biomarkers to track the transition to higher levels of consciousness and statistically test them. Finally, the physicians and medical researchers need to test these in clinical bedside applications while public health and health education experts need to find techniques to promote these to the masses. The public health education research has progressed from knowledge-based interventions to skill-based interventions to theory-based interventions to present day fourth generation multi-theory models (Sharma, 2017b). Multi-theory models help develop brief and precise interventions which would be indispensable in promoting yoga techniques to unravel the mysteries of the mind and consciousness. Such a participatory cognitive influence of multidisciplinary professionals based on "collective intelligence" approach would help popularize techniques to unravel "collective unconscious" and "super consciousness."

ACKNOWLEDGMENT

This research received no specific grant from any funding agency in the public, commercial, or not-for-profit sectors. The author is thankful to the following for their astute comments on this chapter: Dr. John Christensen, Ph.D., Chancellor Emeritus, University of Nebraska at Omaha, Omaha, NE; Dr. R. Srinivasa Murthy, MD, Retired Professor of Psychiatry, India; Dr. Amar Kanekar, Ph.D., MPH, MBBS, MCHES®, CPH, FRSPH, Associate Professor and Graduate Program Coordinator, Health Education/Health Promotion, University of Arkansas at Little Rock, Little Rock, AR; Dr. Traci Hayes, MBA, MS, Dr.PH, Director, Communications & Marketing, RTRN Data Coordinating Center, Mississippi e-Center @ Jackson State University, Jackson, MS; Jacqueline Wilber, E-RYT 500, AYS, M.Ed., Yoga Now, Omaha, NE; & Pamela McNally, KEY practitioner for over 20 years, Omaha, NE.

REFERENCES

Adams, M. V. (2002). African American dreaming and the beast of racism. The cultural unconscious in Jungian analysis. *Psychoanalytic Psychology, 19*(1), 182–198. doi:10.1037/0736-9735.19.1.182

Bandura, A. (1986). *Social foundations of thought and action*. Englewood Cliffs, NJ: Prentice-Hall.

Bisht, D. B. (1985). *The spiritual dimension of health*. New Delhi: Directorate General of Health Services, Government of India.

Brennan, L., Binney, W., Parker, L., Aleti, T., & Nguyen, D. (2014). *Social marketing and behavior change. Models, theory and applications*. Cheltenham, UK: Edward Elgar Publishing.

Carroll, S. (2016). *The big picture: On the origins of life, meaning, and the universe itself.* New York: Dutton.

Carvalho, R. (2014). Synchronicity, the infinite unrepressed, dissociation and the interpersonal. *The Journal of Analytical Psychology, 59*(3), 366–384. doi:10.1111/1468-5922.12085 PMID:24919629

Cashford, J. (2018). 'Who is my Jung?' The progressive, though sometimes ambivalent, expansion of Jung's idea of the collective unconscious: From an 'unconscious humanity' to - in all but name - the soul of the world. *The Journal of Analytical Psychology, 63*(3), 322–335. doi:10.1111/1468-5922.12413 PMID:29750347

Dinan, T. G., Stilling, R. M., Stanton, C., & Cryan, J. F. (2015). Collective unconscious: How gut microbes shape human behavior. *Journal of Psychiatric Research, 63*, 1–9. doi:10.1016/j.jpsychires.2015.02.021 PMID:25772005

Fishbein, M. (2009). An integrative model for behavioral prediction and its application to health promotion. In R. J. Diclemente, R. A. Crosby, & M. C. Kegler (Eds.), *Emerging theories in health promotion practice and research* (2nd ed.; pp. 215–234). San Francisco: Jossey-Bass.

Flay, B. R., & Schure, M. (2012). The theory of triadic influence. In A. C. Wagenaar & S. C. Burris (Eds.), *Public health law research: Theory and methods* (pp. 169–192). San Francisco: Jossey-Bass.

Freeman, M. (2016). From the collective unconscious to the narrative unconscious: Re-imagining the sources of selfhood. *Europe's Journal of Psychology, 12*(4), 513–522. doi:10.5964/ejop.v12i4.1325 PMID:27872663

Freud, S. (1894). The neuro-psychoses of defence. SE, 3, 41-61.

Freud, S. (1923). The ego and the id. SE, 19, 1-66.

Gullatz, S. (2010). Constructing the collective unconscious. *The Journal of Analytical Psychology, 55*(5), 691–714. doi:10.1111/j.1468-5922.2010.01878.x PMID:20961321

Health for All. (2001). *Kundalini energy yoga: Asanas, pranayama, relaxation, and meditation* [DVD]. Omaha, NE: Health for all, Inc. Available from https://www.youtube.com/watch?v=VH1UFVzQwkg https://www.youtube.com/watch?v=XdvwVz-qI1w&t=9s https://www.youtube.com/watch?v=Y3qnDDlSydE&t=16s

Hunt, H. T. (2012). A collective unconscious reconsidered: Jung's archetypal imagination in the light of contemporary psychology and social science. *The Journal of Analytical Psychology*, *57*(1), 76–98. doi:10.1111/j.1468-5922.2011.01952.x PMID:22288542

Jacobi, J. (1943). *The psychology of Jung*. New Haven, CT: Yale University Press.

Jung, C. G. (1968). *Part 1: The archetypes and the collective unconscious* (2nd ed.; Vol. 9; G. Adler & G. F. C. Hull, Trans. & Eds.). Princeton, NJ: Princeton University Press.

Khoury, B., Sharma, M., Rush, S. E., & Fournier, C. (2015). Mindfulness-based stress reduction for healthy individuals: A meta-analysis. *Journal of Psychosomatic Research*, *78*(6), 519–528. doi:10.1016/j.jpsychores.2015.03.009 PMID:25818837

Krishnamurti, J. (n.d.). *Truth is a pathless land*. Retrieved from https://www.jkrishnamurti.org/about-dissolution-speech

Le Bon, G. (2002). The crowd: A study of the popular mind. Mineola, NY: Dover Publications (Originally published 1896)

Mahan, G. P. (1961). *I-God (Secret Power of Kundalini)*. Madras, India: Universal Peace Sanctuary.

Mahan, G. P. (1981). *Kundalini yoga*. Tamil Nadu, India: Universal Peace Sanctuary.

Maharishi, Y. V. (1986). *Karma yoga the holistic unity*. Madras, India: Veththiri Publications.

Maharishi, Y. V. (1989). *Yoga for modern age* (3rd ed.). Erode, India: Vethathiri Publications.

Maharishi, Y. V. (1992). *Journey of consciousness*. New Delhi: Macmillan India Limited.

Maharishi, Y. V. (1998). *Logical solutions for the problems of humanity*. Erode, India: Vethathiri Publications.

Maslow, A. H. (1943). A theory of human motivation. *Psychological Review*, *50*(4), 370–396. doi:10.1037/h0054346

Mehta, P., & Sharma, M. (2012-13). Evaluation of a social cognitive theory based yoga intervention to reduce anxiety. *International Quarterly of Community Health Education*, *32*(3), 205–217. doi:10.2190/IQ.32.3.d PMID:23353562

Meredith-Owen, W. (2011). Winnicott on Jung: Destruction, creativity and the unrepressed unconscious. *The Journal of Analytical Psychology*, *56*(1), 56–75. doi:10.1111/j.1468-5922.2010.01890.x PMID:21241294

Myers, F. W. H. (1903). *Human personality and its survival of bodily death* (Vols. 1–2). London: Longmans.

Romas, J. A., & Sharma, M. (2017). *Practical stress management. A comprehensive workbook*. London: Academic Press.

Sharma, M. (2001). Pilot test of a Kundalini yoga intervention developing mind-body connection. *International Journal of Yoga Therapy*, *11*, 85–91.

Sharma, M. (2005). Effects of a yoga intervention as a supportive therapy in arthritis. *Yoga Studies*, *4*, 12–16.

Sharma, M. (2014). Potential efficacy of yoga in various diseases. *Journal of Symptoms and Signs*, *3*(3), 194–199.

Sharma, M. (2015). Multi-theory model (MTM) for health behavior change. *WebmedCentral Behaviour*, *6*(9), WMC004982.

Sharma, M. (2017a). *Theoretical foundations of health education and health promotion* (3rd ed., pp. 250–262). Burlington, MA: Jones and Bartlett.

Sharma, M. (2017b). Trends and prospects in public health education: A commentary. [Editorial]. *Social Behavior Research & Health*, *1*(2), 67–72.

Sharma, M. (2018). *Introspective meditations for complete contentment (Santosha)*. Omaha, NE: Health for All.

Sharma, M., & Corbin, D. E. (2006). A preliminary study of the effects of an innovative social cognitive theory driven yoga-based behavioral intervention on smoking cessation. *Californian Journal of Health Promotion*, *4*(4), 63–76.

Shor, J. (1944). Review of the psychology of Jung. *Psychological Bulletin*, *41*(7), 496–497. doi:10.1037/h0050824

Singh, K. (1983). *Hinduism. Religions of India*. New Delhi: Clarion Books.

Sivananda, S. S. (1974). *Mind. Its mysteries and control*. Tehri-Garhwal, India: The Divine Life Society.

Smith, J. B. (1994). *Collective intelligence in computer-based collaboration*. Hillsdale, NJ: Lawrence Erlbaum.

Sotirova-Kohli, M., Rosen, D. H., Smith, S. M., Henderson, P., & Taki-Reece, S. (2011). Empirical study of Kanji as archetypal images: Understanding the collective unconscious as part of the Japanese language. *The Journal of Analytical Psychology*, *56*(1), 109–132. doi:10.1111/j.1468-5922.2010.01893.x PMID:21241297

Vivekananda, S. (1973). *Raja yoga or conquering the internal nature*. Calcutta, India: Advaita Ashram.

Weschsler, D. (1971). Concept of collective intelligence. *The American Psychologist*, *26*(10), 904–907. doi:10.1037/h0032223

Yogananda, P. (1998). *Autobiography of a yogi* (13th ed.). Los Angeles, CA: Self-Realization Fellowship.

Section 2
Alchemical Sentience

Chapter 6
Human Cultural Evolution Is Completely Immersed in Natural Evolution

Jan Holmgren
Independent Researcher, Sweden

ABSTRACT

Human culture is presently setting a disastrous mark on planet Earth, overexploiting its resources, causing global warming, causing mass extinctions. We appear already to be in a new geological period, the Anthropocene. The natural sciences are crucial for the understanding of those processes. Equally important are decisions in human culture to act on the large challenges in humane and rational ways. Whitehead's evolutionary Philosophy of Organism, with the introduction of atomic microfeels, inspires a coherent cosmology characterized by the basic position of human consciousness. The humane aspects (pre-physical, metaphysical) of human culture (e.g. the humanities, the religions, the arts) are shown to be importantly causal in the universe. Human empathy and altruism (loving attitudes) combined with rationalism and truthfulness must be enhanced as central ambitions in human societies, not least in areas like politics and economy. The suggested worldview, accepting complementarity, is named multi-aspect monism (MAM).

INTRODUCTION

A somewhat odd characteristic of contemporary consciousness studies is that much of what is studied is such that is most often considered to be unconscious. Mechanisms and chemistry as revealed in neuroscience, for example, are thought to lead up to or underpin consciousness. Consciousness itself then is thought to appear miraculously as a phenomenon distinct from "matter". Through history countless attempts have been made to avoid Cartesian dualisms in order to reach a trustworthy monism. Interesting suggestions for a monist cosmology grounded in physical science, e.g. suggesting "Scintillae of Subjectivity", are presented in Pereira Jr *et al.* (2018). Many have instead attempted to start within consciousness itself, as kinds of idealisms grounded in phenomenology, which is the main attitude adopted below.

DOI: 10.4018/978-1-5225-9065-1.ch006

BACKGROUND

The animal brain is evolved in nature and consciousness certainly appears there for functional reasons (Holmgren 2014). The Philosophy of Organism formulated by Alfred North Whitehead in *Process and Reality* (Whitehead 1978) sees that clearly. Full acknowledgment of Whitehead's repeated and very clear statement that we always ultimately can refer only to our human conscious experiences leads to a somewhat radical conclusion: Whitehead`s cosmology is in fact a meticulously ordered statement of what he could see in his own brain. Of utmost importance for the following is the insight that the same predicament applies to us all. This is an attempt to express some of what I see within my own brain.

MAIN FOCUS OF THE CHAPTER

Central to this chapter is the understanding of human consciousness as inspired by its treatment in Whitehead's philosophy. Hinted at are some consequences for human agents in the universe, with a few implications for the future of human culture and a sustainable planet Earth.

Whitehead´S Metaphysics in Natural Evolution

Whitehead accepts that everything is process in a continuous universe. Human conscious experiences emerge in that universe in the form of large multitudes of apparently discrete atomic feels within human brains. They can loosely be thought to correspond to the "event particles" of his earlier work *The Principle of Relativity* (Whitehead 1922), with an illustration very similar to a light cone. A modified and extended version is shown in Figure 1. The light cone is here defined orthodoxly in Minkowski spacetime and, as suggested in *Hypothesis on consciousness* (Holmgren 1981), with inclusion of the causal effects of a human conscious experience.

Figure 1 shows principles of the human predicament. All human conscious experiences are built of microprocesses in the continuous universe. In processes of actualization they provide seemingly discrete atomic microfeels. Seemingly, because the microfeels are fully integrated in the continuous universe.

Figure 1. The light cone with inclusion of a human conscious experience. The actualized microfeels in themselves and the inter- and intradynamics of their corresponding microprocesses are ultimately not attainable for physics. Those aspects are below physical resolution and will always have to be treated as metaphysical.

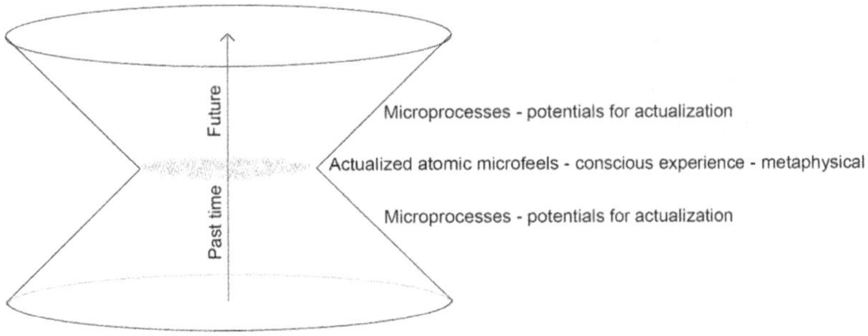

Really individually atomic, because they cannot be split into discernible parts. The microfeels are evolved in natural evolution and correspond to what Whitehead calls "eternal objects", loosely qualia. In order to be felt, each microfeel must have a certain "volume/duration" in the continuous spacetime. For giving us the enormous richness and precision we experience, the microfeels must be huge in numbers and extremely "small" in spacetime. Thanks to our long history of extremely refined natural evolution, very large numbers of combinations of very large numbers of microfeels, often in stably repeated actualizations in extended spacetime, give us fairly exact and reliable inputs from surroundings that can be stored and reactualized in memory engrams (Tonegawa *et al.* 2015). The function of Whitehead's God in The Philosophy of Organism (Whitehead 1978) is replaced here by self-organization in natural evolution.

The natural sciences always aim towards precise mathematical accuracy. Since observations and mathematics all are built in the form of distributions of microfeels, the achievable resolution will irremediably remain limited (Holmgren 2014). The inter- and intradynamics of microfeels and their corresponding microprocesses are necessarily below the resolution of mathematical physics and they will always have to be treated as pre-physical, i.e. metaphysical. There is nothing mysterious in this. The explanation is simply the rich occurrences of extremely small and local processes in the human brain, below the resolution of mathematical physics. As shown in Figure 1 the effects of consciousness give a widening of possible outcomes compared to the abstract and purely physical light cone. Thus, moods, feelings, intentions, decisions, etc. play precise metaphysical causal roles in the universe.

Figure 2 shows a few main steps in the human understanding of the universe, and ultimately the understanding of human nature and culture. Each step is shortly characterized below.

Newtonian Universe (See Figure 2a)

In Newton's times the natural conception of force was immediately observable causal relationships: a horse had to be physically attached to a wagon, a bag of wheat had to be grabbed to move it. Causality had a time aspect, a horse had to be attached before the wagon was moved. Newton's laws on the other hand were based on the idea of instantaneous action at a distance, an idea Newton himself was uncomfortable with. Because of its practical success this new way of thought gradually became a new common sense.

Figure 2. The light cone used to illustrate different human aspects applicable to the universe. a. Newtonian universe – instantaneous action at a distance. b. Einsteinian universe – the speed of light in vacuum is a universal constant – Minkowski space and continuous universe. c. Quantum-mechanical universe – instead of points there are quanta (in fact actualized discrete microfeels in human brains). d. Universe containing relatively simple or static forms of life – Bacteria, Archaea, Fungi, and Plants of Eukariota – effects of coding e.g. in DNA. e. Universe containing advanced mobile forms of life requiring conscious brains – Animalia of Eukariota – effects of microfeels and their interactions. f. Universe containing Homo Sapiens with advanced selfconsciousness and human language – human culture causing Anthropocene.

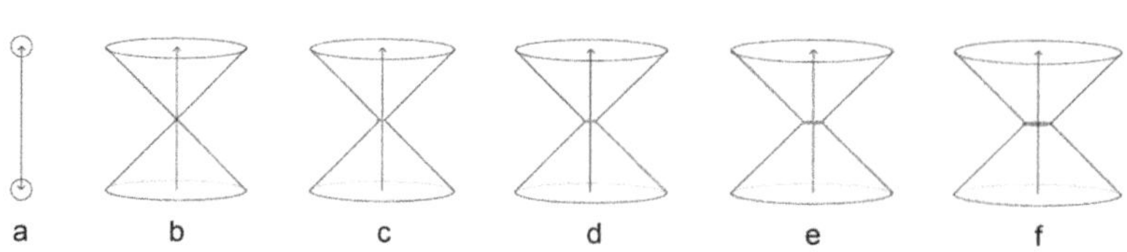

Einsteinian Universe (See Figure 2b)

By recognizing the limit for the speed of light as a cosmological constant, Einstein in the theories of relativity in a way reintroduced a more intuitively natural view, getting rid of instantaneous action at a distance. For each point in the universe there is a limit for which other points can have had an effect on the point (i.e. those within the past light cone) and which other points the point itself can have an effect on (i.e. those within the future light cone). Einstein saw limits for the applicability of physics and he explicitly referred to psychology matters that could not be treated within mathematical physics (Einstein 1959).

Quantum Physics Universe (See Figure 2c)

Whitehead's Philosophy of Organism (Whitehead 1978) clarifies that we never can have direct access to our surroundings, the mediating factor being our organisms. When turning to the minutest discernible structures in the universe, the limits for human resolution cannot be ignored. Human consciousness is evolved for survival in nature and it is not a perfect tool for the understanding of everything. Extremely refined mathematical theories and equally refined experimental methods, with use of advanced IT-technology, may appear to have the potential to extend understanding without limits. However, since the seemingly atomic microfeels that build our conscious experiences necessarily have "volume/duration" in spacetime, they in fact are a serious ultimate limit for the applicability of mathematical physics.

The continuous universe is dimensionless and timeless. The discretization of the continuous in human consciousness is the ground for quantum physics. When each of us looks around in the brain, we will always deal only with large amounts of atomic microfeels. The microfeels will always remain really atomic since they can be felt only when the actualization process is active. Microfeels appear in great diversity, sometimes describable like colors in the color space, sometimes in abstract structures like mathematics. In the generalizations applied by physics the microfeels become quanta, characterized not by their feels but by certain very abstract concepts, like strings, charge, and spin dynamics (Klose 2009). What is known of theories and experiments is solely and entirely actualized in conscious experiences in human brains. This does not entail solipsism. Very much of the input from what happens in surroundings and in the brain, thanks to the enormous refinements achieved in natural evolution, is consistent and trustworthy.

In quantum theory the smallest possible influence of human consciousness is introduced in mathematical physics in the form of the abstract quantum. The real metaphysical moods, feelings, desires, etc., in living organisms have much more wide-ranging influences as potent agents in the universe.

Universe Containing Relatively Simple or Motionless Forms of Life (See Figure 2d)

Life forms entering the universe are novelties and natural products of continuous gradual evolution. Details of their evolutionary histories are often impossible to explore. One trend is a move towards complex structures. There is also frequent loss of complexity or stasis. Coding in hereditary structures, as self-organization, are the main drivers in evolution. In the early or relatively stationary forms of life, Bacteria, Archaea, Fungi and Plants, the demands for coordinated complex behavior may seem of rather little importance. However, researchers have found coevolution and very advanced cooperation in many apparently autonomous organisms.

A central nervous system is not well established in these organisms. Many microprocesses potential for microfeels should be functional as chemical events. Microfeels may have a rather minimal role but still be important actors beyond that envisaged in quantum physics. Quantum physics on the other hand may provide language apt for the study of certain aspects of functions in live organisms (Jedlicka 2017). As shown in the steps of Figure 2, in live organisms a gradual significant widening of potential possibilities for functioning, for action, and for future evolution should be acknowledged.

Universe Containing Advanced Mobile Forms of Life Requiring Conscious Brains (See Figure 2e)

Autonomous mobility in animals meant much enlarged demands for very finetuned coordination and very precise detection of environmental facts. This could only be achieved by immense discretization in comprehensive neural networks with a central organ, brain, where very diverse input could be compared, handled, evaluated, and put into action in various holistic manners. The dynamics within brains is local, the locality enhanced by insulation from the surroundings. Since it all takes place in the continuous universe the binding problem does not occur.

The diversity in nature achieved by self-organization in natural evolution is enormous. There is also remarkable stability in many coded traits, e.g. many genes and body plans are common to a wide variety of species (Bae *et al.* 2015). Many metaphysical moods, feelings, lust, fear, pain, etc., that we humans often share with other animals, are forceful drivers and constraints in animal and human behaviors (Darwin 1871). That such moods etc. really are metaphysical, i.e. not possible to formalize within some language similar to that of mathematical physics, with no mystical implications, should be acknowledged. Their causal roles must be taken seriously and understood in non-reductionist aspects of biology and human individual and social behaviors.

Universe Containing Homo Sapiens With Advanced Self-Consciousness and Human Language (See Figure 2f)

In us, *Homo sapiens*, selfconsciousness and language allowed for creativity beyond that achieved by other animals. Natural evolution is always gradual and completely continuous, even though there may have been times of apparent fast change or stasis. The apparently large step between animals and humans is filled out by extinct *Homo* species, like *erectus*, *habilis*, *floresiensis*, and *neanterthalensis*. In humans advanced selfconsciousness, sociality, mastery of complex languages, and many other finetuned cognitive abilities, evolved into extremely successful complex behavior. The causal impetus of the metaphysical in humans, Whitehead's creativity, is a lot more forceful than in other animals. An evolutionary argument supporting "causal efficacy of consciousness", based in Whiteheadian metaphysics, is suggested in Gaudry (2018).

HUMAN CULTURE IS WHOLLY IMMERSED IN NATURE

Whitehead (1978) recognized two poles in human conscious experiences: one more objective registering physical input from surroundings, and one more subjective providing internal input. In contemporary neuroscience more detailed reflexivities in brains are observed. Strategies similar to trial and error in various aspects, in cooperation with creative fantasy, moved humans into increasingly complex group dynamics, creating novel discoveries and inventions, novel scientific theories, novel religious ideas, novel art, etc. (Holmgren 2017). Much of this certainly somehow has "material" existence outside of human conscious experiences, as partially described by physics. It all is known to us humans only within our full human conscious experiences, built of seemingly discrete microfeels actualized in the continuous universe, intensely playing on metaphysical strings. Human cultural evolution is not distinct from natural evolution, it is natural evolution in full swing (Dobzhansky 1973).

The full acceptance of the metaphysical aspects of human consciousness as causal in the universe, far beyond any physical "Theory of Everything", raises questions about possibilities of inspiring cooperative behavior on a global scale. There should be no doubt that we humans need to unite in strong, durable and effective resolutions for actions against climate change and for conservation of biodiversity on our beautiful planet. The scientific evidences, and even much of the needed remedies, are reasonably well known and are all the time growing in scope and trustworthiness. The large challenges have been known since long time. Why is human action still so diverted, so weak and slow?

Foundational for the understanding of all life on earth is Darwin's theory of natural evolution, with all contemporary and future amplifications, corrections and improvements. The theory is elegant and comprehensive, enormously successful in biology generally, for laymen perhaps most clearly seen in human medicine. It was blatantly misused and corrupted in Nazi quasi-theory, which should be left behind now. To get things right in human affairs on earth there is urgent need for full and enthusiastic acceptance of the theory, and there is great need for ample general education and understanding of it.

Still, one can see only limited interest for natural evolution in the general public, and unashamed denial and even ban on education in certain extremist groups. The reason is probably that natural evolution often mistakenly is understood to deny the metaphysical aspects of life, as if it all would be reduced to physics and chemistry. The full acknowledgement of the reality of human microfeels as basic for all our experiences and cognitions, will instead place the metaphysical aspects as central in human lives. For example, anxiety about climate change, as well as denial about it, can be thought of and dealt with as very concrete structures in the form of patterns of microfeels in human individuals' minds. Consequently, arguments should not only be presented as scientific and statistical hard facts, but also in forms where feelings like empathy and love for others are given attractive expressions.

In such a society, violent fight and fierce competition should not rule. The culture should generally be dominated by good will and strive for beauty. Expressions of violence and destruction would be restrained. The natural sciences, natural evolution included, must be fully acknowledged as the best thought structures achieved in the present epoch (cf. Whitehead's epochal theory of time). It will be generally understood that they always will remain incomplete and must be valued accordingly. Eventually, even substantial disarmament and widespread peace may become achievable.

EDUCATION, RELIGION, AND MULTI ASPECT MONISM (MAM)

The success of humans in conquering and exploiting our planet is causing large dangers, such as global warming and mass extinctions. Global political decision-making is difficult but not impossible, as shown e.g. by successful actions against ozone depletion. The challenges now facing humanity call for radical changes in life-styles, challenges that involve most aspects of human societies.

Metaphysical aspects, feelings, values, decisions, and their consequent behaviors in individual behavior, politics, economy, are crucial for the success in establishing a sustainable situation on earth. Rational decisions need to be grounded deeply in empathy and love for present and future human generations, but also in empathy and love for all remaining and future wildlife and ecological systems.

Each individual person has an organism and brain structure which is determined by individual hereditary structures and enormously varied and evolving experiential structures stored in the body and in memories. Like animals, we tend to form rather stable personalities, that in humans to some extent can be classified, e.g. according to the five-factor model (FFM: openness to experience, conscientiousness, extraversion, agreeableness, neuroticism).

Education is the most powerful means we have to help us learn to cooperate in meaningful ways, in spite of all differences. Traditional education however tends to reinforce the reigning forces in a society, and therefore may need radical adjustment when serious lifestyle issues are at stake. We are all in want of some stability in a world where insecurity and anxiety for death are constantly present. Many turn to religions in the hope to find answers to the deepest questions in life.

In his book *Religion in the making* (1996) Whitehead stated: "When religion ceases to seek for penetration, for clarity, it is sinking into its lower forms. The ages of faith are the ages of rationalism." and (on Christianity and Buddhism) "The self-sufficient pedantry of learning and the confidence of ignorant zealots have combined to shut up each religion in its own forms of thought. Instead of looking to each other for deeper meanings, they have remained self-satisfied and unfertilized."

There is certainly need to question static dogmas. Religious teachings that bluntly contradict each other or violate well established scientific insights should be avoided in schools. Democratic secular and rational political systems should be preferred, while religious freedom is guaranteed on a personal private level. Too fanciful and scientifically unfounded beliefs tend to corrupt rational thought. They should be confined to private spheres, and should not be allowed to interfere in secular life. There is urgent need for human societies to become responsible and in rational ways change their life-styles.

All human aspects, involving metaphysical aspects as well as the natural sciences, cannot be expected easily to merge into one common harmonious worldview. A realistic cosmology, accepting unavoidable complementarity and probably many commonsense paradoxes, can be named Multi Aspect Monism (MAM) (Holmgren 2014).

CONCLUSION: DIGITAL UNIVERSE VS. MICROFEEL UNIVERSE

A theory of everything based on physics must presuppose that the universe is fundamentally restricted to being Einsteinian or quantum-physical in accordance with Figure 2b or 2c. A digital or mathematical universe is basic e.g. for scary ideas about robots acquiring superintelligence and dominating humans (Tegmark 2014).

A universe containing microfeels in human metaphysics as in Figure 2f cannot be reproduced and hijacked by computers. What humans want decides our future. Digitalization, AI, and robotics are not threatening but will be used as subordinate tools. Rational scientific arguments will probably not suffice for the needed change of human behaviors. Human societies need to relocate large resources from technical to non-technical and humane nature-friendly endeavors in order to achieve rational and common decisions on goals for a future universe. One major challenge is to limit human population growth on earth in humane ways. Do we want a few humans to survive in extremely inhospitable environment on planet Mars, while life on earth is ruined? Or do we want to care for life in continued natural evolution on our planet, with enormous potential for happy lives in peace and prosperity?

REFERENCES

Bae, B.-I., Jayaraman, D., & Walsh, C. A. (2015). Genetic Changes shaping the Human Brain. *Developmental Cell*, *32*(4), 423–434. doi:10.1016/j.devcel.2015.01.035 PMID:25710529

Darwin, C. (1871) The Descent of Man, and Selection in Relation to Sex. New York: Appleton.

Dobzhansky, T. (1973). Nothing in biology makes sense except in the light of evolution. *The American Biology Teacher*, *35*(3), 125–129. doi:10.2307/4444260

Einstein, A. (1950). *Out of My Later Years*. New York: Philosophical Library.

Gaudry, J. D. (2018). Liking what's good for you: Evolution, Subjectivity and purpose. *Cosmos and History: The Journal of Natural and Social Philosophy*, *14*(3), 359–375.

Holmgren, J. (1981). Hypothesis on consciousness. Academic Press. doi:10.13140/RG.2.1.1611.5049

Holmgren, J. (2014). Natural Evolution and Human Consciousness. *Mens Sana Monographs*, *12*(1), 127–138. doi:10.4103/0973-1229.130322 PMID:24891802

Holmgren, J. (2017). Human consciousness - a parsimonious approach. Principles for concept formation, for abstraction, and for the foundations of logic, mathematics, physics, and biology. Academic Press. doi:10.13140/RG.2.2.35901.67041

Jedlicka, P. (2017). Revisiting the Quantum Brain Hypothesis: Toward Quantum (Neuro)biology? *Frontiers in Molecular Neuroscience*, *10*, 366. doi:10.3389/fnmol.2017.00366 PMID:29163041

Klose, J. (2009). Process Ontology from Whitehead to Quantum Physics. In H. Atmanspacher & H. Primas (Eds.), *Recasting Reality-Wolfgang Pauli's Philosophical Ideas and Contemporary Science* (pp. 151–170). Springer-Verlag. doi:10.1007/978-3-540-85198-1_8

Lewis, S. L., & Maslin, M. A. (2015). Defining the Anthropocene. *Nature*, *519*(7542), 171–180. doi:10.1038/nature14258 PMID:25762280

Pereira, A. Jr, Nunn, C., Nixon, G. M., & Pregnolato, M. (2018). Consciousness and Cosmos: Building an Ontological Framework. *Journal of Consciousness Studies*, *25*(3-4), 151–205.

Tegmark, M. (2014). *Our Mathematical Universe – My Quest for the Ultimate Nature of Reality*. Brockman, Inc.

Tonegawa, S., Liu, X., Ramirez, S., & Redondo, R. (2015). Memory Engram Cells Have Come of Age. *Neuron*, *87*(5), 918–931. doi:10.1016/j.neuron.2015.08.002 PMID:26335640

Whitehead, A. N. (1922). *The principle of Relativity with applications to Physical Science*. Cambridge University Press.

Whitehead, A. N. (1978) Process and Reality: An Essay in Cosmology (D.R. Griffin & D.W. Sherburne, Eds.). New York: The Free Press (Originally published 1929)

Whitehead, A. N. (1996). *Religion in the making: Lowell lectures 1926*. New York: Fordham University Press.

Chapter 7
Relativity and Cognitive Ethics

Richard Sieb
Independent Researcher, Canada

ABSTRACT

Ethics originate from conscious experience. All categories of ethics (meta-ethics, normative ethics, applied ethics, descriptive ethics) are knowable only through conscious experience. Hence, conscious experience might be considered a meta-ethic (the origin and basis of all ethics). Conscious experience appears to us as a unified four-dimensional space-time continuum or field. A neural correlate for conscious experience modeled by Einstein's special theory of relativity has been found in the human brain. Conscious experience can be described and understood using relativistic physics. The principles of relativistic physics therefore influence ethics. Three universals emerge from relativity which mediate conscious experience and ethics: the laws of physics, the speed of light, and space-time intervals. The presence of these universals suggests that conscious experience (observed physical reality) is determinate (predictable). We do, however, have free will (choice), and this free will appears to be governed by ethics.

INTRODUCTION

Ethics are the moral principles which govern a person's behavior. They make up a branch of philosophy involved in systematizing, defending, and recommending concepts of right and wrong conduct. Ethics investigates such questions as "What is the best way for people to live?" and "What actions are right or wrong in particular circumstances?" Ethics seeks to resolve questions of human morality by defining concepts such as good and evil, right and wrong, virtue and vice, justice and crime. There are four basic categories of ethics: meta-ethics (ethical philosophy related to universal law, the origin and nature of ethics), normative ethics (contextual behavioral standards), applied ethics (application of ethics to particular situations or issues), and descriptive ethics (empirical neurophysiological research on ethics). The purpose of this chapter is to utilize descriptive ethics to describe how the other three categories of ethics emerge.

DOI: 10.4018/978-1-5225-9065-1.ch007

BACKGROUND

Ethics has historically evolved mainly through vigorous debate. There has been little, if any, application of actual physics to the understanding of ethics, as our understanding of physics has not reached an appropriate level. Through the ages, ethics has been profoundly influenced by prevailing scientific, metaphysical, cultural, and religious ideas (Koch,2016). Many of these were based on assumptions which were clearly articulated, others not so much. Terms such as "consciousness", "mind", and "soul" have great relevance for ethics. The meaning and conceptualization of these terms has greatly changed over the centuries, with major consequences for ethics. For instance, it was perfectly ethical in the middle ages to accuse people of witchcraft or heresy and burn them alive, not so much today. Philosophers, theologians, scholars, scientists, and physicians have defined, conceptualized, reified, denied, and redefined these terms in an effort to come to grips with the mystery of our inner life (Koch,2016) and to come up with an enduring system of ethics. By conceptualizing consciousness in relativistic terms, it is hoped that a major step in the understanding of consciousness and ethics has been taken.

In humans, knowing the world occurs through spatiotemporal experiences and interpretations (Maniadakis & Trahanias,2011). All knowledge is defined, conceptualized, and understood only through conscious experience. We would not have knowledge of anything, including ethics, if it was not gained through conscious experience. Science is the creation of knowledge. It is dependent on the observer's subjectivity or conscious experience (Hossenfelder,2015;Sieb,2016b). Many of the world's greatest scientists (Einstein, Newton, etc.) used their conscious experiences (imagination, thought experiments) to generate their greatest discoveries (such as special and general relativity, Newton's Laws, description of quantum phenomena, etc.). Scientific discovery is also colored by ethics. Research is increasingly accepting the subjectivity (conscious experience) of the individual and the role of values and ethics in every phase of research (from the choice of what to research and why, how to carry out the research, and recognizing that the ethical stance of the researcher influences the nature of the knowledge that is created). Quantum mechanics, for instance, provides evidence that the presence of the observer, and the questions they ask, influence the nature of reality that emerges (Sieb,2016b). Utilization of animals and uninformed individuals in research is strictly controlled. There is a realization that you cannot separate conscious experience, ethics, and knowledge. Ethics plays a major role in the way the process unfolds, and the course of evolution may well be influenced by the nature of the ethical choices that are made. There are huge possibilities and opportunities for our evolutionary journey that lie (currently latent) within consciousness and ethics. Since the physics behind human conscious experience have never before been understood, it has never before been possible to describe in scientific terms the workings of the mind and the origin of ethics. This is the first paper in which actual laws of physics are utilized to understand conscious experience and ethics. By doing so, it is hoped a better understanding of ourselves, the workings of the mind, reality, and ethics is achieved. This endeavor is based on an ever-increasing amount of empirical research, some of which is described in the following pages.

CONSCIOUSNESS

The reticular activating system originates in the upper brainstem reticular core and projects through the intralaminar thalamic nuclei to the cerebral cortex to mediate arousal, attention, and consciousness, ie. the waking state (Sieb,2013). The noradrenergic locus coeruleus, the mesencephalic reticular formation, the cholinergic nucleus basalis of Meynert, the dorsal hypothalamus, the intralaminar thalamic nuclei, and the tegmentum make up the reticular activating system (Sieb,2013). The reticular activating system and the cerebral cortex are necessary to maintain consciousness. Damage to the reticular activating system results in coma. Partial destruction or inhibition of the reticular activating system results in a partial loss of consciousness known as the acute confusion state or delirium. Ethical judgment and conduct goes out the window when the reticular activating system is damaged.

Conscious Experience

Conscious experience makes up the content of consciousness, ie. what is in consciousness (Sieb,2004, 2013,2015). Conscious experiences are self-generated internal explicit states, arise rapidly in recurrent systems, take time to form, have a prolonged duration, are seamless, structurally-complex, ineffable, transparent, bounded, unified and coherent, informative, serial, limited in capacity, subject to interference, new or novel, variable, flexible, project outwards, can gain access to other systems, and arise from attention (Sieb,2004,2013). Conscious experience appears as a unified field. It is a shared illusion of an interactive Metaverse (considered here as the experienced universe); where every thought, word, or action has implications for the entire Metaverse (because conscious experience is a unified field). Every thought, word, or action has agency (the capacity to act). Conscious experience may be defined as the direct observation of conscious events. Hence the observer is central to conscious experience. Every thought, word, or action appears to emanate from the observer. Individuality, independence, autonomy, agency, and free will (choice) are the hallmarks of the conscious observer. Agency may be unconscious and involuntary or purposeful and goal-directed (intentional actions). An agent typically has some sort of immediate awareness (conscious experience) of their physical activity and the goals that the activity is aimed at. In goal-directed action, an agent implements a kind of direct control or guidance over their behavior. This may be due to ethics (morality). Morality is the differentiation of decisions and actions into those that are proper or improper (Long & Sedley,1987). Moral agency is an individual's ability to make judgments about their behavior based on some notion of right and wrong and to be held accountable for their actions (Angus,2003). A moral agent is a being who is capable of acting with reference to right and wrong. As moral agents, we use morality to govern our free will. Every individual develops their own particular ethics. It is the individual, in the first part, that determines "right thought", "right feeling", or "right action". Society regulates and conditions individual conduct so as to keep it appropriate and acceptable to the society. Non-compliance with societal control results in ostracism and swift retribution. Society determines "right thought", "right feeling", or "right action" in the second part.

How do subjective states like conscious experiences arise in the physical brain (the "explanatory gap" or "hard problem of consciousness")? What is the physics behind these subjective states? Why does conscious experience appear to us the way it does? How does the brain build a system of ethics from conscious experience? Is there a meta-ethic? For hundreds of years, scientists have pondered these problems. Many consider them unsolvable. The relativistic concept of consciousness provides answers to all these questions. It outlines the physics behind conscious experiences, closes the explanatory gap, and solves the hard problem of consciousness. It explains why conscious experience appears to us the way it does. Relativistic consciousness shows how the brain can build a system of ethics from conscious experience and that conscious experience might be considered a meta-ethic.

The Relativistic Concept of Consciousness

Conscious experience is observed to have four dimensions (three dimensions of space, one dimension of time). If you look around you, some conscious events are observed simultaneously, but at different three-dimensional spatial positions; other conscious events are observed at the same position, but at different times. For example, when I examine my conscious experience of this room, many conscious events (door, window, light, light switch, walls, ceiling, floor, desk, chair, computer, monitor, keyboard, hands, mouse, gloves, etc.) are observed simultaneously, but in different three-dimensional spatial positions. These events are separated only by space (since they are observed at different spatial positions), but not by time (since they are observed simultaneously). When events are observed simultaneously, but in different places, the frame of reference is known as space-like (events separated only by space). I observe other conscious events (my finger and the power button on the computer, my hand and the mouse, my hands and the gloves, a glass and a cup) at the same spatial positions (when I press the button, move the mouse, put the gloves on, or replace the glass with a cup), but at different times. These events are separated only by time (since they are observed at different times), but not by space (since they are observed at the same spatial positions). When events are observed at the same place, but at different times, the frame of reference is known as time-like (events separated only by time). Notice that the conscious events separated by time also have cause-effect and past-future relations. Separation in time gives rise to causality and past-future relationships. Conscious experience therefore appears to have a space-like and time-like organization. This organization is typical and characteristic of conscious experience in general (all conscious experiences are organized in this manner) and indicates that conscious experience is organized in space and time. Since there are three dimensions of space (length, width, height) and one dimension of time, conscious experience can be said to have four dimensions. We observe (experience) the world in four dimensions.

Space and time play a critical role in the organization of conscious experiences. Space is the boundless three-dimensional extent in which "events" occur and have relative position and direction; time is a continuum in which "events" succeed one another from past through present to future (any dictionary). Space is often conceived in three linear dimensions, but modern physicists usually consider it with time, as part of a boundless four-dimensional continuum called space-time. Space-time is any mathematical model that combines space and time into a single continuum (Petkov,2010). Conscious experience may be considered a space-time continuum. Einstein's special theory of relativity is the most successful model of space-time and is a viable model for conscious experience. In fact, what Einstein appears to have developed in his special theory of relativity is a model of human conscious experience. This is reasonable as conscious experience is our observed reality and Einstein's theory of relativity was conceived to explain our observed reality. Einstein used his own conscious experiences (imaginings, thought experiments-Hossenfelder,2015) to develop his theories of relativity (special and general). Hence it is not surprising that his theories actually model his conscious experience. Emmanuel Kant also prophetically concluded that space and time are not discovered by humans to be objective features of the world, but are part of a systematic framework for the organization of our experiences (Lucas & Hodgson,1985).

SPECIAL THEORY OF RELATIVITY

Einstein's special theory of relativity depends on frames of reference (observational perspectives of space described using coordinate systems). Special relativity is based on two postulates: the laws of physics are invariant (the same) in all inertial (constant velocity) frames of reference and the speed of light is the same for all observers, regardless of the motion of the observers (Einstein,2001). Special relativity predicts a wide range of consequences (which have been experimentally verified). These include *length contraction, time dilation, relativistic mass, mass-energy equivalence ($E=mc^2$, where c is the speed of light in a vacuum), a universal speed limit (c), and relativity of simultaneity* (DiSalle,2009;Einstein,2001; Feynman,1998;Roberts and Schleif,2007). Time dilates or lengthens (slows) and lengths contract (shorten) at higher speeds of a reference frame (observer) relative to another; this keeps the laws of physics, *c*, and space-time intervals *invariant* (the same) in all *inertial* frames of reference. Three universals arise from relativity: the laws of physics, *c*, and space-time intervals. In relativity, time cannot be separated from space, because the observed rate at which time passes depends on the relative velocity of the observer. Time and space are interwoven into a single continuum (the space-time continuum). Since conscious experience is a space-time continuum, Einstein's special theory of relativity is a viable model for conscious experience. The presence of the universals suggests that our observed physical reality (conscious experience) is deterministic (predictable). The determinant nature of conscious experience is evident in that its dynamics are well-described by classical physics (Jaeger,2014;Morin, 2008;Sieb,2016a,b).

Space-Time Intervals

An event is the fundamental entity of observed physical reality represented by three coordinates of space and one coordinate of time in the space-time continuum postulated by the theory of relativity (Einstein, 2001;Petkov,2010). In relativistic terms, a conscious event is the fundamental entity of conscious experience (observed physical reality) represented by three coordinates of space and one coordinate of time in the space-time continuum postulated for conscious experience. In relativistic physics, an event has a unique position specified by four coordinates (x,y,z,t). The unification of space and time is exemplified by the common practice of selecting a metric such that all four dimensions are measured in terms of units of distance (x,y,z,ct), where "*c*" represents the speed of light in a vacuum and "*ct*" represents distance along the time axis (Petkov,2010;"Wikibooks",2015). Since *c* is equal to 1, *ct* reduces to *t*. This enables mathematical description of space-time, the positions of events in space-time, and most importantly, space-time intervals.

In three-dimensional space, the separation between two objects is measured by the distance between them (the distance is purely spatial and always positive). In space-time, the separation between two events is measured by the invariant space-time interval between the events, which takes into account not only their spatial separation, but their temporal separation as well (Petkov,2010;"Wikibooks",2015). Space-time interval may be formulated as follows ("Wikibooks",2015): the space-time interval "s" is equal to the difference between the space coordinates "Δr" of two events minus the difference between the time coordinates "$c\Delta t$" of the two events (in practice the square is utilized as the sign of the space-time interval "s" is indefinite-positive, negative, or zero; Δr is a displacement vector).

(space-time interval) $s^2 = \Delta r^2 - c^2 \Delta t^2$ or $s^2 = \Delta r^2 - \Delta t^2$, if c=1

The Light Cone

Space-time intervals are also defined by the light cone. A pulse of light emitted from a point spreads outwards in all directions at *c*. If light is confined to a two-dimensional plane, the light from a flash spreads out in a circle (Figures 1 and 2). If the growing circle is drawn with the vertical axis representing time (Figures 1 and 2), the result is a light cone (Penrose,2005; "Wikibooks", 2015).

In reality, there are three space dimensions, so light would actually form an expanding sphere and the light cone would be a 4D version of a cone. The concepts are easier to visualize with the number of space dimensions reduced to two. A light cone represents the path that a flash of light, emanating from a single event and travelling at the speed of light in all directions, would take through space-time. Light cones are the same for all events and for all observers (since *c* is a universal constant). Given an event A, all events that can be reached by a light pulse from A form a future light cone, while all events that can send a light pulse to A form an inverted past light cone (Figures 1 and 2). Three types of space-time intervals are defined by the light cone and the formula for space-time interval. Space-time intervals play

Figure 1. Space-time intervals (Wikipedia, 2016)

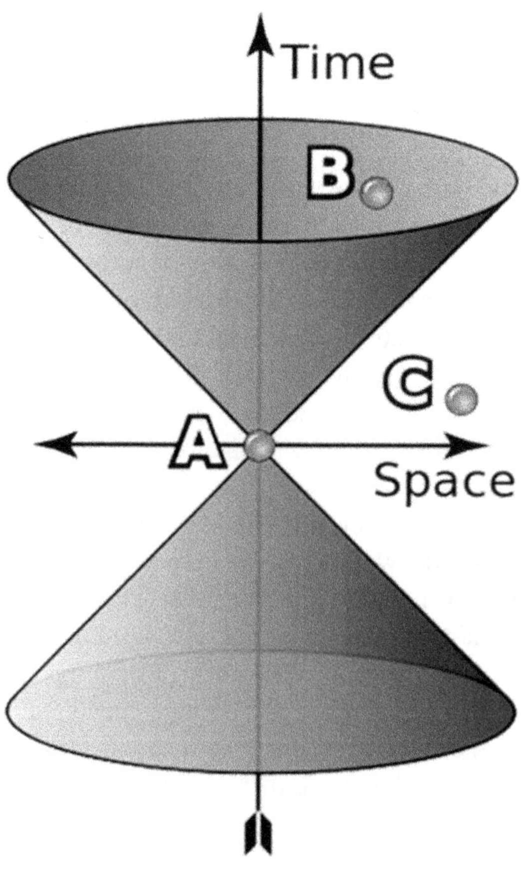

Figure 2. Past and future light cones (Wikipedia, 2016)

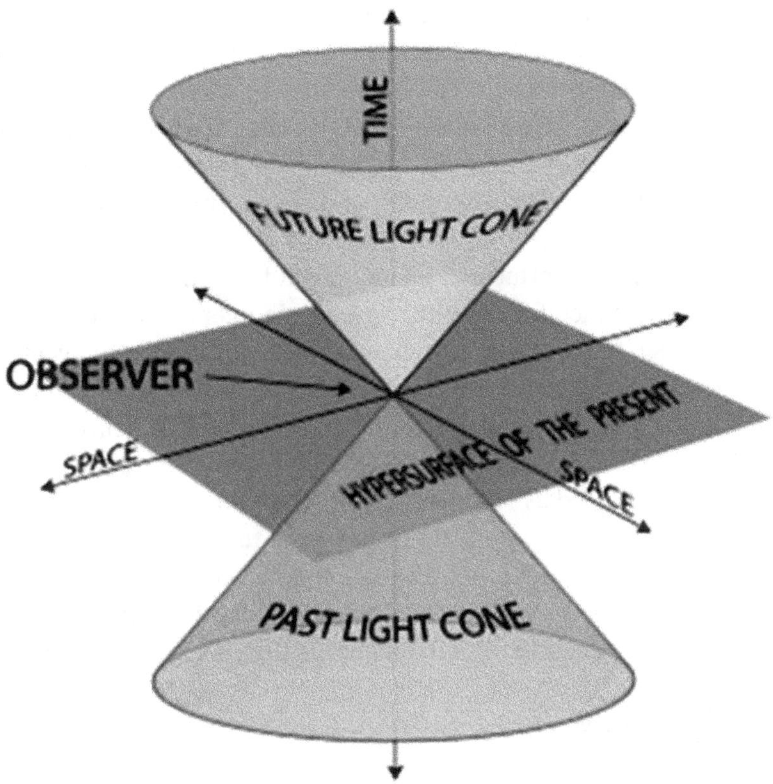

a crucial role in the organization of all conscious experiences, including those of ethics. Space-time intervals explain why conscious experiences appear to us the way they do.

Light-Like Space-Time Intervals

When light travels between two events, the spatial separation of the events (Δr^2)is equal to the distance light travels (Δt^2) and $\Delta r^2=\Delta t^2$. The separation of the events in space is equal to their separation in time. Hence the space-time interval equals zero ($s^2=0$). Events occurring to a photon, as it travels along its path, and events *on* a light cone are of this nature. The difference in the space coordinates of two events *on* a light cone is exactly equal to the difference in the time coordinates of the events (Figure 1). This separation of events is called a light-like or "null" space-time interval. This occurs when you directly observe a light source or an event (the event is visible because of reflected light). Light-like space-time intervals may help determine the nature of conscious events through the conveyance of information about the events to the eyes.

Time-Like Space-Time Intervals

If the difference in the time coordinates of two events is greater than the difference in the space coordinates of the events ($\Delta t^2>\Delta r^2,\ s^2 < 0$), the spatial separation of the events is less than the distance light travels. Hence there is enough time between the events that signals or information can travel between the events at less than the speed of light. This separation is called a time-like space-time interval and the events fall inside a light cone (A and B in Figure 1). A reference frame exists where two events may be separated only by time, but not by space. The events may be observed at the same spatial position, but not at the same time. The events may be observed at the same spatial position, but at different times (the events are separated only by time). Time-like space-time intervals could account for the experience of conscious events at the same spatial positions, but at different times (this can occur only with time-like separation of the events).

Because signals and other causal influences cannot travel faster than the speed of light, light cones define the concept of causality. Since for time-like space-time intervals (inside light cones), signals or information can travel between the events at less than the speed of light, one event can influence or be influenced by the other event, by signals or information that does not need to travel faster than the speed of light. One event could be the cause or effect of the other event. If B causes A, B exists in the past history (in the past light cone) of A (Figures 1 and 2). If B is caused by A, B exists in the future (in the future light cone) of A (Figures 1 and 2). The past light cone of an event represents the boundary of its causal past and the future light cone the boundary of its causal future. Events with time-like space-time interval separation may be said to have a past or future relation. This can be observed in conscious experience. Events which occur at the same position, but at different times (time-like space-time interval separation) often have a cause-effect relationship and necessarily a past-future relationship.

Space-Like Space-Time Intervals

If the difference in the space coordinates of two events is greater than the difference in the time coordinates of the events ($\Delta r^2 > \Delta t^2, s^2 > 0$), the spatial separation of the events is greater than the distance light travels and the events fall outside a light cone (A and C in Figure 1). This separation is called a space-like space-time interval. For space-like space-time intervals (outside light cones), there is not enough time between the occurrence of the events that signals or information can travel between the events at less than the speed of light. There exists a reference frame where two events may be separated only by space, but not by time. The events may be observed at the same time, but not at the same spatial position. The events may be observed simultaneously, but at different spatial positions (the events are separated only by space). Space-like space-time intervals could account for the experience of conscious events simultaneously, but at different spatial positions (this can occur only with space-like separation of the events).

When a space-like space-time interval separates two events, not enough time passes between their occurrences for there to exist a cause-effect relationship crossing the spatial distance between the events at the speed of light or slower. Generally, the events are considered not to have a cause-effect or past-future relation. This can be observed in ordinary conscious experience. Conscious events which occur simultaneously, but in different spatial positions (space-like separation of events), do not have cause-effect or past-future relationships.

ORGANIZATION OF CONSCIOUS EXPERIENCE

Space-time intervals account for the organization of conscious experience. Space-time intervals in effect link (associate) conscious events to form a coherent four-dimensional conscious experience (a coherent four-dimensional observation of conscious events). Space-time intervals explain why conscious experience appears to us the way it does. They also explain cause-effect and past-future relationships. Conscious experience may be described as an *orientation* in space and time-an awareness of the existing situation with reference to space, time, and identity. It is an *understanding* of the position of the observer in space and time. This is important for human cognition. There is a substantial amount of supporting evidence for a neural correlate of human conscious experience. This neural correlate is modeled by Einstein's special theory of relativity and appears to utilize space-time intervals for the organization of four-dimensional conscious experiences.

Neural Correlate of Conscious Experience

A fundamental feature of conscious experience is the *spatial organization of events. "Place cells"* have been discovered in the hippocampus (O'Keefe,1976;O'Keefe & Dostrovsky,1971). Place cells fire at particular places in a spatially-structured environment. A place cell fires when the subject passes through a specific small region of space called the place field. Place fields are considered allocentric rather than egocentric because they are defined with respect to the outside world, rather than the body. By orientation based on the environment, place cells can work effectively as neural maps of the environment (Jeffery, Anderson, Hayman, & Chakraborty,2003). The hippocampus essentially encodes a neural representation of surrounding space-a neural representation of the layout of the environment. The discovery of place cells led to a series of investigations that culminated in a book entitled *"The Hippocampus as a Cognitive Map"* which argued that the hippocampal neural network instantiates cognitive maps for spatial memory function (O'Keefe & Nadel,1978). This motivated hundreds of experimental studies aimed at clarifying the role of the hippocampus in spatial memory and spatial navigation (Maurer, Vanrhoads, Sutherland, Lipa, & McNaughton,2005;Moser & Moser,1998). Considerable data indicate the hippocampus is critical to episodic memory in humans (Rolls,2013;Sieb, 2015;Steinvorth, Levine, & Corkin,2005). Episodic memory is memory of personal past *experiences* of a particular time and place with reference to the person as a participant-observer (Munoz-Lopez, Mohedano-Moriano, & Insausti,2010;Schacter, Gilbert, & Wegner,2011;Sieb,2015). Conscious recall is required to demonstrate episodic memory (Munez-Lopez et al.,2010;Tulving,1983;Ullman,2004); after all, episodic memory is memory of personal past conscious experiences. A fundamental feature of episodic memory is the *spatial representation of events* composing a unique experience. An allocentric representation of events in the environment is encoded in the hippocampus and stored as episodic memory and hippocampal place cells play a crucial role in the establishment and maintenance of this representation (Serino & Riva,2014;Sieb,2015). Place cells are part of a complex circuit that informs memory and awareness (Jeffery,2007;Jeffery et al.,2003).

Another fundamental feature of episodic memory is the *temporal organization of serial events* that compose a unique experience (MacDonald, Lepage, Eden, & Eichenbaum,2011). The hippocampus is essential to encoding unique sequences of events as well as disambiguating sequences that share common events. Studies on humans have shown that the hippocampus is critical to remembering the flow of events in distinct experiences and, in doing so, bridges temporal gaps between non-contiguous events. "There is a robust hippocampal representation of sequence memories, highlighted by *time cells*, that encode successive moments during empty temporal gaps between key events, while also encoding *location* and *ongoing behavior* (MacDonald et al.,2011)." The hippocampal representation of time is quite similar to its representation of space (MacDonald et al.,2011): a large proportion of neurons are engaged in both; time cells fire at discrete moments during "empty" periods in a temporally-organized memory, much as place cells fire at discrete locations, devoid of specific stimuli; time cells signal the nature and timing of salient events, just as place cells signal the nature and spatial location of salient events; time cells disambiguate overlapping sequence memories, just as place cells disambiguate overlapping routes; and time cells partially "retime" when key temporal parameters are altered, just as place cells partially "remap" when critical spatial cues are altered. These findings suggest that hippocampal neuronal ensembles segment temporally-organized memories much as they segment spatial memories. MacDonald et al. (2011) conclude that "place cells and time cells reflect fundamental mechanisms by which hippocampal neural networks parse any spatiotemporal context into quantal units of where and when important events occur and bridge, and thereby organize elements, in a conceptual organization of events." Eichenbaum (2014) concludes that "the firing properties of time cells parallel the properties of place cells and provide an additional dimension that is integrated with the spatial dimensions." *Both investigators conclude that "hippocampal neurons differentially encode key events in space and time (space-time) and compose unique spatially and temporally-organized representations of specific experiences." "The representation of time and space in the hippocampus is a fundamental mechanism for organizing the elements of experience* (MacDonald et al.,2011)."

Space-time intervals appear to be the fundamental units utilized by the brain to organize the elements of experience. There is increasing supporting evidence suggesting that space-time intervals are represented in the hippocampus and extracted by the posterior parietal cortex (PPC) and prefrontal cortex (PFC) for the organization of conscious experiences. In 2014, John O'Keefe, May-Britt Moser, and Edvard Moser were awarded the Nobel Prize in Physiology or Medicine for the discovery of *"grid cells"*, cells that constitute a *"positioning system"* in the brain. A grid cell is a type of neuron found in the brains of many species that allows them to *"understand"* their position in space (Doeller, Barry, & Burgess,2010;Hafting, Fyhn, Molden, Moser, & Moser,2005;Sieb,2016a,b). Grid cells have been identified in rats, bats, monkeys, and humans. A grid cell fires when a freely-moving animal traverses a set of small regions (firing fields) roughly equal in size and arranged in a periodic triangular array that covers the entire available environment (Hafting et al.,2005). Firing fields are equally spaced apart, such that the distance from one firing field to all six adjacent firing fields is approximately the same. Firing fields are positioned such that the six neighboring fields are located at approximately 60 degree increments (Figure 3). Doeller et al. (2010) provided the first evidence for grid-cell-like representations in humans that implicates a network of regions supporting spatial cognition and autobiographical (episodic) memory. This network consisted of the **entorhinal/hippocampal/subicular, posterior and medial parietal, lateral temporal, and medial prefrontal cortical areas**. The signal was greatest in the right entorhinal cortex. *Doeller et al. (2010) concluded that "grid cells provide a strikingly periodic representation which is suggestive of very specific computational mechanisms* (such as of space-time intervals)."

The hippocampus has reciprocal connections with the entorhinal cortex and many other brain areas (Serino & Riva,2014). Besides output to the entorhinal cortex, additional output goes to other cortical areas, including the **PFC and PPC**. A key function of PPC is the implementation of visuospatial attention and of spatial processing in general (Wendelken,2015). The intraparietal sulcus (IPS), which separates the inferior and superior parietal lobes (IPL, SPL), has been shown to contribute to the maintenance of spatial location information and the IPL is a locus of spatial-relational processing. Lara and Wallis (2015)

Figure 3. A hexagonal lattice; the dots are firing fields of a grid cell (Wikipedia, 2016)

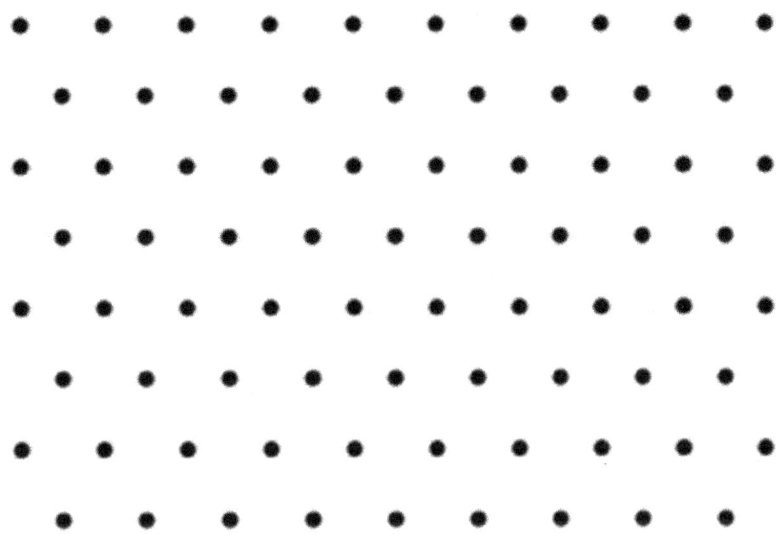

found that "the strongest activity of the overwhelming majority of PFC neurons reflect *the spatial location of stimuli and the passage of time*." Both PPC and PFC have been implicated in time-experiencing (Wendelken,2015): the cerebellum, right PPC, right PFC, fronto-striatal circuits, and insular cortex for *duration perception*; the inferior frontal and superior temporal lobes, hippocampus, medial PFC, medial parietal and posterior cingulate cortex for *past–future distinction and mental time travel*; the PFC, IPC, superior colliculus, and insular cortex for *synchronous and asynchronous event distinction*; and the posterior sylvian regions, PPC, and temporo-parietal networks for *temporal order judgment*. Wendelken (2015) observed "stronger PPC activation for processing *inequalities*, than for processing equalities, and argued that this was due to representation of the more specific *inequality relationships* in PPC." This is interesting because *space-like and time-like space-time intervals are inequalities*. Wendelken (2015) found that "current evidence is most consistent with accounts that involve *estimation and probabilistic computation for the PPC and PFC*." This is just what would be expected if the PPC and PFC processed space-time interval inequalities. PPC is a primary contributor to mathematical (numerical) cognition. In addition to these parietal regions, regions of the frontal lobe are also active in calculation tasks (Wendelken,2015). There is evidence that numerical cognition is intimately related to spatial cognition (Wendelken,2015). Regions of the parietal cortex show shared activation for both spatial and numerical processing, and as mentioned above, for temporal processing. These various lines of research suggest a *strong, but flexible, connection between numerical, spatial, and temporal cognition in PPC and PFC*. These studies support the contention that the PPC and PFC may preferentially extract space-time interval inequality information from the hippocampus in a process of relational integration. Wendelken (2015) found that "*the extraction of mental relations from the hippocampus by the PPC and PFC initiates an important process of relational integration*."

Perception is the identification, organization, and interpretation of sensory information in order to represent and understand the environment (Schacter et al.,2011); an *awareness (experience)* of the elements of the environment through physical sensation; physical sensation interpreted in the light of *experience*. Perception is shaped by learning, memory, expectation, and attention (Bernstein,2010;

Gregory,1987), processes which the hippocampus, PPC, and PFC are engaged in (Sieb,2013,2015). Perception also has features which may be conditioned by the hippocampus-PPC-PFC: constancy (the ability to recognise the same event from widely varying sensory inputs-Atkinson, Atkinson, & Smith, 1990;Bernstein,2010), grouping (humans naturally perceive events as organized sets of qualia-Goldstein, 2009;Gray,2006;Wolfe et al,2008), contrast (qualia can be affected by the qualities of context-Corsini, 2002;Kushner,2008;Popper,2010), experience (with experience, organisms can make finer perceptual distinctions and learn new kinds of categorization-Sumner,2009), motivation (Coon & Mitterer,2008; Sieb,2013;Weiten,2010), and expectation (a predisposition to perceive things in a certain way-Coon & Mitterer,2008;Sieb,2013;Weiten,2010). Perception appears to involve the hippocampus-PPC-PFC. Since conscious experience is such an intimate part of perception, the hippocampus-PPC-PFC is probably also involved in the organization of conscious experiences.

Reasoning (the capacity to reach novel conclusions on the basis of existing premises) is among the most complex of cognitive processes (Wendelken,2015). Reasoning is an integral part of the establishment and application of ethics. Bilateral PPC activation was found during relational reasoning and left PPC activation during propositional reasoning. Within PPC, reasoning is most strongly associated with activation of middle to posterior IPL, and to a lesser extent, with neighboring regions of SPL. Left PPC demonstrated greater involvement than right PPC. Selectivity for higher-order visuospatial reasoning, but not semantic reasoning, was found in right PPC. There were notable similarities between reasoning activations and activations associated with visuospatial processing and attention, particularly on the right, and between reasoning and phonological processing, particularly on the left. The mid-IPL appears to be unique for relational reasoning. Wendelken (2015) observed stronger PPC activation for reasoning with inequalities, than for reasoning with equalities and that current evidence points away from logical rule-following as a primary mechanism for reasoning and is more consistent with accounts that involve estimation and probabilistic computation. He found that it was clearly indicated that the pattern of activation in PPC associated with reasoning is most closely related to that for mathematical cognition. Wendelken (2015) found that rostrolateral PFC extracts mental relations from the hippocampus and is specialized for second-order relational reasoning. He found that current results are consistent with the possibility that rostrolateral PFC may share this duty with a sub-region of mid-IPL. Although direct anatomical connections between rostrolateral PFC and mid-IPL have not been reported, it is noteworthy that these two regions demonstrate strong functional connectivity during task execution and even at rest. The PPC and PFC also contribute to episodic memory. Parietal activation is most commonly associated with the endorsement of stimuli as having been previously encountered (Nelson, McDermott, Wig, Schlagger, & Petersen,2013;Wagner, Shannon, Hahn, & Buckner,2005), though associations with memory encoding (Uncapher & Wagner,2009) and memory confidence (Johnson, Suzuki, & Rugg,2013) have also been noted. The PFC (particularly in the left hemisphere) is involved in the formation of new episodic memories. Patients with damage to the PFC can learn new information, but tend to do so in a disordered fashion (Janowsky, Shimamura, & Squire,1989). The PFC may be essential for remembering the contextual details of episodic memory (Gabriele & Kao,2007), help organize information for more efficient storage, or underlie semantic strategies which enhance encoding (Gabriele & Desmond,1998). Hence the PPC and PFC are intimately involved in complex cognitive functioning (such as involved in ethics) and space-time interval inequalities extracted from the hippocampus appear to play a key role in this processing.

Causality

Causal inference is a fundamental component of cognition and perception, binding together conceptual categories, imposing structures on perceived events, and guiding decision-making (Cummins,2014), including that of ethics. Causality is distinct from mere contingency or covariation (Cummins,2014). In causality, one event has the power to bring about another event. In covariation and contingency, two events are simply statistically dependent on one another. Neuro-imaging studies show that the brain distinguishes causal events from non-causal events (Cummins,2014). There are significantly higher relative levels of activation in the right middle frontal gyrus and the right IPL for causal relative to non-causal events. Causal judgments, beyond associative judgments, generated distinct activation in left dorsolateral PFC and right precuneus (part of the SPL), substantiating the particular involvement of these areas in assessments of causality. Perceptual causality can be distinguished from inferential causality (Cummins,2014). Inferential causality activates the medial frontal cortex, with particular left hemispheric involvement. Perceptual causality activates the right parietal lobe suggesting that the right parietal lobe is involved in the processing of the spatial attributes of causality. The PPC and PFC therefore appear to be involved in the assessment of causality. Since they also appear to be involved in the processing of space-time interval inequalities, assessments of causality could arise in conjunction with the processing of space-time interval inequalities by the PPC and PFC. This supports the contention that space-time interval inequalities are represented in the hippocampus and extracted by the PPC and PFC for cognition.

Working Memory

Working memory is the holding in mind of multiple pieces of information for a brief period of time and its manipulation. Working memory has been linked to attention, perception, conscious experience, learning, cognitive development, cognitive function, and memory (Lara & Wallis,2015;Sieb,2015, 2016a,b). The concept of working memory describes a process of short-term storage of information to support ongoing or upcoming actions, and is considered a crucial component of the executive control of goal-directed behavior (Mansouri, Rosa, & Atapour,2015). The retention of task-relevant information is essential for complex behaviors which evolve in time, in order to maintain the perception and actions in a coherent and goal-directed framework. Working memory is crucial for the temporal organization of behavior, linking processes across delays. Working memory is considered an essential intermediate stage enabling further manipulation and integration of information involved in perceptual and mental functions. Functionally, working memory may be considered the provisional retention of perceptual information for prospective action, a type of focal attention whereby perception is reorganized and re-represented, becoming explicit, functional, and **conscious** (Sieb,2004,2015,2016a,b). The PFC, PPC, thalamus, and parts of the basal ganglia (caudate, globus pallidus) are crucial for working memory function (Sieb,2013, 2015,2016a,b). There is an emerging consensus that most working memory tasks recruit a network of prefrontal and parietal cortical areas. An increasingly large number of investigations have found that the feedback connectivity between the PFC and PPC is required for working memory and conscious experience. Positive feedback appears to be involved (Sieb,2004,2013,2015). Positive feedback supports synchronized neural oscillations in the theta range (Enriquez-Geppert, Huster, Figge, & Herrmann, 2014;Sieb,2016a,b), which have been implicated in perception, conscious experience, cognition, and memory. The PPC and PFC appear to play a substantial role in the functioning of working memory and the organization of conscious experience. As will be seen, these areas also are involved in the making of moral decisions and judgments.

EMOTIONS, MOTIVATION, EXPECTATIONS, DECISIONS, AND SOCIAL BEHAVIOR

The orbitofrontal cortex (OFC) and anterior cingulate cortex (ACC) are intimately involved in emotions, motivation, expectation, decision-making, and social behavior (Sieb,2013). The OFC and ACC are both subdivisions of the prefrontal cortex (John,2009). The OFC is primary for emotion, social behavior, and reinforcement-expectancies for decision-making (Sieb,2013). The OFC is considered the central executive for emotional and social control (John,2009). The OFC has been found to utilize affective valence information from the amygdala to influence emotional reactions (Sieb,2013). The OFC continuously extracts affect-laden recognition patterns or cues (stimulus-reinforcement associations) from sensory input, associates these with prior experiences with general categories, and is responsible for a subjective sense of perceptual coherence of environmental information (Sieb,2013). Complete destruction of the OFC may produce the complete loss of emotional and social behavior (Sieb,2013). Less extensive damage produces a loss of emotional control, leading to emotional and social behavioral dysfunction: the disinhibition or pseudo-psychopathic syndrome (John,2009;Sieb,2013). This syndrome is characterized by disinhibition, diminished self-supervision of behavior, distractibility, impulsiveness, tactlessness and loss of interpersonal sensitivity/empathy, utilization and imitation behavior, inappropriate jocularity, sexual preoccupation and jesting, hypomanic symptoms, neglect of personal care, poor hygiene, decreased social judgment, antisocial acts, and limited insight. Damage to the OFC produces a breakdown of morality and ethical conduct. The key functions of the ACC appear to revolve around the detection of errors or shortfalls from some standard, anticipation and preparation for action, and regulation of emotion (John,2009;Sieb,2013). Lesions of the ACC result in an inability to detect errors, severe difficulty resolving conflict, emotional instability, inattention, and akinetic mutism (Sieb,2013). Lesions of the ACC may result in the apathetic or pseudo-depressive syndrome (John,2009;Sieb,2013). This syndrome is characterized by apathy (a loss of motivation or drive): emotional apathy-absence of interest, excitement, and intensity of emotional responsiveness to positive or negative events; affective apathy-flat, unchanging emotional expression; cognitive apathy-decreased generative thinking, curiosity, engagement with usual activities, interest in learning and new experiences, and lack of concern with one's health, family, or future; motor apathy-lack of effort, productivity, ability to sustain activities, and initiation of new activities, increased dependence on others to structure activities, difficulty in initiating motor acts, hesitation when starting a new movement, and tendency to not gesture when speaking. The OFC and ACC definitely appear to be involved in the regulation of emotion, social behavior, and the individual's morality and ethical conduct.

Morality

The brain areas that are consistently involved when humans reason about moral issues have been investigated by a large-scale quantitative meta-analysis of the brain activity changes reported in the moral neuroscience literature (Bzdok et al.,2011). Bzdok et al. (2011) found that the neural network underlying moral decisions overlapped the neural network representing others' intentions (theory of mind) and the neural network representing others' (vicariously experienced) emotional states (empathy). This study supports the hypothesis that moral reasoning is related to both seeing things from other peoples' point of view (theory of mind) and to grasping other peoples' feelings or emotions (empathy). "These results provide evidence that the neural network underlying moral decisions is probably domain-global and might be dissociable into cognitive and affective sub-systems (Bzdok et al.,2011)." The explicit making of moral right and wrong judgments coincides with activation in the ventromedial prefrontal cortex (OFC and ACC) while intuitive reactions to situations containing implicit moral issues activates the temporo-parietal junction area ((Harenski, Antonenko, Shane, & Kiehl,2010;Young, Camprodon, Hauser, Pascual-Leone, & Saxe,2010a;Young et al.,2010b). These are the same areas that are involved in perception, working memory, the organization of conscious experiences, reasoning, assessing cause-effect relations, emotion, and social behavior. The ability to recognize and vicariously experience what another individual is undergoing may be a key step forward in the evolution of social behavior, and ultimately, morality (Rizzolatti, Fadiga, Gallese, & Fogassi,1996;Vedantam,2010). The inability to feel empathy is one of the defining characteristics of psychopathy (De Wied, Goudena, & Matthys,2005;Fernandez & Marshall, 2003).

FUTURE RESEARCH DIRECTIONS

Parents, friends, teachers, and other associates play a powerful role in the acquisition of ethical standards and principles and in the making of ethical decisions and judgments. Research in education and psychology should continue to search for insights into how this occurs and how it may be modified and controlled. There are certain general behavioral standards (normative ethics) of personal ethics that may be instilled in a person at an early age: concern for the well-being of others, respect for the autonomy of others, trustworthiness, honesty, willing compliance with the law, justice, fairness, not taking unfair advantage of others, benevolence, doing good, preventing harm, and reverence for life. Certain general behavioral standards of professional ethics may be acquired: impartiality, objectivity, openness with full disclosure, confidentiality, due diligence, professional responsibility, and avoiding conflict of interest. Certain general behavioral standards of global ethics may be instilled: global justice, society before self, environmental stewardship, interdependence and responsibility for the whole, and reverence for place. Personal ethics are at the first checkpoint in any given situation and often override professional and global ethics. Research and instruction into how these general standards may be acquired and applied in various situations and scenarios (applied ethics) may help individuals be more successful and happy.

Since society exerts such a powerful control over our ethical conduct, we should conduct our lives with dignity and seek out positive societal influences. The best possible empirical knowledge should be utilized in ethics. Speculation and unsubstantiated assumptions should be carefully considered (like assumptions taken on faith, spiritual, or metaphysical inclinations). Participating ideas should have appropriate weight. Harmful outside influences (drugs, people, groups) should be avoided. An individual should be free to enhance or modify their ethics through study, discussion, meditation, or other beneficial endeavors (cognitive liberty). Making good ethical decisions and judgments seems to be self-sustaining, leading to good health and satisfying relationships.

Research on the ability of individuals to recognize and vicariously experience what other individuals are undergoing (theory of mind and empathy) may provide insight into how the ethical constitution of a person is established and how ethical decisions and judgments are elaborated. This could entail study of the activity and function of "mirror neurons" in the brain, particularly those found in the prefrontal cortex and posterior parietal cortex. Such study might also help in the understanding, diagnosis, and treatment of psychopathy. Targeting the ventromedial prefrontal cortex and posterior parietal cortex with various clinical techniques (electrostimulation, electromagnetic stimulation) might be beneficial. It seems like our regard of others is very important in our development of ethics and the making of ethical decisions and judgments. Educational and psychological research on the ways we relate to and regard others may be very practical and important in developing ethics.

CONCLUSION

Human conscious experience appears to be a meta-ethic. In humans, knowing the world occurs through spatiotemporal experiences and interpretations. All knowledge, including ethics, is defined, conceptualized, and understood only through conscious experience. Human conscious experience is organized in four dimensions. Einstein's special theory of relativity is a model for human conscious experience and a neural correlate based on this model has been found in the human brain. Special relativity outlines the physics behind how we view the world (our observed reality) and ultimately ethics. Hence the explanatory gap is closed and the hard problem of consciousness solved (how something subjective like conscious experience can arise in something physical like the brain). Relativistic physics also explains why conscious experience appears to us the way it does. The positions of events in space and time (space-time) are encoded in the hippocampus. Grid cells in the entorhinal cortex provide a reference grid in the hippocampus for the computation of space-time interval relations. Space-time interval relations are extracted from the hippocampus by the prefrontal cortex and posterior parietal cortex for perception, the organization of conscious experience, memory, assessment of cause-effect and past-future relations, and cognition (including the acquisition of ethics and the making of ethical judgments and decisions). Ethics is deduced and interpreted through conscious experience. Hence conscious experience might be considered a meta-ethic, the origin and basis of all ethics. The same brain areas (prefrontal cortex, posterior parietal cortex) that are involved in working memory, the organization of conscious experiences, assessment of causality and past-future relations, episodic memory, and cognition are also involved in the making of ethical decisions and judgments. Hence there is an intimate connection between conscious experience and ethics. There is increasing awareness of the importance of conscious experience (subjectivity) and ethics in scientific research. Hence the creation of human knowledge is guided ty ethics. Three universals emerge from relativity: the laws of physics, the speed of light, and space-time intervals. These universals suggest that our conscious experience (our observed physical reality) is deterministic (predictable). This is supported in that the laws of classical physics provide an accurate description of the dynamics of our observed physical reality (conscious experience). It appears like the function of ethics is to govern and guide our free will (choice) in this deterministic world.

It appears like we acquire ethics and make ethical decisions through our vicarious experience (observation) of the intentions (theory of mind) and feelings and emotions (empathy) of others. Our experience of the behavior and feelings of other people is paramount to the acquisition of ethics and the making of ethical decisions and judgments. Hence the people we interact with (mainly parents, friends, teachers, authorities, and other associates) play a huge role in the acquisition of ethics and our ethical conduct. How we observe and interpret the behavior of other people is the origin and basis of our ethics. The ethics we develop govern and guide our free will (the goal-directed behaviors we choose to make). Hence ethics play a large role in how our lives unfold and evolve. Ethics may help determine our occupations, relationships, leisure activities, and general behavior. Hence ethics are very important for the growth and success of the individual. Ethics may determine the growth and evolution of whole societies and civilizations. For example, the Romans believed that it was ethical to kill, enslave, or conscript outsiders. This led to the large growth of Roman civilization, but also to its eventual decline. Ethics is controlling the growth and dynamics of our present society, to what end is still to be determined. Ethics are important in scientific research and the growth of our knowledge base. Ethics appears to be influencing human evolution at all levels. How this may play out in the future should be of great concern to everyone.

REFERENCES

Atkinson, R. L., Atkinson, R. C., & Smith, E. E. (1990). *Introduction to Psychology*. Orlando, FL: Harcourt Brace Jovanovich.

Bernstein, D. A. (2010). *Essentials of Psychology*. Boston: Cengage Learning.

Bohr, N. (1976). Niels Bohr, Collected Works: Vol. 3. *The Correspondence Principle (1918-1923)*. Amsterdam: North-Holland.

Bzdok, D., Schilbach, L., Vogeley, K., Schneider, K., Laird, A. R., Langner, R., & Eickhoff, S. B. (2012). Parsing the neural correlates of moral cognition: ALE meta-analysis on morality, theory of mind, and empathy. *Brain Structure & Function*, *217*(4), 783–796. doi:10.100700429-012-0380-y PMID:22270812

Coon, D., & Mitterer, J. O. (2008). *Introduction to Psychology: Gateways to Mind and Behavior*. Boston, MA: Cengage Learning.

Corsini, R. J. (2002). *The Dictionary of Psychology*. Florence, KY: Psychology Press.

Cummins, D. D. (2014). Neural correlates of causal power judgments. *Frontiers in Human Neuroscience*, *8*(1014). PMID:25566033

De Wied, M., Goudena, P. P., & Matthys, W. (2005). Empathy in boys with disruptive behavior disorders. *Journal of Child Psychology and Psychiatry, and Allied Disciplines*, *46*(8), 867–880. doi:10.1111/j.1469-7610.2004.00389.x PMID:16033635

Disalle, R. (2009). Space and Time: Inertial Frames. In E. N. Zalta (Ed.), *The Stanford Encyclopedia of Philosophy*. Retrieved from http://plato.stanford.edu/archives/sum2002/entries/spacetime-iframes/#Oth

Doeller, C. F., Barry, C., & Burgess, N. (2010). Evidence for grid cells in a human memory network. *Nature*, *463*(7281), 657–661. doi:10.1038/nature08704 PMID:20090680

Eichenbaum, H. (2014). Time cells in the hippocampus: A new dimension for mapping memories. *Nature Reviews. Neuroscience*, *15*(11), 732–744. doi:10.1038/nrn3827 PMID:25269553

Einstein, A. (2001). Relativity: The Special and the General Theory (R. W. Lawson, Trans.). Florence, KY: Routledge. (Originally published 1920)

Enriquez-Geppert, S., Huster, R. J., Figge, C., & Herrmann, C. S. (2014). Self-regulation of frontal-midline theta facilitates memory updating and mental set shifting. *Frontiers in Behavioral Neuroscience*, *8*(420). PMID:25538585

Fernandez, Y. M., & Marshall, W. L. (2003). Victim empathy, social self-esteem, and psychopathy in rapists. *Sexual Abuse*, *15*(1), 11–26. doi:10.1177/107906320301500102 PMID:12616926

Feynman, R. P. (1998). *Six Not-So-Easy Pieces: Einstein's Relativity, Symmetry, and Space-Time*. New York: Basic Books.

Gabriele, J. D., & Desmond, J. E. (1998). The role of left prefrontal cortex in learning and memory. *Proceedings of the National Academy of Sciences of the United States of America*, *95*(3), 906–913. doi:10.1073/pnas.95.3.906 PMID:9448258

Gabriele, J. D., & Kao, Y. (2007). Development of the declarative memory system in the human brain. *Nature Neuroscience*, *10*(9), 1198–1205. doi:10.1038/nn1950 PMID:17676059

Goldstein, E. R. (2009). *Sensation and Perception*. Boston, MA: Cengage Learning.

Gray, P. O. (2006). *Psychology* (5th ed.). New York: Worth.

Gregary, R. (1987). Oxford Companion to the Mind. New York: Oxford.

Hafting, T., Fyhn, M., Molden, S., Moser, M.-B., & Moser, E. I. (2005). Microstructure of a spatial map in the entorhinal cortex. *Nature*, *436*(7052), 801–806. doi:10.1038/nature03721 PMID:15965463

Harenski, C. L., Antonenko, O., Shane, M. S., & Kiehl, K. A. (2010). A functional imaging investigation of moral deliberation and moral intention. *NeuroImage*, *49*(3), 2707–2716. doi:10.1016/j.neuroimage.2009.10.062 PMID:19878727

Hossenfelder, S. (2015). Lost in thought. How important to physics was Einstein's imaginings? *Scientific American*, *313*(3). PMID:26455101

Jaeger, G. (2014). What in the (quantum) world is macroscopic? *American Journal of Physics*, *82*(9), 896–905. doi:10.1119/1.4878358

Janowski, J. S., Shimamura, A. P., & Squire, L. R. (1989). Source memory impairment in patients with frontal lobe lesions. *Neuropsychologia*, *27*(8), 1043–1056. doi:10.1016/0028-3932(89)90184-X PMID:2797412

Jeffery, K. (2007). Integration of sensory inputs to place cells: What, where, why, and how? *Hippocampus*, *17*(9), 775–785. doi:10.1002/hipo.20322 PMID:17615579

Jeffery, K., Anderson, M., Hayman, R., & Chakraborty, S. (2003). A proposed architecture for the neural representation of spatial context. *Neuroscience and Biobehavioral Reviews*, *28*(2), 201–218. doi:10.1016/j.neubiorev.2003.12.002 PMID:15172764

John, P. J. (2009). Frontal-temporal dysfunction in schizophrenia: A selective review. *Indian Journal of Psychiatry*, *51*(3), 180–190. doi:10.4103/0019-5545.55084 PMID:19881045

Johnson, J. D., Suzuki, M., & Rugg, M. D. (2013). Recollection, familiarity and content sensitivity in lateral parietal cortex: A high-resolution fMRI study. *Frontiers in Human Neuroscience*, *7*(209). PMID:23734122

Koch, C. (2015). Two natural philosophers, centuries apart, converse about the mind. *Scientific American Mind*, *26*(2), 28–31. doi:10.1038cientificamericanmind0315-28

Koch, C. (2016). Constructing the modern mind. *Scientific American Mind*, *27*(3), 22–27. doi:10.1038 cientificamericanmind0516-22

Kushner, L. H. (2008). *Contrast in judgements of mental health*. Ann Arbor, MI: ProQuest.

Lara, A. H., & Wallis, J. D. (2015). The role of prefrontal cortex in working memory: A mini review. *Frontiers in Systems Neuroscience*, *9*(173). PMID:26733825

Long, A. A., & Sedley, D. N. (1987). *The Hellenistic Philosophers: Translations of the Principal Sources with Philosophical Commentary*. New York: Cambridge University Press.

Lucas, J. R., & Hodgson, P. E. (1985). *Space, Time and Causality: An Essay in Natural Philosophy*. New York: Oxford University Press.

MacDonald, C. J., Lepage, K. Q., Eden, U. T., & Eichenbaum, H. (2011). Hippocampal "time cells" bridge the gap in memory for discontiguous events. *Neuron*, *71*(4), 737–749. doi:10.1016/j.neuron.2011.07.012 PMID:21867888

Maniadakis, M., & Trahanias, P. (2011). Temporal cognition: A key ingredient of intelligent systems. *Frontiers in Neurorobotics*, *5*(2). PMID:21954384

Mansouri, F. A., Rosa, M. G. B., & Atapour, N. (2015). Working memory in the service of executive control functions. *Frontiers in Systems Neuroscience*, *9*(166). PMID:26696841

Maurer, A. P., Vanrhoads, S. R., Sutherland, G. R., Lipa, P., & McNaughton, B. I. (2005). Self-motion and the origin of differential spatial scaling along the septo-temporal axis of the hippocampus. *Hippocampus*, *15*(7), 841–852. doi:10.1002/hipo.20114 PMID:16145692

Morin, D. (2008). *Introduction to Classical Mechanics*. New York: Cambridge University Press. doi:10.1017/CBO9780511808951

Moser, M.-B., & Moser, E. I. (1998). Functional differentiation in the hippocampus. *Hippocampus*, *8*(6), 608–619. doi:10.1002/(SICI)1098-1063(1998)8:6<608::AID-HIPO3>3.0.CO;2-7 PMID:9882018

Munoz-Lopez, M. M., Mohedano-Moriano, A., & Insausti, R. (2010). Anatomical pathways for auditory memory in primates. *Frontiers in Neuroanatomy*, *4*(129). PMID:20976037

Nelson, S. M., McDermott, K. B., Wig, G. S., Schlagger, B. I., & Petersen, S. E. (2013). The critical roles of localization and physiology for understanding parietal contributions to memory retrieval. *The Neuroscientist*, *19*(6), 578–591. doi:10.1177/1073858413492389 PMID:23778789

O'Keefe, J. (1976). Place units in the hippocampus of the freely moving rat. *Experimental Neurology, 51*(1), 78–109. doi:10.1016/0014-4886(76)90055-8 PMID:1261644

O'Keefe, J., & Dostrovsky, J. (1971). The hippocampus as a spatial map. Preliminary evidence from unit activity in the freely-moving rat. *Brain Research, 34*(1), 171–175. doi:10.1016/0006-8993(71)90358-1 PMID:5124915

O'Keefe, J., & Nadel, L. (1978). *The Hippocampus as a Cognitive Map*. London: Oxford University Press.

Penrose, R. (2005). *The Road to Reality*. London: Vintage Books.

Petkov, V. (2010). *Minkowski Spacetime: A Hundred Years Later*. Berlin: Springer. doi:10.1007/978-90-481-3475-5

Popper, A. N. (2010). *Music Perception*. Berlin: Springer.

Rizzolatti, G., Fadiga, L., Gallese, V., & Fogassi, L. (1996). Premotor cortex and the recognition of motor actions. *Brain Research. Cognitive Brain Research, 3*(2), 131–141. doi:10.1016/0926-6410(95)00038-0 PMID:8713554

Roberts, T., & Schleif, S. (2007). *What is the experimental basis of special relativity?* Usenet Physics FAQ.

Rolls, E. T. (2013). The mechanisms for pattern completion and pattern separation in the hippocampus. *Frontiers in Systems Neuroscience, 7*(74). PMID:24198767

Schacter, D. L., Gilbert, D. T., & Wegner, D. M. (2011). *Semantic and episodic memory. In Psychology* (2nd ed.; pp. 240–241). New York: Worth Incorporated.

Serino, S., & Riva, G. (2014). What is the role of spatial processing in the decline of episodic memory in Alzheimer's disease? The "mental frame syncing" hypothesis. *Frontiers in Aging Neuroscience, 6*(33). PMID:24653696

Sieb, R. A. (2004). The emergence of consciousness. *Medical Hypotheses, 63*(5), 900–904. doi:10.1016/j.mehy.2004.04.005 PMID:15488666

Sieb, R. A. (2013). The emergence of emotion. *Activitas Nervosa Superior, 55*(4), 115–145. doi:10.1007/BF03379732

Sieb, R. A. (2015). Memory in four dimensions. In M. Sakakibara & I. Etsuro (Eds.), *Memory Consolidation* (pp. 259–313). New York: Nova Science Publishers, Inc.

Sieb, R. A. (2016a). *Four-Dimensional Consciousness*. Retrieved from relativisticconsciousness.com

Sieb, R.A. (2016b). *Relativistic Consciousness*. Retrieved from relativisticconsciousness.com

Steinvorth, S., Levine, B., & Corkin, S. (2005). Medial temporal lobe structures are needed to re-experience remote autobiographical memories: Evidence from H.M. and W.R. *Neuropsychologia, 43*(4), 479–496. doi:10.1016/j.neuropsychologia.2005.01.001 PMID:15716139

Sumner, B. (2009). *The effect of experience on the perception and representation of dialect variants. Journal of Memory and Language*.

Taylor, A. (2003). *Animals & Ethics: An Overview of the Philosophical Debate*. Peterborough, Canada: Broadview Press.

Tulving, E. (1983). *Elements of Episodic Memory*. New York: Oxford University Press.

Ullman, M. T. (2004). Contributions of memory circuits to language: The declarative/procedural model. *Cognition*, *92*(1-2), 231–270. doi:10.1016/j.cognition.2003.10.008 PMID:15037131

Uncapher, M. R., & Wagner, A. D. (2005). Posterior parietal cortex and episodic encoding: Insights from fMRI subsequent memory effects and dual-attention theory. *Neurobiology of Learning and Memory*, *91*(2), 139–154. doi:10.1016/j.nlm.2008.10.011 PMID:19028591

Vendantum, S. (2007). If it feels good to be good, it might be only natural. *The Washington Post*.

Wagner, A. D., Shannon, B. J., Hahn, I., & Buckner, R. L. (2005). Parietal lobe contributions to episodic memory retrieval. *Trends in Cognitive Sciences*, *9*(9), 445–453. doi:10.1016/j.tics.2005.07.001 PMID:16054861

Weiten, W. (2010). *Psychology: Themes and Variations* (8th ed.). Boston: Wadsworth/Cengage Learning.

Wendelken, C. (2015). Meta-analysis: How does posterior parietal cortex contribute to reasoning? *Frontiers in Human Neuroscience*, *8*(1042). PMID:25653604

Wikibooks, The Free Textbook Project. (n.d.). Retrieved 11:18, June 24, 2016 from https://en.wikibooks.org/w/index.php?title=Special_Relativity/Spacetime&oldid=2766972

Wolfe, J. M., Kluender, K. R., Levi, D. M., Bartochuk, L. M., Herz, R. S., Klatsky, R. L., & Lederman, S. J. (2008). *Gestalt grouping principles. Sensation and Perception* (2nd ed.; pp. 78–80). Sunderland, MA: Sinauer Associates.

Young, L., Bechara, A., Tranel, D., Damasio, H., Hauser, M., & Damasio, A. (2010b). Damage to the ventromedial prefrontal cortex impairs judgment of harmful intent. *Neuron*, *65*(6), 845–851. doi:10.1016/j.neuron.2010.03.003 PMID:20346759

Young, L., Camprodon, J. A., Hauser, M., Pascual-Leone, A., & Saxe, R. (2010a). Disruption of the right temporoparietal junction with transcranial magnetic stimulation reduces the role of beliefs in moral judgments. *Proceedings of the National Academy of Sciences of the United States of America*, *107*(15), 6753–6758. doi:10.1073/pnas.0914826107 PMID:20351278

Chapter 8
A Mathematico-Physical Understanding of the Not-Being Potential and Creation of Parallel Universes by the Energy of Consciousness in Life Forms

Changsoo Shin
Seoul National University, South Korea

Nami Lee
Seoul National University Hospital, South Korea

ABSTRACT

This chapter approaches the subject of the not-being potential that created the universe from a mathematical and physical point of view. It postulates that the not-being potential existed prior to the Big Bang, the energy that was produced from the not-being potential to the being became a highly dense cosmic egg. This cosmic egg of infinite temperature exploded to create the current universe. This assumption is demonstrated by mathematical formulas and physics laws, which provide a solid scientific foundation for the aforementioned theory. Consequently, the authors show that the transformation of life forms' unconsciousness into consciousness creates parallel universes in a similar way by which the nor-being potential creates the universe.

INTRODUCTION

According to the recent big bang theory that explains the beginning of life in the universe, humans with egos appeared on earth only 13.8 billion years after the universe was created (Hawking & Penrose, 1970). Humans have shown great interest in matters of birth and death, as well as our place of origin and destination upon death, while religion continuously attempts to provide an answer. Therefore, it is

DOI: 10.4018/978-1-5225-9065-1.ch008

possible to say that the origin of religion can be traced as far back as the birth of Homo sapiens as a species (King, 2009).

[The field/study of] physics has evolved from Newton's classical mechanics to Einstein's general theory of relativity, quantum mechanics and particle physics, and made great progress in understanding both the macro and micro sides of the material universe. Accompanying these developments was another attempt to explain the structure of the universe using the superstring theory (Zweibach, 2009). From the 1970s to 1980s, John H. Schwarz and Michael Green argued that all matter in the universe was composed of tiny strings of energy that vibrate in various dimensions (Green, Schwartz, & Witten, 1988). Furthermore, physicists came to accept the cosmic background radiation theory as it relates to the primeval universe. The theory explains that the universe began to form from a single point with a very high density and temperature. The membrane theory, a big bang theory that explains how the universe came to be, also attempts to describe how the big bang occurred. In his membrane theory, E. Witten brought five different string theories together and argued that the big bang was created by a collision of branes between two parallel universes (Witten, 2012). However, the question of who initialized the vibration of so many tiny strings remains unresolved.

In this paper, we put forward a hypothesis that assumes time started when the supernatural potential that had existed as the not-being became the being, and a gigantic amount of high-density energy was released in the process. According to our hypothesis, the ultra-high density energy that was released during the not-being-to-being transformation process is assumed to be the big bang. We argue that the not-being can be expressed using an infinite recursive Heaviside step function, and assume it to be the not being's potential. We claim that the differentiation of the not-being with respect to time is the process of becoming being, and in a world where only time exists, the energy produced thereafter became a highly dense cosmic egg. Similarly, the consciousness of human beings is also a kind of ardent wish which arises from the process of differentiating a lump of the unconsciousness with respect to time, and has a considerable amount of energy whose magnitude and strength are sufficient to create multi-universes even though they are insignificant compared to the energy of the big bang.

In 1957, H. Everett extended the micro-scale Schrödinger's cats example to macro scale multi universes (Everett, 1957), but, discouraged by contemporary physicists, he died at the young age of 52. Multi universes, which we will discuss, are similar to his famous multi-universe idea but are not directly linked to this paper's main idea of generating the energy for multi universes during the transformation from unconsciousness to consciousness in life forms.

In treating distributions in distribution (recursive Heaviside step function), we suggest how to eliminate the issue of the chain rule in obtaining derivatives of recursive Heaviside functions by introducing a hybrid ontology. As the main objective of this paper, we try to attribute our unconsciousness to a special solution of the multi-dimensional advection partial differential equation, as well as to prove that it can be represented by the finite recursive Heaviside step function similar to the not-being potential during the unconscious to conscious transformation process. Consequently, we exist within a certain kind of awareness system originating from our consciousness and beliefs.

POTENTIAL, FORCE, AND ENERGY

In this paper, we attempt to postulate the existence of the not-being potential through mathematical theories. Let it be clear that by trying to express the existence of the not-being potential using simple mathematical formulas, this paper does not intend to display any disrespect for the not-being potential. This is one of many attempts to define the not-being potential as a concept comprehensible to humans.

Before dealing with numerical formulas related to the not-being potential, we first looked at a potential theory which is often cited in the study of physics. The potential theory is applied to the studies of gravitational fields and electrostatic fields, and to derive Green's functions for elastic wave equations in an infinite homogeneous medium (Pilant, 1979), which is a set of hypothetical concepts in physics and mathematics.

We define concepts such as the electromagnetic field, the gravitational field, strong force, weak force, power and energy in four-dimensional time and space. In this paper, we assume that the unconsciousness is only defined in time, with current time t and past times $\tau_1, \tau_2, \tau_3, \cdots, \tau_n$ because it is impossible to point out the exact locations of human consciousness and unconsciousness on a space coordinate system (Davis, 1983). If we open up a brain, for instance, we will be able to observe the flow of blood, polarization through ion channels, electric currents and other phenomena but not be able to say that consciousness is located inside the brain. If consciousness or unconsciousness exists in the brain, it should be possible to observe it when we open up the brain.

A concept of unconciousness widely accepted in psychology has wide spectrums. In here, we define the unconciousness as a potential that saves incidences life forms of an ego have expereienced, in overlapped order or in mixed order or in disordered way. We borrowed a concept of the potential defined in classical physics, for instance, the electric potential, and a force resulted by taking derivatives of the potential with respect to distance, for example, the electrical field, the energy defined by multiplication of force with distance. Then we define a conciousness as a psychokinetic force that comes by differntiating of the potential(unconciousness) with respect to time, the psychokinetic energy is defined as a multiplication of psychokinetic force with unit time.

In summary, we treat here the unconciousness as the psychokintic potential, a force resulted by taking derivatives of the psycokinectic potential with respect to current time, as the psychokinectic force and a multiplication of psychokitectic force with unit time as the psychokinectic energy. Furthermore, we assume that the psychokinetic energy potential,the psychokinetic force and the psychokinectic energy cannot be measured in ways of classical physics in the universe in where we(material bodies) exit. Material bodies may exist as illusion to the material universe whereas the material universe may be illusion to the soul world. We assumed that time can be treated as a vector instead of scalar. However, our assumption cannot be proven nor verified. It is only a hypothesis in deriving both the advection equation of the supernatural potential and the multidimensional advection equation of the human mind.

A HYBRID ONTOLOGY

In our existing mathematics, we have learned that discontinuous functions exist. However, if our universe is an analogue world, then discontinuous functions cannot exist. The Heaviside sequence function, supposing k is a finite number close to an infinitely large number, can be expressed in the following way.

$$H_{like}\left(t - t_0\right) = \lim_{k\to\infty} \frac{1 + \tanh k\left(t - t_0\right)}{2} \tag{1}$$

The delta function can be similarly defined.

$$\delta_{like} \equiv H_{like}\left(t-t_0\right) = \lim_{k\to\infty}\frac{d}{dt}\frac{1+\tanh k\left(t-t_0\right)}{2} = \lim_{k\to\infty}\frac{k}{2}sech^2 k\left(t-t_0\right) \tag{2}$$

If our universe consists only of a continuous function, the Heaviside function (discontinuous) is not computationally feasible by approaching k to infinity. However powerful computers may become accessible in the eternal future, the limit cannot be computed numerically as in the case of π. But mathematicians can imagine that the Heaviside sequence can converge on the discontinuous step function by approaching k to infinity. we again stress that e limit of the Heaviside sequence cannot be realized computationally. Therefore, we exclude discontinuous functions and use continuous function space which is dense in l_2 space. With this approach, we cannot define an exact Heaviside step function given by

$$H\left(t\right) = \begin{cases} 1 \ \text{ when } t \geq 0 \,\text{where} t \text{is a real number} \\ 0 \ \ when t < 0. \end{cases} \tag{3}$$

However, both the Fourier series representation of the Heaviside function and the limit of the Heaviside sequence can be expressed as

$$H\left(t\right) = \begin{cases} 1 \ \text{ when } t \geq 0 \,\text{where} t \text{is a real number} \\ \frac{1}{2} \ \text{ when } t = 0 \\ 0 \ \ when t < 0. \end{cases} \quad . \tag{4}$$

we discussed plural derivatives of several Heaviside functions that have the same shape in (Shin, 2016a).

On the other hand, we consider the Heaviside function as a digital ontology. For example, consider a digital world consisting of infinite bits (Zahedi, 2015). In such a case, Δt is discretized in infinitely small intervals, but not 0, and then normalized by Δt. Then, we assume that the analog world is interpolated exactly as a linear interpolation in the digital world since Δt is so small. This defines a hybrid ontology.

$$H\left(t\right) = \begin{cases} 1 \ \text{ when } t \geq 0 \,\text{where} t \text{is an integer} \\ 0 \ \ when t < 0. \end{cases} \tag{5}$$

Accordingly, the Heaviside function given by Equation (5) can be displayed as shown in Figure 1. Applying the analog ontology, it can be expressed as a piecewise function shown in Figure 2. This Heaviside Sequence is continuous, but the limit is a discontinuous function, so it is not complete in terms of l_1 norm (Kreyszig, 2011). However, if we use a continuous function belonging to C^0, we can create a Heaviside-like function (binary function) with the first differential.

Figure 1. The Heaviside like function given by Equation (3) within the digital ontology

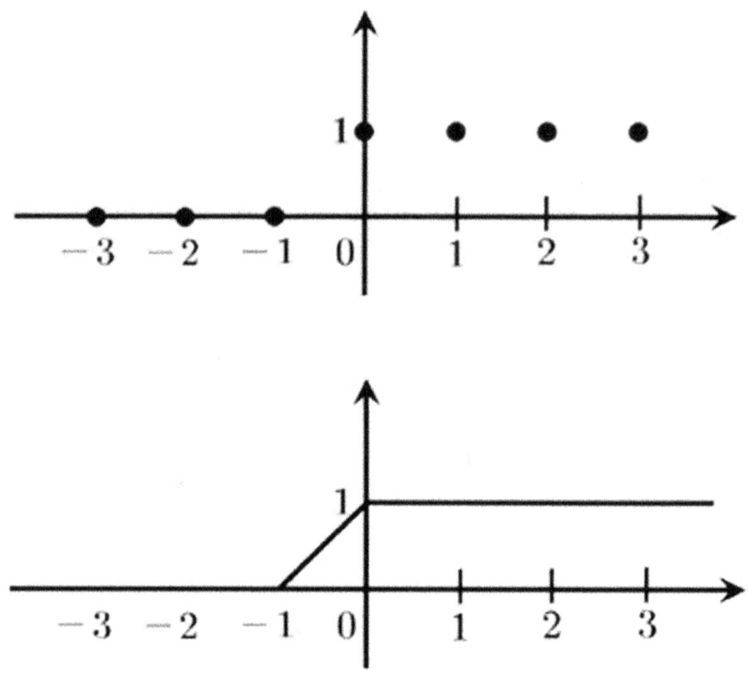

Figure 2. Derivative of Heaviside like function at $t = 0$ and $t = -1$ by following the definition of derivatives of central difference formula

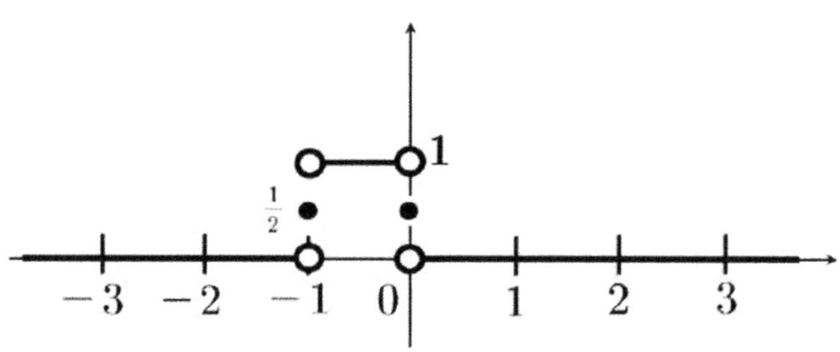

$$\left.\frac{dH(t)}{dt}\right|_{t=0} = 1 \tag{6}$$

The differential given by Equation (6) is shown in Figure 2.

Note that we take the differentials of Heaviside like functions by averaging values obtained from the definition of derivatives of both forward difference formulas and backward difference formulas, which are equivalent to taking the definition of the derivative of central difference formulas. Then, we have

1/2 at both $t=0$ and $t=-1$, as shown in Figure 2. By the reduction of the Fourier series for discontinuous functions, we can have a derivative defined at $t=0$ and $t=-1$, as shown in Figure 3.

Even if the world consists of infinite bits, the differential of the Heaviside function has a unit value. The recursive binary Heaviside function is a piecewise continuous function and the finite difference value is also defined. If we denote the recursive binary Heaviside function's differential by $\delta_{like}(0)$, a delta-like function, then its value assumes 1. Then we can apply the chain rule without difficulty, since the infinity of the original delta function becomes both finite and the unit value . Therefore $\left.\frac{dH_{like}(t)}{dt}\right|_{t=0}=\delta_{like}(t)=1$. Moreover, although $\left.\frac{dH_{like}(t)}{dt}\right|_{t=-1}=1$ is possible, it is ignored. This is because $t\geq 0$, and thus exceeds the range we have defined. As such, we can express this as

$$\left.\frac{dH_{like}(t)}{dt}\right|_{t=0}=1.$$

Consider, for example, a hybrid ontology for the shape and differentials of two recursive binary Heaviside-like step functions;

$$u_1=H_{like}\left(t+\tau_1 H_{like}(t+\tau_2)\right) \quad \text{and} \quad u_1=H_{like}\left(t+\tau_1 H_{like}(t+\tau_2)\right)\left.\frac{dH(t)}{dt}\right|_{t=0} \tag{7}$$

Suppose that $\tau_1=1$ and $\tau_2=2$. Both functions are represented in the same shape, as shown in Figure 4 and Figure 5.

The differential of u_1 is expressed as follows.

$$\frac{\partial u_1}{\partial t}=\delta\left(t+H_{like}(t+2)\right)(1+\delta_{like}(t+2). \tag{8}$$

Figure 3. A restored derivative of the Heaviside step function by reduction of the Fourier series of a discontinuous function

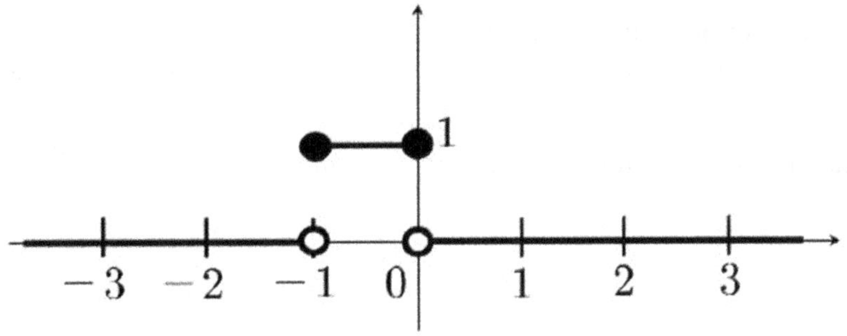

Figure 4. The first Heaviside-like step function in Equation (7), $u_1 = H_{like}\left(t + H_{like}(t+2)\right)$

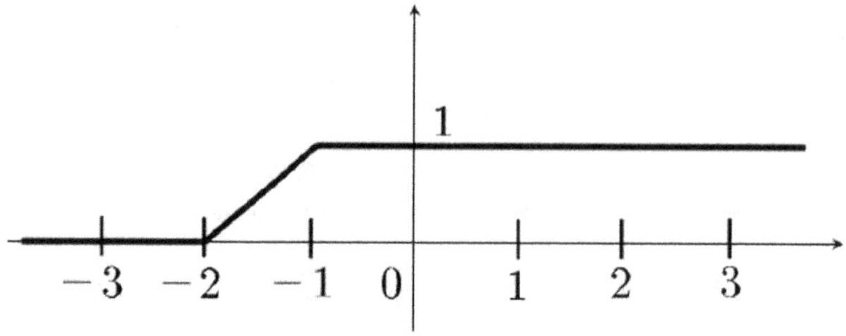

Figure 5. The second Heaviside-like step function in Equation (7), $u_2 = H_{like}\left(t + 2H_{like}(t+1)\right)$

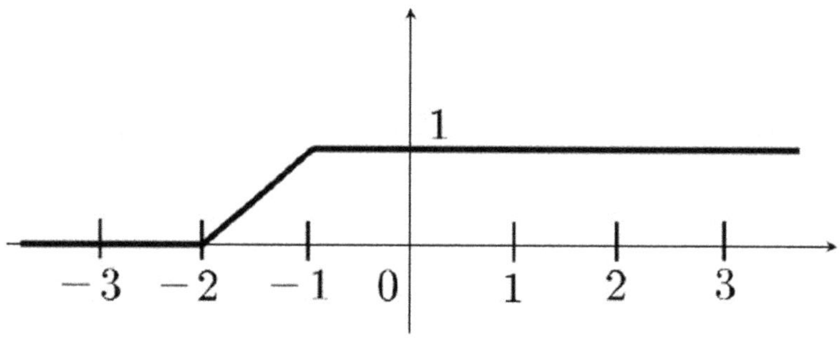

Here,because $H_{like}\left(t+2\right)$ isnon-zeroonlywhen $t \geq -2$,itbecomes $\delta_{like}\left(t + H_{like}\left(t+2\right)\right) = \delta_{like}\left(t+1\right)$ and consequently $\left.\frac{\partial u_1\left(t\right)}{\partial t}\right|_{t=-1} = \delta_{like}\left(t+1\right)$. Also the differential for u_2 is expressed as follows.

$$\frac{\partial u_2}{\partial t} = \delta\left(t + 2H_{like}\left(t+1\right)\right)(1 + 2\delta_{like}\left(t+1\right). \tag{9}$$

Here, because $H_{like}\left(t+1\right)$ is non-zero only when $t \geq -1$, it becomes $\delta\left(t + 2H_{like}\left(t+1\right)\right) = 0$, and consequently $\left.\frac{\partial u_2}{\partial t}\right|_{t=-2} = 0$. Therefore, u_1 and u_2 are represented by functions with the same shape but different differentiated function values, as shown in Figure 6 and Figure 7.

Thus, even if the shapes are the same, there is a function in which the differential values are expressed in distinct functions (Shin, 2016a). In this paper, H means H_{like}, δ means δ_{like}, and ∞ denotes ∞_{like}. It is assumed that the differential can be obtained by applying the chain rule.

Figure 6. The derivative of u_1 *given by Equation (8), reduced to* $\frac{\partial u_1}{\partial t} = \delta_{like}\left(t+1\right)$

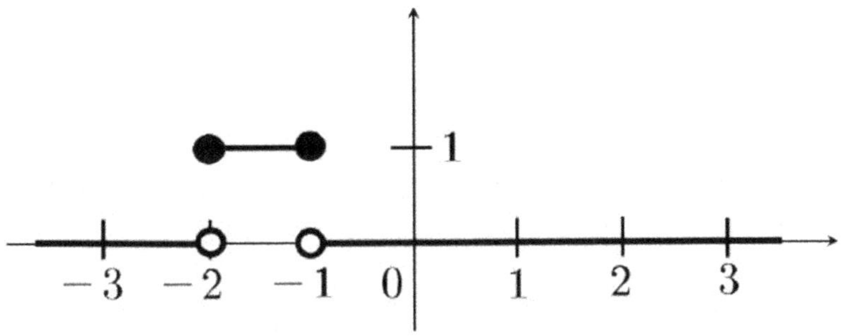

Figure 7. The derivative of u_2 *given by Equation (9), reduced to* $\frac{\partial u_2}{\partial t} = 0$

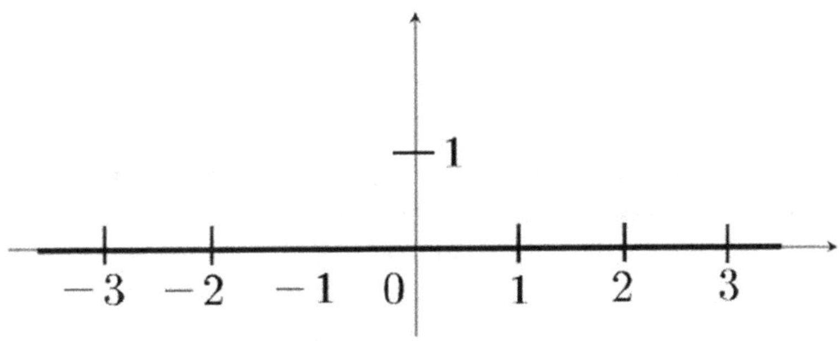

THE SUPERNATURAL POTENTIAL'S ADVECTION EQUATION

We presume that $U=$ constant and represents the not-being potential. By denoting the time vectors as t and τ (t represents current time and τ represents past time), the differential dU becomes

$$dU = \frac{\partial U}{\partial t}dt - \frac{\partial U}{\partial \tau}d\tau = 0. \tag{10}$$

Equation (10) is derived by mimicking the exact differential equation as done by Kreyszig (11). In Equation (10), we can regard the total derivative of potential U as $\nabla U \cdot dT$, where $\nabla U = \frac{\partial U}{\partial t}\hat{i} + \frac{\partial U}{\partial \tau}\hat{j}$, ($\hat{i}$ and $\hat{j}$ are unit vectors, t and τ are in the current time direction and the past time direction). After setting both dt and $d\tau$ equal to unit time, we obtain the one-dimensional advection equation

$$\frac{\partial U}{\partial t} - \frac{\partial U}{\partial \tau} = 0. \tag{11}$$

Here, U is the ultra-unconscious potential, t is the present time and τ is the past time. Equation (11) can be reorganized into

$$\frac{\partial U}{\partial t} = \frac{\partial U}{\partial \tau} \tag{12}$$

by transposing the past time term on the left-hand side to the right-hand side. The general solution of Equation (12) is $U = f\left(-t - \tau\right)$, and the particular solution becomes

$$U_n = H\left[-t - \tau H\left[-t - \tau H\left[-t - \tau H\left(-t - \tau H\left(-t - \tau \cdots\cdots\right)\right)\right]\right]\right]. \tag{13}$$

Let's call this particular solution an n th degree recursive Heaviside step function. For example, a quadratic recursive Heaviside step function is written as

$$U_2 = H\left[-t - \tau H\left(-t - \tau\right)\right]. \tag{14}$$

From the fact that Equation (13) meets the requirements to be a solution for the homogeneous advection equation, we postulate that the supernatural potential has always existed in single dimensional time. After differentiating with respect to t, Equation (14) becomes

$$\frac{\partial U_2}{\partial t} = \delta\left[-t - \tau H\left(-t - \tau\right)\right]\left[-t - \tau\delta\left(-t - \tau\right)\right] \tag{15}$$

and when $t = -\tau$, it becomes

$$\frac{\partial U_2}{\partial t} = \delta\left(-t - \tau\right)\left[-1 + \tau\delta\left(-t - \tau\right)\right] = -\delta\left(-t - \tau\right) + \tau\delta^2\left(-t - \tau\right) = -\delta\left(0\right) + \tau\delta^2\left(0\right). \tag{16}$$

Also, when we differentiate Equation (13), the n th degree recursive formula, with respect to t, the result becomes

$$\frac{\partial U_n}{\partial t} = -\sum_{k=1}^{n} \tau^{k-1}\delta^{(k)}\left(-t - \tau\right), \tag{17}$$

and for the infinite recursive formula, the result becomes

$$\frac{\partial U_\infty}{\partial t} = -\sum_{k=1}^{\infty} \tau^{k-1}\delta^{(k)}\left(-t - \tau\right) \tag{18}$$

Where the absolute value of A should be a number greater than 1 both in the hybrid ontology as well as in the analog ontology. Note that $\delta\left(-t-\tau\right)$ becomes infinity in the analogue ontology. Both τ and δ function contribute to a generation of psychokinesis energy, whereas in the hybrid ontology, $\delta\left(-t-\tau\right)=1$ at $t=-\tau$, the τ only contribute to generate the psychokinesis energy. The not-being potential, which is expressed as a potential of the unconscious, is a form of psychokinetic energy and exists only in time. However, such not-being potential is beyond human imagination at the moment of differentiation. When multiplied by the unit time at the instant when the not-being potential began to take effect, an infinite amount of energy is released which is defined only in time (Shin, 2016b). This is equivalent to the big bang in modern physics. The hypothesis postulates that there was only time and no space before the big bang. Therefore, if we multiply the function by Δt, which is the unit time, and psychokinesis defined in time, we obtain energy as a result as we define work by $W=\vec{F}\cdot\vec{S}$ where $\vec{F}$ is a force and $\vec{S}$ is a displacement vector.

THE SUPERNATURAL POTENTIAL

If $n=3$, for example, the supernatural potential is expressed as

$$P_{the\ not-being\ potential}=U_3\left(t,\tau\right)=H\left[-t-\tau H\left[-t-\tau H\left(-t-\tau\right)\right]\right]. \tag{19}$$

In Figure 8, we plot the Heaviside step function $H\left(-t-\tau\right)$. The not-being potential given by Equation (19) obtained from Equation (13) with $n=3$, assumes the same shape as given by Figure 8 regardless of the order n of the recursiveness in the recursive Heaviside step function given by Equation (13).

Figure 8 and Equation (19) both demonstrate the potential of the not being and it is necessary to take a closer look at Equation (19) to examine the process of the potential transforming itself into energy. The recursive Heaviside step function that represents the not-being potential in this paper includes an infinite number of discontinuous functions. However, the shape of the graph remains the same even if there are multiples recursive Heaviside step functions. While it is possible to carry out differentiation using the chain rule in the case of a continuous function, it is impossible to do the same thing with the recursive

Figure 8. The plots of the Heaviside step function $H\left(-t-\tau\right)$

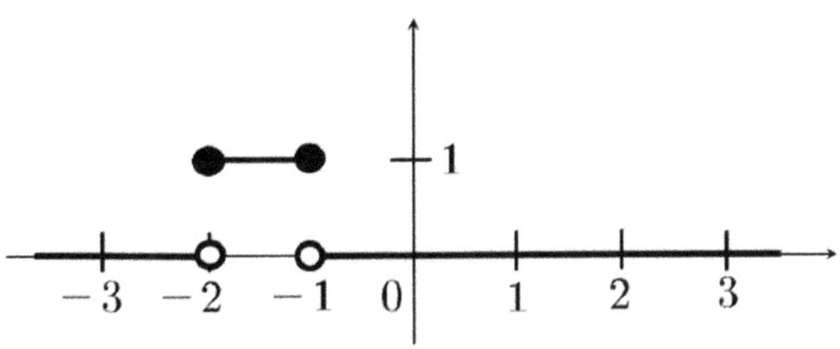

Heaviside step function because it contains numerous discontinuities. In this paper, as introduced in the hybrid ontology section, we take the differentiation of recursive Heaviside functions by the chain rule as we do in taking the derivative of continuous functions.

For instance, consider a function U_2 given by

$$U_2 = H\left[-t - \tau H\left(-t - \tau\right)\right]. \tag{20}$$

When Equation (20) is differentiated with respect to t, we obtain

$$\frac{\partial U_2}{\partial t} = \left[-t - \tau\delta\left(-t - \tau\right)\right]\delta\left(-t - \tau\right) = -\delta\left(-t - \tau\right) + \tau\delta^2\left(-t - \tau\right). \tag{21}$$

If we hypothesize that the not-being potential is represented by the infinite recursive Heaviside step function and was conscious at a certain point of time $-\tau$, it is possible to conclude that the not-being to the being (unconscious-to-conscious) energy of the not-being conscious state was sufficient to create the universe. There is no way of knowing how recursive the function was, but we can say that the supernatural potential was an ultra-unconscious being, and that the current universe was created at the moment when the supernatural being became conscious—at the point in time when the potential produced energy of an infinitely high temperature and an infinite density that is the foundation of explaining the universe in modern physics.

It appears, however, that the not-being potential has not gone through another process of coming into consciousness since the birth of the universe. An enormous amount of energy would have been produced if there was another unconscious-to-conscious transformation process, and the universe would have collapsed. In this sense, Friedrich Nietzsche's famous statement that God, being shown as the supernatural potential to us, potential is dead might hold some truth. The supernatural potential created the Universe but did not work further on it. At least, the infinite recursive not being did not become being.

INDUCTION OF THE MULTI-DIMENSIONAL ADVECTION EQUATION FOR HUMAN UNCONSCIOUSNESS

Let us denote the potential of human unconsciousness by U. In addition, let us call the first moment that an individual gains consciousness τ_1 and the moments from unconsciousness to consciousness $\tau_2, \tau_3, \cdots, \tau_n$, respectively. If the current time of being unconscious is τ_0 and t is the current time, the potential U satisfies the equation

$$dU = -\frac{\partial U}{\partial \tau_1} d\tau_1 - \frac{\partial U}{\partial \tau_2} d\tau_2 - \cdots - \frac{\partial U}{\partial \tau_{n-1}} d\tau_{n-1} - \frac{\partial U}{\partial \tau_n} d\tau_n + \frac{\partial U}{\partial t} dt = 0 \tag{22}$$

where it has been assumed that $t \geq \tau_k$ and $\tau_k < 0$ for $k = 1,2,\cdots,n$. As we did in deriving the nnot -being potential's exact differential equation, if we set $d\tau_1, d\tau_2, \cdots, d\tau_n$ and dt to the unit time, Equation (22) becomes

$$\frac{\partial U}{\partial t} = \sum_{k=1}^{n} \frac{\partial U}{\partial \tau_k}. \tag{23}$$

Since humans have existed in a finite amount of time, the equation that defines human consciousness and unconsciousness can be expressed as

$$\frac{\partial U}{\partial t} = \left(\frac{\partial U}{\partial \tau_1} + \frac{\partial U}{\partial \tau_2} + \cdots + \frac{\partial U}{\partial \tau_n} \right). \tag{24}$$

General solutions of Equation (24) can be given as:

$$U\left(t\right) = f\left(t + \frac{\tau_1 + \tau_2 + \cdots + \tau_n}{n} \right) \tag{25}$$

and a special solution is given by:

$$U_p\left(t\right) = F\left(t + \tau_1 + \tau_2 + \cdots + \tau_n\right) = f\left(t + \tau_1\right) + f\left(t + \tau_2\right) + \cdots + f\left(t + \tau_n\right). \tag{26}$$

After an individual is born, if his *n*th unconsciousness turns into consciousness, this can be expressed as a finite recursive function such as (Shin & Cha, 2015; Shin, Ha, Chung, & Park, 2012)

$$U = H\left[-t - \tau_1 H\left[-t - \tau_2 H\left[-t - \tau_3 H\left[-t - \tau_4 H\left(\cdots\right)\right]\right]\right]\right] + t. \tag{27}$$

We add constant value of unit to the recursive Heaviside step function,that leads to express default conciousness state of life forms. Hereafter,we omit default conciousness(t) state and its derivative (1).

Since human memories often fall into oblivion, the value of the time derivative of the potential U without current time t, which represents unconsciousness, can sometimes assume zero, while the past times $\tau_1, \tau_2, \cdots, \tau_n$ continue to coexist. Even when time derivatives of human consciousness do not vanish, even though the amount of released energy is tiny compared to that of the not-being potential, it becomes possible to create a universe with its own time axis separated from others. Let us investigate a simple example where the potential is given by

$$U = H\left[-t - \tau_1 H\left(-t - \tau_2\right)\right]. \tag{28}$$

When differentiated with respect to time t, Equation (28) becomes

$$\frac{\partial U}{\partial t} = \delta\left[-t - \tau_1 H\left(-t - \tau_2\right)\right]\left[-1 + \tau_1 \delta\left(-t - \tau_2\right)\right]. \tag{29}$$

and its values can be grouped into three cases as (i) $\tau_1 > \tau_2$, (ii) $\tau_1 = \tau_2$, and (iii) $\tau_1 < \tau_2$. When (i) $\tau_1 > \tau_2$, the time derivative becomes

$$\frac{\partial U}{\partial t} = -\delta\left(-t - \tau_1\right) = -\delta\left(0\right) \tag{30}$$

which occurs at $t = -\tau_1$. Similarly, when (ii) $\tau_1 = \tau_2$, the time derivative becomes

$$\frac{\partial U}{\partial t} = -\delta\left(-t - \tau_1\right) + \tau_1 \delta^2\left(-t - \tau_1\right) = -\delta\left(0\right) + \tau_1 \delta^2\left(0\right). \tag{31}$$

which also occurs at $t = -\tau_1$. Finally, when (iii) $\tau_1 < \tau_2$, the time derivative can be

$$\frac{\partial U}{\partial t} = 0 \text{ or } \frac{\partial U}{\partial t} = -\delta\left(0\right) \tag{32}$$

where the first value appears when $t = -\tau_1 \neq 0$, while the second value appears when $t = -\tau_1 = 0$. The power of human consciousness can be expressed with a delta function, with many powers of the delta function, or 1. In other words, human beings can transfer themselves to another universe by changing the time axis along which they exist according to their level of consciousness.

THE SIMILIARITIES AND DIFFERENCES BETWEEN HUMAN CONSCIOUSNESS AND THE NOT-BEING CONSCIOUSNESS, AND THE MULTI-UNIVERSE

We made an assumption that the not-being potential's unconscious state is expressed by an infinite recursive Heaviside function, and that an infinite amount of energy is produced during the process of becoming fron the not-being potential to being conscious at instant time In humans, we assume that unconsciousness is expressed by a finite recursive Heaviside function that includes everything we see, hear and feel with our senses at each time $\tau_1, \tau_2, \cdots, \tau_n$ overlapping with one another in disordered manner. Human beings also produce an finite amount of energy, albeit a very small amount compared to that of the not-being potential, at the moment of becoming the being instantly. This energy creates a universe which is called a parallel universe created by the mind. As such, the energy produced by a person's unconscious-to-conscious transformation process is significant as well. However, human energy is never expressed with an infinite recursive Heaviside step function because humans are born and exist in a finite amount of time. Human energy, which is expressed with recursive Heaviside step functions of each different time frame, is close to zero when compared to that of the not-being potential. Still, a person's conscious energy is enormous yet to be determined as well, and can be used to create one's own universe based on one's imagination. Our brain is an organ to take derivatives of the unconciousness instantly and becomes conciousness state plus default conciousness state.

However, humans have a particular mechanism for remembering and forgetting. For example, we sometimes forget what we had for lunch last Saturday but still remember our first kiss with a previous significant other. In the case of forgetting, if there is an incorrect combination of time frames τ_1, τ_2, τ_3, τ_4, and τ_5 in the recursive Heaviside step function, a value which is obtained by differentiating with respect to the current time t becomes zero, therefore resulting in the default stage (Shin et al., 2012). In this case, a person's conscious energy is a unit constant unit value and he is not able to change the universe in which he exists.

Take, for example, how three people's reactions to a lost dog can differ according to their unconsciousness. We express these three different reactions using the recursive Heaviside step function and differentiate them with respect to the current time t.

Thus we see the reasons behind the different reactions in a mathematical way. For instance, among Jane, Paul, and Bill, Paul would want to take the dog home because he lost his own dog, and Bill may be terrified because of a childhood memory of being bitten by a dog, while Jane may feel nothing at all towards the dog, as shown in Figure 9 (Shin & Cha, 2015).

The human reactions, unlike those of the not-being potential, depend on past experiences. Even though there is a heavy and infinite overlapping of unconsciousness at times τ_k, it must be remembered that humans have a unique characteristic of forgetting if the unconciousness of humans exited in disordered sequences. For example, if humans become unconscious ten trillion different times, there are ten trillion squared different unconscious states including instances where various states coexist. Even if we attempt psychoanalysis, it may be possible to observe only a very small part of the unconsciousness.

In Figure 10, we look at Schrödinger's cat. The cat remains both alive and dead from cyanide poisoning until someone looks into the box. Regarding such ideas, H. Everett proposed the theory of multiple

Figure 9. Functions indicating the situation and mental states of Jane, Paul, and Bill

Name	Mental State
Jane	 $U_1 = H(-t-\tau_1^J)$ $\frac{\partial U_1}{\partial t} = -\delta(-t-\tau_1^J)$
Paul	 $U_2 = H[-t-\tau_2^P H(-t-t_1^P)]$ $\frac{\partial U_2}{\partial t} = \delta[-t-\tau_2^P H(-t-\tau_1^P)][-1+\tau_2^P \delta(-t-\tau_1^P)]$
Bill	 $U_3 = H[-t-\tau_3^B H[-t-\tau_2^B H(-t-\tau_1^B)]]$ $\frac{\partial U_3}{\partial t} = \delta[-t-\tau_3^B H[-t-\tau_2^B H(-t-\tau_1^B)]]$ $\times[-1+\tau_3^B \delta[-t-\tau_2^B H(-t-\tau_1^B)][-1+\tau_2^B \delta(-t-\tau_1^B)]]$

Figure 10. The parallel universe of the two Schrödinger's cats

universes (Byrne, 2010) which are deemed selective parallel universes as shown in Figure 11. According to the scenario of the parallel universe created by the mind, the universe of the live cat or that of the dead cat is developed depending on one's perception of the cat poisoned by cyanide. When one sees the cat, if the derivative of one's unconsciousness does not vanish, then the former or the latter universe will develop depending on if one thinks the cat is alive or dead.

The two parallel universes are similar to one another. It is a matter of whether observation comes first or consciousness comes first. Generally, it would be more valid to say that we observe with a conscious mind. As a result, there is a different time zone for each universe depending on our consciousness, so the parallel universes that we argue exist would be similar to those in Figure 11. The existing parallel universe theory does not provide a clear explanation of the energy source for the creation of an infinite universe. For this, we argue that the finite amount of energy that is produced when human unconsciousness transforms into consciousness creates new universes in which one belongs. We also believe that although the energy is tremendous, it is impossible to be aware of the scale of that energy unless we move into the future and observe it ourselves.

In the microscale Schrödinger's cat, a six month-old baby who has no knowledge or no experience of cyanide (no memory of the danger of cyanide, unaware of what cyanide is) observes a cat with another chance to live. However, all grown-up males and females who know of the danger of cyanide will assume that the cat dies.

Schrödinger's cats on a macro scale are primarily influenced by the observer's experience, unconsciousness, memory and other factors. For example, if a cat's possibility of survival is indicated on the x -axis, the probability will show a normal distribution. An observer with little knowledge of the effects of cyanide will have a normal distribution that is slightly skewed toward a lower probability of survival. If you observe many cats dying from cyanide, you will naturally expect a more skewed probability distribution to the left. On the other hand, if there is no knowledge of the effects of cyanide, or if a large number of live cats are observed, the probability distribution is skewed to the right. If you take the positive view that the cat will live, a more skewed probability distribution to the right will occur. A similar case would result in a parallel universe similar to Schrödinger's cat in the macro-world, depending on the will of a cancer patient. In fact, many clinical doctors have reported cases where a patient's immune system stopped working properly once he or she lost the will to fight the disease.

Figure 11. The idea of selective parallel universes excerpted from (Zahedi, 2015)

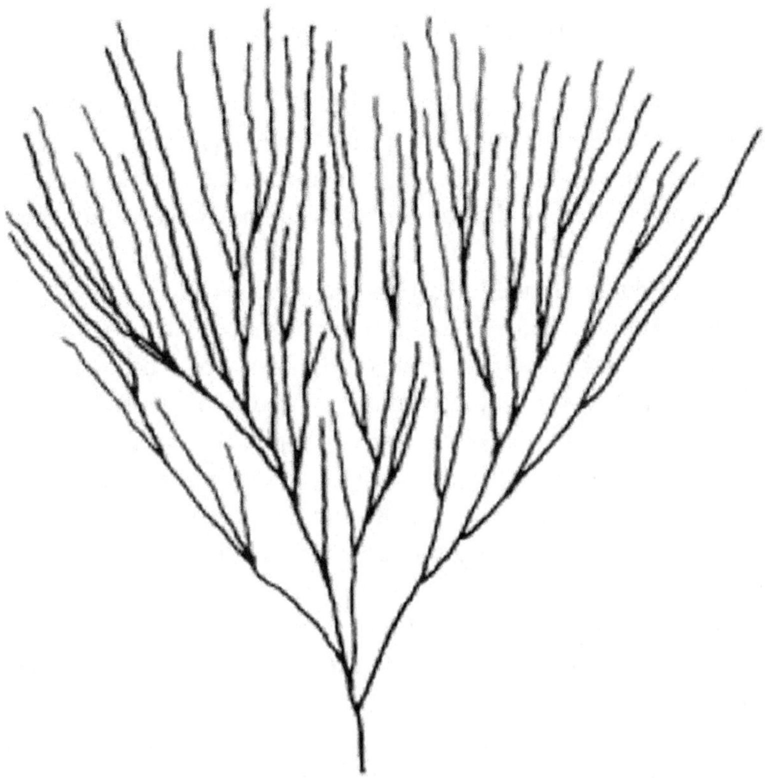

A HYPOTHESIS ON NON-REPLICABLE DYNAMIC STATE OF HUMAN UNCONSCIOUSNESS

In the above section, the human consciousness can be expressed into the overlapped, mixed and disordered sequence recursive Heaviside step function as almost unlimited past time incidences. When we recall something in past, we could remember it at any moment we like to recall. But we cannot recall something important as we want to recall it, tomorrow we can recall it or we can recall it after a week or after a month or after several years or never remember it. Our unconsciousness can be re-ordered, re-overlapped and or re-mixed at any moment, to some extent, as a function of past incidence times. This assumption explains why we cannot recall something as soon as we want to recall. This hypothesis will result in making any future device to record human unconsciousness or consciousness impossible, although we advance both science and technology in long term future, that means we cannot be replicated by any devices forever and our soul is unique.

CONCLUSION

Human beings have fostered great civilizations through the development of mathematics and science. However, there is as yet no clear explanation of the spiritual side of humankind. We have so far attempted to provide an explanation for the not-being potential's creation of the universe using an infinite recursive Heaviside step function of an not being potential and numerical differentiation. The human unconsciousness, and also that of an animal with an ego, is similar to that of the not-being potential. Our conscious process is tremendous, as an infinite number of universes can be created and determined by conscious activities and self-reflection.

Human unconsciousness can be defined using different values such as the square or the n powers of past time τ, which is a differentiated value of a finite recursive Heaviside step function. This is an idea that can only exist in time and not in a general coordinate space. The sizes of the values are generally expressed as infinite in an hybrid ontology, but they could vary when expressed in a sequence. Human unconsciousness differs from that of the not-being potential in that it is not concentrated at a specific time τ but mixed or placed or disordered in a time sequence and has the distinct characteristic of forgetting. The human energy produced during the unconscious-to-conscious process is small compared to that of the supernatural potential, but it is sometimes enough to make a change in the universal space and thus create another universe.

We have also discussed the various images an individual can embody in different people's unconsciousness. The individual can exist in many different forms in the unconsciousness of others, but his unconscious mind is unconscious of these different forms. So the individual can be locked in many different universes depending on how people think of him in a conscious way. Depending on whether the images sealed in people's unconsciousness are good or bad, the individual may experience different fates.

We proved that the some energy from life forms's unconsciousness to consciousness can make our universe diverge into the parallel universes by assuming that human's unconsciousness is some sort of potential. Our brain is the computing machine organ of taking derivatives of the human's potential(unconsciousness) instantly. However, the process from the not-being potential to the being occurred instantly and disappeared, resulting into generation of both our universe and life forms whose potential resemble the not being's potential.

The analogue ontology and the hybrid ontology, the past time τ is a critical factor to determine the size of psychokinesis energy. However, in the analog ontology, a size of δ function matters, whereas in the hybrid ontology, δ function did not contribute in estimating the size of psychokinesic energy. The recursive Heaviside step function representing life forms's unconciousness can have same shapes but their derivatives are different to each other. This nature can be applied to the not –being potential. However many times the not-being's potential is nested, the shape maintains the same shape.

REFERENCES

Byrne, P. (2010). *The Many Worlds of Hugh Everett III: Multiple Universes, Mutual Assured Destruction, and the Meltdown of a Nuclear Family*. Oxford University Press.

Davis, P. (1983). *God and the New Physics*. Simon and Schuster.

Everett, H. (1957). 'Relative state' formulation of quantum mechanics. *Reviews of Modern Physics*, *29*(3), 454–462. doi:10.1103/RevModPhys.29.454

Green, M., Schwarz, J. H., & Witten, E. (1988). *Superstring Theory* (Vol. 1). Cambridge University Press.

Hawking, S. W., & Penrose, R. (1970). The Singularities of Gravitational Colllapse and Cosmology. *Proceedings of the Royal Society of London. Series A, Mathematical and Physical Sciences*, *314*(1519), 529–548. doi:10.1098/rspa.1970.0021

King, B. 2009. *Evolving God: A provocative View on the Origins of Religion*. Doubleday Publishing.

Kreyszig, E. (2011). *Advanced Engineering Mathematics* (10th ed.). John Wiley and Sons Inc.

Pilant, W. (1979). *Elastic Waves in the Earth*. Elsevier Scientific Publishing Company.

Shin, C. (2016a). Introducing a function with plural derivatives. *Zeitschrift für Angewandte Mathematik und Physik*.

Shin, C. (2016b). *God before the Big Bang*. International Education & Research Journal.

Shin, C., & Cha, D. (2015). An Attempt towards Mathematical Description of the Human Mental State. *ADAM*, *5*(1), 1–8.

Shin, C., Ha, W., Chung, W., & Park, S. (2012). Unresolvable Human Mental States Based on a Parallel Universe Theory. *ADAM*, *1*(1), 18–29.

Westen, D. (1999). The Scientific Status of Unconscious Process. *Journal of the American Psychoanalytic Association*, *47*, 1061–1106. doi:10.1177/000306519904700404 PMID:10650551

Witten, E. (2012). Fivebranes and Knots. *Quantum Topology*, *3*(1), 1–137. doi:10.4171/QT/26

Zahedi, R. (2015). *On Discrete Physics (Digital Philosophy/Digital Cosmology) and the Cellular Automaton: A Perfect Mathematical Deterministic Structure for Reality - as a Huge Computational Simulation*. rXiv:1501.01373v3

Zwiebach, B. (2009). *A First Course in String Theory*. Cambridge University Press. doi:10.1017/CBO9780511841620

Section 3
Mediated Sentience

Chapter 9
The Media-Dream Model:
Science Fiction as Archetypal Representation

Stephen Brock Schafer
Pacific Rim Enterprises, USA

ABSTRACT

We live in a world not of science but of science fiction. Like pixel patterns from unconscious software is projected onto a monitor, unconscious archetypal quantum patterns are projected as what Carl Jung called archetypal representation (AR). Projected images are then subject to the vagaries of personal perception, so it may be stipulated that no absolute reality exists for humans. Rather, each person lives in a perceptual fiction. According to Carl G. Jung, dreams are projections from quantum-level unconscious dimensions into the cognitive dimension of "consciousness." In the language of dream analysis, Jung would have described the science fiction genre as a prospective (future-oriented) dreamscape of archetypal representations. In the media-dream model, quantum patterns are derived from research in cognitive neuroscience and physics. Contemplated as AR, the sci-fi genre is predictive of cultural futures and formats psychological motives and morality. Sci-fi has the potential to detect the psychological dynamics at work during the paradigm shift into a dreamscape of illusion.

INTRODUCTION

According to Jungian principle, dreams can be analyzed with the language of symbolism in-order to access and heal incoherent individual and collective psychic states. This healing process can also be understood as a learning dynamic that is based on biofeedback. Jung called the biofeedback "Compensation", and Jungian Compensation is a process leading to the achievement of a coherent state of being that he called "Individuation". Therefore, achievement of coherent states that result from meaningful insight (learning) as to the unconscious significance of dream symbols becomes the healing dynamic for individuals and cultures. Science fiction films can be analyzed as if they were collective media-dreams. They provide prophetic insights that foster coherent states and throw light on alternative cultural life-paths that are being explored during the paradigm shift. Remembering that the same dynamic of dream

DOI: 10.4018/978-1-5225-9065-1.ch009

(image) analysis and healing potential can be applied to all mediated illusions including news coverage, the science of pixel pattern projection becomes the key to creating a coherent culture.

It may be stipulated that science fiction has had a profound influence on culture. Notwithstanding groundbreaking cognitive research on the "unconscious", according to the same psychological dynamics (association, correlation, & repetition) used successfully for decades by advertisers and, now, by neuro-marketers, we understand why science fiction becomes reality. It can be demonstrated that the human reality is a product of the dynamics of dreams as researched by Carl Jung. Though the importance of dreams has been recognized since time immemorial, it is only now—in the grasp and with the evidence provided by the virtual reality of the technologically mediated dreamscape—that a majority of humans can realize, indeed, humans create their own reality of dreams. Taking full responsibility and accountability for this creative agency will change everything. The award-winning science fiction writer, A. A. Attanasio (2016), says:

We live in a world not of science but of science fiction. Consider gravity. Newton's law of universal gravitation took people to the moon and back several times – more than a half century after Einstein's general theory of relativity made that law fiction.

Even Einstein's view of gravity becomes fiction inside black holes and at the first instant of the Big Bang. And the enigmas of dark matter and dark energy in this century have moved physicists to suggest that gravity may "leak" from other dimensions (String Theory) or perhaps emerge from the thermodynamics of spacetime (Entropic Gravity).

Observational experimentation, even with something as fundamental as gravity, can only ever offer us fictions. We're biological, after all, and most of the universe is not. Categories like time and volume, which define the parameters of our world, turn out to be illusions [mathematical constructs] in cosmic reality.

Correlated research by Carl Jung and George Lakoff provides the cognitive reasons why science fiction—that always takes the form of some sort of mediation—becomes reality. The same cognitive dynamic applies to the critical problems of the media—political hacking, surveillance, the creation of reality with repetitive lies, and subliminal coercion with advertising and neuromarketing. The premise of this paper is that science fiction dramas that heal cultural collectives with coherent mediated biofeedback can be designed as *Psych*ecology video game (PEG) research instruments.

Notwithstanding such sci-fi films as *Jacob's Ladder, Artificial Intelligence* and *Vanilla Sky* that have actually explored cognitive science and modern issues based on the illusions of physics and psychology, perhaps the best illustrations of scientific sagacity are the classic science fiction film, *Blade Runner* (BR) and its sequel, *Blade Runner 2049* (BR49). Both BR and *Blade Runner 2049*—based on the prescient stories by Phillipe Dick—evoke robotic worlds of "Replicants" that surpass their human creators in all functional areas—perhaps even spiritual morality. These are worlds wherein the *Singularity* has become real. With great skill, both films portray future man-made simulated cultures inhabited by advanced AI organisms. Jung would categorize sci-fi films as prospective dreams of the future. Both films raise critical questions about the morality of what current science is doing right now. The premise (statement of the dramatic problem) of BR asks the questions: "What is the nature of human responsibility in a sentient universe, and what are the responsibility for creating life—both organic and mechanical?" These are profound questions that nobody is asking with the exception of a few who are stimulated by sci-fi. BR

2049 extends the moral-ontological premise to the issues and alternatives raised by evolution within a sentient cosmos. The context of BR 2049 is transitional from a human-dominated culture to a culture that is characterized by super-human AI-sentience in which robots are evolving to the stage of becoming Souls that are "begotten, not made".

Already, abundant scientific research is contemplating this evolution, professing to have created AI that is smarter than its human creators (Anonymous, 2017; How we build machines that can think for themselves, 2016; National Institutes of Natural Sciences, 2016; Wiltshire, 2016; Anonymous, 2016; Garling & Cully, 2015; Lewis, 2014; McCarthy, Minsky, Rochester, 1955; Geril, n.d.) and that should be considered a new species (Anonymous, 2014). It is also important to remember the research of Reeves and Nass who verified the Media Equation; i.e., media images are considered real. (Reeves & Nass, 1996)

Contemplated as (real) archetypal representations, the Media-sphere-as-dreamscape constantly divulges important information relative to negative incoherent patterns in the collective unconscious of cultural personae. As to their significance, Jung defined three types of dreams (Jacoby, 1973, p. 74):

1. Dreams in which a reaction of the subconscious supplements or complements impressions of recent experience—"impressions of the day". The conscious component is stronger and its energy potential flows toward the unconscious in order to reestablish coherence.
2. Dreams are unprovoked by consciousness, but arise from a spontaneous action of the unconscious. This creates conflict between consciousness and unconsciousness and the energy potential is balanced.
3. "But when the contrary position of the unconscious is the stronger, the gradient runs from the unconscious to consciousness. It is then that we have the significant dreams which can sometimes utterly change and even reverse the orientation of consciousness." (Jacoby, 1973, p. 74-75) These are the strangest dreams.

Because they have no relationship to consciousness, this third type is archetypal, hard to interpret, overpowering, and often oracular. The interpretation of such dreams must proceed slowly and with great caution because, "Beneficial as an archetypal dream may be if its content can be understood correctly and integrated at the proper moment, it can also prove extremely dangerous and may lead to psychosis if the dreamer's ego is too narrow to confront it and come to terms with it." (p. 75) The current Media-sphere includes all three types of Jungian dreams, but as an illustration of the third type, the contextual USA Media-sphere may be understood as a projection of traumatic cultural failure to address problems of the cultural unconscious. These failings are so dire that if they cannot be positively integrated into consciousness, democratic culture will devolve into totalitarian form. Science fiction may be the oracles necessary to understand the content of the Media-dream and to correctly integrate the message into consciousness.

Prospective Dreams

According to Jung, dream contents can exert a reductive or prospective function—both forms may be salutary, but "They may compensate negatively, reducing the individual to his human nullity and to his dependence on physiological, historical, and phylogenetic conditions (this material was brilliantly investigated by Freud), or else positively by providing a kind of 'guiding image' which corrects a self-devaluating attitude and brings about a 'better' direction of consciousness. Also, the dream's prospective function must be distinguished from its compensatory function." (p. 78-79) In other words, compensation is *purposive* and brings to consciousness elements that are repressed or unappreciated. "The prospective function, on the other hand is *an anticipation* in the unconscious of future conscious achievements, something like a preliminary sketch, or a plan roughed out in advance." (Jacoby, p. 79; Jung, General Aspects of Dream Psychology, pp. 255, 258) Therefore sci-fi as dream analogs can serve multiple cognitive purposes.

Jung's methodology for interpreting dreams is fundamentally different from the "reductive" methodology employed by Freud. This is critically important because Jung's amplification method is based on an essentially metaphorical exegesis of symbolic patterns. As the reader will see, such exegesis will be first used to research contextual psyche and next to design PEG. With these design elements in mind it becomes possible to re-design the global Media-sphere with coherent mediated biofeedback to contextual cultural personae.

Carl Jung's primary focus was on the harmonization (compensation) of incoherent patters expressed symbolically in dreams. Both Blade Runner stories can be understood as compensational. They provide compensation for the current cultural pathology, but they are primarily prospective. They provide valuable anticipations of a simulated AI future.

Prospective cognitive patterns locked within "archetypes" of the unconscious can be used to prepare for the future. In 21st century language, Jungian dream analysis is based on the projection of quantum archetypes—as quantum electrodynamic (QED) patterns—into various mediated states—all of which have the same unified energy dynamics. These mediated states include the waking dream (so-called "real life"), memories, fantasies, television, and now the Internet, social media, political advertising, and neuromarketing. All these "dreams" have the same rhetorical (dramatic) structure. The factor that links such diverse fields as physics, dream analysis, and cognition is the common denominator of ***dramatic structure***. Once the hypothesis of dream-ontology is accepted by science, Jung's contribution to understanding dream dynamics becomes central to problem-solving in an age of mediated illusion.

BACKGROUND

Computational Types

Variances in types of computation including analog, digital, and quantum are relevant to the theory of Psychecology video games (PEGs) that use sci-fi game stories. The dynamics of real-life experience are the same as the experience of playing video games, and both are interactive in the same way as dreams. Therefore, PEGs can be designed as dream analogs. An analog computer is a form of computer that uses the continuously changeable aspects of physical phenomena such as electrical, mechanical, or hydraulic quantities to model the problem being solved. In contrast, digital computers represent varying quantities symbolically, as their numerical values change. Both types can be used to model PEGs, but because analog computers use continuous values rather than discreet values, processes cannot be reliably repeated with exact equivalence. However, this characteristic of "continuous values" may provide a more authentic way to model cognitive states than digital modeling. As to quantum modeling, at present too little information exists. Logically, quantum computing should be completely consistent with quantum field theory as applied to cognitive research with PEGs.

Setting up an analog computer[1] required scale factors to be chosen, along with initial conditions; i.e., starting values for-the purpose of creating the required network of interconnections between computing elements. In theoretical PEG design, initial conditions would include dramatic architecture, cognitive linguistics, Jungian dream dynamics (projection) as represented by Myers-Briggs personality types, neurological connectionism, coherence as measured by heart rate variability, and the mathematics of Fourier transforms and quantum field theory. In some instances, after a certain running time interval, the analog computer could repeatedly return to the initial-conditions state to reset the problem and to run it again. Backpropagation is a common method for training a neural network, so perhaps a return to initial conditions could be relevant to cognitive training in PEGs, but only if computer capacity allows for nearly instantaneous recovery of initial-conditions.

Archetypal Representations

According to Jung's definition, archetypal Representations are pattern projections from the unconscious—much like pixel patterns are projected from software. In his study of the quantum nature of Psyche, Carl G. Jung used the term to define a syntax for the language of dream symbolism as communication between conscious and unconscious cognitive dimensions. In the psychic framework, Jung defined quantum patterns that exist in the personal and collective unconscious as *Archetypes* and this term can be applied to current research such as George Lakoff's *Family Metaphors*. Jung's lifelong research on the dynamics of dreams clearly affirms that human perceptions of reality are projections of those unconscious patterns. If Jungian research is not considered valid by some, Jungian principles have been verified by current cognitive research and translated into neurobiological frameworks. (Schafer, 2018a; Schafer, 2018b; Schafer, 2016a; Schafer, 2016b; Schafer, 2016c; Schafer, 2016d; Schafer, 2012e)

Everything unconscious is projected; i.e., it appears as a property or activity of an object. Only by an act of self-knowledge are these contents integrated with the subject, detached from the object and recognized as psychic phenomena (Alchemy, par. 346) The phenomenon of projection is an integral part of the mechanism of the unconscious, and since the unconscious plays a considerable part in every psyche, there can be no psychic life without a certain amount of projection. In dreams or waking, in individuals or groups, in relation to persons, things, or conditions it stands wholly outside the conscious will. "Projection," says Jung, "Is never made; it happens." (Unconscious, par. 150) He defines it as the "expulsion of a subjective content into an object", in contrast to introjection, which "consists in taking an object into the subject." (Types, par. 346)

In mathematical and programming terms, Archetypes of the unconscious are projected into consciousness as recursive *Archetypal Representations*—like mirror images of the unconscious patterns. For example, mathematical pixel programs are projected onto the computer monitor as images to which a game player responds. Jung differentiated archetypes from their projections mostly on-the basis of whether they were conscious or not. However, from the cognitive viewpoint, most anything that manifests as an image—not only dreams, but memories, fantasies, hopes, so-called "real" life, and media—are perceptual illusions that qualify as Archetypal Representations. In terms of researching the modern Media-sphere, this dreamlike sci-fi dynamic of illusion takes on unprecedented importance.

Add to this the major research by Reeves & Nass—published (1996) as *The Media Equation*—the verification that, at unconscious levels, all human demographics perceive media images as "real", we can begin to understand how the medium of science fiction not only derives from the collective unconscious as defined by Jung, but acts as a biofeedback loop to reframe quantum cognitive patterns.

The last twenty years of cognitive research—especially that of George Lakoff—has explored the neuro-dynamics of communication between conscious and unconscious dimensions. The salient finding of this research is that both conscious and unconscious cognitive dimensions are structured as drama (narrative architecture), and these dramas provide the framework for reality perceptions as-well-as self-perceptions. As Lakoff says, "We live our stories":

In the language of the brain—its very building blocks—words and thoughts are defined relative to narrative frames and conceptual metaphors. The very fact that we recognize these cultural narratives and frames means that they are instantiated physically in our brains. We are not born with them, but we start growing them soon, and as we acquire the deep narratives, our synapses change and become fixed. A large number of deep narratives can be activated together. We cannot understand other people without such cultural narratives. But more important, we cannot understand ourselves—who we are, who we have been, and where we want to go—without recognizing and seeing how we fit into cultural narratives." (2008, pp. 33-34)

Extensive research has confirmed the dramatic-quantum-structural fact with a combination of linguistic theory and technological imaging procedures; i.e., MRI images tell stories about the condition of physical organs and EEG images tell us stories about the veracity of a subject's answers to questions. The research of Lakoff and others confirms a basic tenet of Jungian research on dream dynamics. Jung observed that dreams have dramatic structure. Therefore, dramatic structure provides the fundamental framework for researching media-dreams as Archetypal Representations and for exploring their meanings with exegesis of narrative architecture.

Jung found that most dreams show a certain similarity of structure. Unlike Freud, he believes that they form a self-contained whole, a dramatic action which can meaningfully be broken down into the elements of a Greek play. (1) Place, time, dramatis personae: this is the beginning of the dream, which often indicates the scene of action and the cast of characters (2) Exposition or statement of the problem. Here is presented the central content: the unconscious frames the question to which it will reply in the course of the dream; (3) Peripety this is the "backbone" of the dream; the plot is woven, the action moves toward a climax, transformation, or catastrophe; (4) Lysis, the solution, the outcome of the dream, its meaningful conclusion and the disclosure of its compensatory message. This rough pattern, on which most dreams are constructed, forms a suitable basis for interpretation. (Jacoby, 1973, p. 83)

Because the stages of Jungian Compensation that lead to meaningful insight and healing are relatively subjective, literary exegesis provides a relatively subjective framework for development of interpretive algorithms. The stages are as follows:

The Jungian technique of analyzing a dream may be divided into the following stages: description of the present [contextual] situation of consciousness; description of preceding events; investigation of the subjective context and, where archaic motives appear, and comparison with mythological parallels--finally, in complicated situations, comparison with objective data obtained from third persons. Relative to the narrative architectural model of exegesis, these stages might be equated to (1) statement of the story's premise, (2) backstory, (3) investigation of the themes and narrative symbolism of the story, (4) investigation of mythic symbolism, and (5) analysis of how various characters in the story function as mirrors for the protagonist's personality profile.

On the other hand the course taken by the contents of the unconscious consists roughly of the following seven stages: (1) lowering of the threshold of consciousness [suspension of disbelief], permitting the contents of the unconscious to emerge; (2) rising up of the contents of the unconscious [view sequence of the plot including flashback and flash-forward] in dreams, visions, fantasies; (3) perceiving the contents and holding them fast by consciousness [understanding of the relationship between plot sequence and view sequence]; (4) investigation, clarification, interpretation, and understanding of the meaning of the contents [repetition of symbolic meaning in different forms]; (5) integration [gradual synchronization] of this meaning with the general psychic situation of the individual; (6) acquisition, incorporation, and elaboration of the meaning thus found [achievement of relatively coherent cognitive realization based on meaningful insights during gameplay]; (7) integration of the "meaning" [realization of lysis as the answer to the problem established with the premise], its organic incorporation in the psyche becoming so complete that it "enters the blood stream" as it were, becomes a knowledge secured by instinct. (Jacoby, 1973, p. 82)

The measurement of Coherent states during these Jungian analytical stages provides a solid basis for quantifiable access to psychological meaning; i.e. Jungian Compensation and Coherence appear to be synonymous. (Schafer, Stephen Brock, 2016) So what is Coherence?

Physiological coherence, also referred to as heart coherence, cardiac coherence or resonance is a functional mode, measured by heart rate variability (HRV) analysis wherein a person's heart rhythm pattern becomes more ordered and sine-wave like at a frequency of around 0.1 Hz (10 seconds). (Ginsberg, Berry, & Powell, 2010) The term physiological coherence embraces several related phenomena—auto coherence,cross coherence, synchronization, and resonance—all of which are associated with increased order, efficiency, and harmony in the functioning of the body's systems. When one is in a coherent state, it reflects increased synchronization and resonance in higher-level brain systems and in the activity occurring in the two branches of the autonomic nervous system (ANS), as well as a shift in autonomic balance toward increased parasympathetic activity.

Psychologically, coherence reflects increased emotional and perceptual stability and alignment among the physical, cognitive, and emotional systems. Coherence and resilience are closely related as each has the quality of being both a process and an outcome as they rely on physiological and psychological processes that create resilient outcomes. In addition, both are states rather than traits that vary over time as demands, circumstances and level of maturity change. (Malin & Srivastava, 1979)

Because Resilience is related to self-management and efficient utilization of energy resources across four domains; physical, emotional, mental and spiritual, it can be correlated with Jungian Individuation or Maslow's stage of self-actualization. Physical resilience is basically reflected in physical flexibility, endurance and strength, while emotional resilience is reflected in one's ability to self-regulate the degree of emotional flexibility, positive emotions and relationships. Mental resilience is reflected in our attention span, mental flexibility, an optimistic world view and ability to integrate multiple points of view. Spiritual resilience is typically associated with our commitment to core values, intuition, and tolerance of others' values and beliefs. These dimensions of resilience can be programmed in-order to increase coherent states relative to the exposition of a story in Psychecology video games (PEGs) and the premise of protagonists in the drama. (Schafer, 2018), in publication) According to Aristotle, Exposition in good drama is the statement of the story problem, while lysis is the resolution of the problem. Character premise is defined as "what a character learns." (Glassner, 2004; Schafer, 2007) If a game-win depends on resolution of the story problem and is associated with exposition and character premise, the game-win would also be associated with increased coherence in the in the framing of the player's cognitive unconscious.

When we are in a coherent state, the increased physiological efficiency and alignment of the mental and emotional systems accumulates resilience (energy) across all four energetic domains. Having a high level of resilience is important for not only recouping from challenging situations, but for preventing unnecessary stress reactions (frustration, impatience, anxiety) that deplete our physical and psychological resources. As explained above, winning a coherently structured DBG would result in energy efficiencies that have an impact on the global field.

As discussed extensively in other papers (Schafer, 2012; Schafer, 2016b; Schafer, 2016c; McCraty & Schafer, 2016; Schafer, 2011b), Jungian Compensation and Coherence are synonymous. Because Jungian dreams–as-drama and Lakoff's principles of neuro-biological drama agree as to the structural common denominator, it may be extrapolated that media-dreams share the dramatic common denominator and can be analyzed accordingly. Therefore, the meaning of such media-dreams as sci-fi PEGs can be correlated with the measurable factor called Coherence using the dramatic premise as key. (Schafer, 2007) Because a dream-story's dénouement leads to meaningful insight, use of Jungian dream analysis for film exegesis should best be applied to a series of films such as the Terminator series. On the other hand, a single film includes a sequence of scenes that could be understood as a complex series of dreams.

MEDIA AND SCIENCE FICTION AS ARCHETYPAL REPRESENTATION

The essence of Jung's healing dynamic is his clinical verification that when the symbolic language imbedded in these projected images is analyzed and supplemented with biofeedback from a psychiatrist, the patient can experience healing insight as to the ***meaning*** of the dream. The Jungian method can be directly applied to learning, so ***learning = healing*** and both can be mathematically quantified in-order to discover the "contextual unknown" in any experience. The simple equation for this learning/healing dynamic is provided by Aristotle's formula for analogy: ***A/B = X/Y*** where any factor can be designated as unknown and calculated accordingly. Of course, this fundamental formula can amplified with complex algorithms-as-scale-factors and applied to dream analysis in-order to resolve the unknown/unconscious source of either ignorance or neurosis. Moreover, the dynamics of quantum field mathematics (QFT) may be instantiated in the theoretical genre called *Psych*ecology video games (PEG's)

In the process of defining his *Analytical Psychology*, Jung spans the bridge that links psychology, language, and physics. Modern simulation modelling has employed diverse algorithms to quantify the bio-rhetorical (dramatic) common denominator and has applied mathematics to AI, robotics, the Media-sphere, and video game design. When the mathematics of quantum field theory (QFT) are added to the computational mix (Brooks, 2016), possibilities for metaphorical exegesis are expanded with realistic pragmatism. Moreover, when coherent states as researched by the *Institute of Heartmath* are recognized to be synonymous with Jungian compensation, a sound basis for contextually meaningful quantification of unconscious psychological states will be established. Science now has the mathematical Keys to the Kingdom in its grasp. All the ducks are in a row, and the scientific problem-solving challenge is to differentiate between complex subjective moral health and psychosis within the fields of illusion.

This is an ongoing challenge that may persist for a thousand years, and humans need all the help they can get. Throughout this paper, the salient point is reiterated: Measurably coherent states constitute frequencies of cosmic order that may be equated with "spiritual guidance"; i.e., robust research verifies the benign effects of harmony on biological, cognitive-sentient systems. (Alabdulgader et al., 2018; McCraty, Rollin & Childre, 2010; McCraty, Rollin, 2005, 2004, n.d.; Papadopolous & Byng-Hall, 2007; Schafer, 2016d; Schafer, 2012; Tomasino, 2007)

Perhaps the most practical way to understand and apply these keys to the Kingdom is with simulated dramatization and mediated biofeedback using "prospective" sci-fi stories.[2] If the science fiction genre can be dynamically equated to Jungian dreams and simulated in PEGs, the genre and the media-at-large become potent and veridical tools for psychological healing on both individual and collective scales.

Formerly inscrutable Jungian dream-energy dynamics are now susceptible to close-scrutiny with coherent QFT dynamics and the media dynamics (based on Fourier mathematics) that enable the Internet. QFT strongly suggests that the media-sphere can be viewed as a Jungian dreamscape, and the QFT dynamics of Jungian Psyche appear to be significantly correlated with the QFT dynamics of the holographic media-sphere. It is likely that this understanding of media-dream dynamics is significant for understanding the reality of illusion intrinsic to understanding moral behavior within the illusion. We are already witness to the mediated scope of immoral behavior in the form of cyber-warfare, neuromarketing, surveillance, and political hacking. Understanding the media-sphere-as-dreamscape could make all the difference between generating psychotic collective behavior and generating healthy behavior.

It may be stipulated that multidisciplinary modeling and simulation technology is very complex, but at the base of modeling is the Aristotelian formula that links numbers with corresponding language. In his explanation of analogy, Aristotle observes that: "Life" (A) is to "Old Age" (B) as "Day" (X) is to "Evening." (Y). Therefore, A is to B as X is to Y can be written as A/B = X/Y. According to this analogy, if the nature of "Old Age" is the unknown, something about "Old Age" can be discovered by comparing it to the evening of a day and may be verbalized as, "the evening of life" or "Day's old age." Also, this linguistic-metaphorical dynamic can be linked to the dramatic common denominator as syntax and formulated into the sophisticated algorithms of our computational reality.

Beyond Aristotle, the Pythagorean aphorisms (Hall, 2003, pp. 191-221) provide a rich framework for correlating numbers with letters and syntax. Such computational dynamics as applied to simulation, AI, robotics, and medicine have become mainstream, and now science is exploring both inner and outer cosmos with the *common denominator* that can be translated—given some insight into ancient so-called metaphysics—into the languages of color (optics) and sound (harmonics).

All stories—drama & narrative—have been shown to have the same cognitive/dramatic dynamics as Jungian dreams (Schafer, 2016a; Schafer, 2016b; Schafer, 2016c), so careful analysis of the multi-functional sci-fi genre can be accomplished using mathematics based on algorithms for literary exegesis. Algorithms already exist for this purpose. (Deep Neural Networks Learns to Judge Books by their Covers, 2016; Anonymous, 2016 a & b; Wu & Xi, 2016) Any such exegesis must include drawing correspondences among QFT dynamics, cognitive framing dynamics, and linguistic dynamics. In-order to do so, mythic, religious, and esoteric sources must be algorithmically integrated as tools for evolutionary insight. It is no longer acceptable to ignore the scientific knowledge of the past or to consider it primitive or anomalous. Rather, the orphaned mystery schools (Egyptian, Zen Buddhist, Pythagorean, Aristotelian, Alchemical, Freemasonic, Jesuit, etc.) must be seriously re-evaluated.

One of the most comprehensive scientific repositories of these mysteries is the research and practice of Carl. G. Jung. In the language of dream analysis, Jung would have described the sci-fi genre as prospective dreams (prophesy) or "that which will become" (Jacoby, Yolande, 1973), and his research can be directly correlated with neurobiological/cosmic patterns of the emergent holographic, mediated, quantum electrodynamic (QED) reality. Such prospective sci-fi dreams can now be designed and programmed as ***Psych*ecology** video games (PEG).

Doing so will shine significant light on fundamental issues of the human experience and condition. In the future-world portrayed by *Blade Runner 2049*, the core story issue has to do with the Soul. The story asks the questions: What is the Soul? What difference does the Soul make? Do advanced robots have souls? Does a soul have anything to do with having procreative potentials? Do correlations exist among pro-*creativity,* sentience, consciousness, evolution, and the Soul? Is evolution related more to quantum electrodynamics or to neuro-biology and what is the meaning of the relationship? Can the relationship be defined in terms of QF sentience? Is evolutionary meaning purposeful? Does God have a plan? The answers to the biogenetic questions have already been affirmed (Lipton, 2005, pp. 49-74), so the gap between prophesy and science is rapidly diminishing.

These questions all trend toward insight as to the core meaning of the human "dream" which is denominated the *"Galations syndrome"*. (Galations 4:4-5) Based on the theme, "born not made", BR 2049 explores a story premise that addresses the ontological issue of "Divine Grace" vs. "Acts", and the difference between freedom and slavery. So, it may be stipulated that science fiction contemplates deep issues relative to the human condition, and—for instance—the difference between the "real" and the "simulated".

In the simulated dream-world of Blade Runner 2049, cyborgs[3] rule and holographic sentient beings that have convincingly human *functions* (physical, emotional, mental, & "spiritual?") are commonplace technologies. The only way in which humans—who have been relegated to a pseudo administrative authority--excel is that they are "begotten not made". In other words, humans are somehow superior only because they have souls, but the story premise is about an evolutionary *pro-creative* event in which a human and a cyborg create offspring. Establishing this narrative, the film points the viewer in the direction of meaningful discovery beyond current imagining and beyond any existing prospective problem-solving.

Another Sci-fi film, *Equilibrium*, portrays a future culture that—based on the hypothesis that emotion (and, therefore the Soul) is the root of all evil—has demonized and nearly eradicated emotion. The false premise of the plot demonstrates that the hypothesis is idiotic because emotion is also the basis for all of life's joy. This is a paradox that humans should understand, but—due to a cultural over-emphasis on rational materialism—they don't. Moreover, reason cannot be separated from emotion by rational surgery. Any attempt to do so renders so-called "reason" psychotic and diabolical. (Equilibrium)

What is it you want?
It is every man's quest to discover who he really is,
but the answers exist in the present—not the past
The past is a construct of the mind. It blinds us. It fools us, into believing it,
but the heart wants to live in the present. Look for your answers there."(Mathias: Total Recall)

Though its significance may be overlooked, this brief monologue in the remake of the science fiction film, *Total Recall*, constitutes an illustration of how science fiction creates the future; i.e., according to QFT, the future is imbedded in the present. It establishes a story "premise" upon which value structures may be built. The monologue sums up morality in a future that is metaphysical in both the ancient and modern scientific sense. It expounds the fundamental challenge of simulation in a quantum-holographic and counter-intuitive unified field of sentience. Essentially—based on both ancient and modern scientific precedents—in a Monad or unified field, everything occurs in the present—even memories and fantasies are unique to a present moment.

In such a field as projected by physics, morality exists in an ongoing moment that is quintessentially defined in terms of human choices and stored as metaphorical resonances rather than data-banks of the Psyche. (Koriat & Goldsmith, 1996) This does not mean that morality is relative. Just the opposite! It suggests that ***morality is absolute.*** Absolute morality can best be understood in terms of the ***coherent morality of the heart***. In the world's philosophy and religion, this concept is reiterated constantly, but—for modern audiences—the concept requires explanation. Ironically, proof of the metaphysical status of the Heart is coming from modern science. As-yet, few understand the practical existential value of "thinking with the heart". Yet, current research by the Institute of Heartmath and its colleagues world-wide has explored the neurobiological dynamics of the heart extensively and in comprehensive detail that has led to such findings as:

- The heart has its own brain.
- The heart has powerful impact on all functional systems (thinking, feeling, sensing, & intuiting).
- The heart's impact is far more powerful than the impact of the brain
- "Thinking with the heart" may be accomplished simply by directing thought to the bodily heart area. (McCraty & 2010; McCraty, 2005)

This is just a beginning, but when it is collectively realized, the coherent morality of the heart will require the sort of discipline advocated—and mostly ignored—by all the great religions. In-order to function within a coherent state, it will be necessary to apply the kind of genius necessary to transcend the illusion of cause-and-effect. Though—like Newton's Law—cause-and-effect—Karma—is an important aspect for understanding the current human condition. On the far side of karma exists a state of being that is Monadic; A reality that can be described as transcendent coherence where—we are told, "I and the Father are one". We have a long way, yet, to travel, but sci-fi is our guide. This state of being can be characterized as collective transcendence—a sort-of super conductivity—but the process can begin now with the focusing of a critical mass of human thought using science fiction as a language to generate mediated dream simulations.

All the new intelligence about current developments in robotics research and AI assumes that herein lies the future. Everyone seems to believe this prophesy, but—if true—this could be the greatest curse the planet and humanity have ever imagined. It is an understatement to say that there is insufficient research addressing the morality of such a future.

Machines are already very intelligent, and researchers are already gifting robots with advanced sensory skill and emotion. (McCraty & Schafer, 2017; McCraty, n.d.; McCraty, 2005) All robotic functions are being addressed with unprecedented rapidity: "Deep Neural Networks Learn to Judge Books by their Covers" (Wiltshire, 2016); "How we build machines that can think for themselves" (Anonymous, 2017); "The Last Mile of loT: Artificial Intelligence (IT)" (Banafa, 2016); "History of AI beating humans"

(Wiltshire, 2016). However, though robots and the AI with which they are programmed seem magical, they are not nearly as miraculous as the humans that designed them.

It is important to remember that machines have no hearts (Perhaps this is a popular fallacy.), so research that presumes to study the human future should carefully study the nature of the heart—not only as an organ but as a cosmic mediator. This message has already become the premise in a growing field of science-fiction that includes *Blade Runner and BR 2049 Terminator Salvation, Total Recall,* and *I Robot.*

The subject of future morality in unified fields or virtual realities has grand scale, the general parameters of which this author has discussed in other publications (Schafer, 2018; Schafer, 2016a; Schafer, 2016b; Schafer, 2016c; Schafer, 2016d; Schafer, 2012; Schafer & Yu, 2011a; Schafer, 2011b; Schafer & Palamides, 2011c; Schafer, 2008; Schafer, April 2007; Schafer, October 2006). However, because of the vast scope of the subject, this chapter attempts to show how specific examples of science fiction as archetypal representations can be subjected to exegesis for-the purpose of discovering unconscious cultural meaning.

As detailed above, literary exegesis provides a practical simulacrum—an insubstantial form or semblance—of Jungian dream analysis. With-regard to exegesis of films as media-dreams, both Jungian principles and current neurolinguistic research can be applied. The narrative structure of dreams is the common denominator. As explained in the Background section, the definition of Aristotelian dreams can be applied both to Jungian exegesis and to the findings of current neurolinguistic research.

The well-known neuro-linguist, George Lakoff observes that 98% of human function is based in dimensions of the (emotional) cognitive unconscious. He says that conscious thought is reflective, like looking at yourself in a mirror. (Lakoff, 2008, p. 9) This is in complete agreement with Jung's statement, "Nothing influences our conduct ***less*** than do intellectual ideas…for such ideas represent forces that are beyond logical justification and moral sanction." (Jacoby, 1973) Instead,

To a considerable extent we can guide and control the contents of consciousness by our will; the unconscious, however has a continuity and order that is independent of us and impervious to our influence. The archetypes are its centers and fields of force. Thus, the contents that sink into the unconscious are subjected to a new, invisible order that is not accessible to conscious knowledge; their path is often refracted, and frequently their aspect and significance are changed in a manner incomprehensible to us. This absolute inner order of the unconscious is a haven and helper amid the upheavals and accidents of life, provided that we know how to deal with it…The archetypes are also akin to what Plato called the "idea". But Plato's idea is a model if supreme perfection only in the "luminous" sense, whereas Jung's archetype is bipolar, embodying the dark side as well as the light. (Jacoby, 1973, p. 42)

So, contrary to the assumption of human rationality, the principles of Jung and Lakoff are in conformity. Lakoff cites estimates that 98% of human mental function resides in the "cognitive [emotional] unconscious"[4]—that our embodied (neurobiological) minds reside below the level of consciousness and our moral and political thought is patterned as narrative-metaphorical frames that determine how we function in both the biological and social worlds. (Lakoff, 2008, p. 11)

In the language of the brain—its very building blocks—words and thoughts are defined relative to narrative frames and conceptual metaphors. "The very fact that we recognize these cultural narratives and frames means that they are instantiated physically in our brains. We are not born with them, but we start growing them soon, and as we acquire the deep narratives, our synapses change and become fixed. A large number of deep narratives can be activated together. We cannot understand other people without such cultural narratives. But more important, we cannot understand ourselves—who we are, who we have been, and where we want to go—without recognizing and seeing how we fit into cultural narratives." (Lakoff, 2008, pp. 33-34) Some stereotypical narratives are as follows:

- Rags-to-Riches
- Pull-Yourself-up-by Your Bootstraps
- The American Dream
- American Redemption: hero fails, then succeeds
- The-Gold-Digger
- How-to-Marry-a-Millionaire
- The Woman's Lot
- Teenage Drug Abuse
- Death of a Child
- Born Again: The Cycle of Death & Rebirth

Relative to semantic "scripts," Lakoff cites Roger Shank and Robert Abelson (Lakoff, 2008, p. 249) in his paraphrase, "Complex narratives—the kinds we find in everyone's life story, as well as in fairy tales, novels, and dramas—are made up of smaller narratives with very simple structures." (Chapter 18 of Political Mind goes into detail regarding cognitive framing of the nervous system including a significant criticism of Chomsky's "New Enlightenment" approach.)

Lakoff says, "The neural circuitry needed to create frame structures is relatively simple [neural binding circuitry, neural signatures, event structures] and so frames tend to structure a huge amount of our thought. Sounding remarkably like Jung who observed that dreams have dramatic action that can meaningfully be broken down into the elements of a Greek play (Jacoby, 1973, p. 83), Lakoff explains that each frame has roles (like a cast of characters), relations between the roles, and scenarios carried out by those playing the roles." (Lakoff, 2008, p. 22; Goffman, 1974) "There is a protagonist, the person whose point of view is being taken. The events are good and bad things that happen. And there are appropriate emotions that fit certain kinds of events in the scenarios." (Lakoff, 2008, p. 23) Citing Goffman (1974), Dr. Lakoff also observes that words are all defined relative to conceptual frames: "Groups of related words, called 'semantic fields,' are defined with respect to the same frame. Thus words like 'cost,' sell,' 'goods,' price,' buy,' and so on are defined with respect to a single frame," (Lakoff, 2008, p 22; Goffman, 1974), and the roles of Buyer, Seller, Goods, and Money, form a narrative field context for the frame. In *Moral Politics*, Lakoff explains in detail how "much of moral reasoning is metaphorical reasoning"

As to how such metaphors were deduced, Lakoff says:

The metaphors have been arrived at independently by looking at two kinds of evidence: (1) how the language of a non-moral domain (e.g., the financial domain) is used to talk about the moral domain, and (2) how the forms of reason used in non-moral domains (e.g., health, strength) are used to reason about the moral domain." (Lakoff, 2002, p. 63)

An example of how narrative-metaphorical resonance can influence morality is discussed by Lakoff in terms of the core metaphor of moral accounting. According to this fundamental metaphor, a syllogism is established: well-being is good; well-being is moral; wealth contributes to well-being; to be wealthy is moral. According to this *Moral Accounting metaphor*, economic words like *owe, debt,* and *pay* are used in a moral context. (Lakoff, 2008, p. 63) The result is a logical framework in which the metaphorical resonance suggests that wealth is better than poverty (good), so being wealthy is moral (good). Lakoff explains:

A conceptual metaphor is a correspondence between concepts across conceptual domains, allowing forms of reasoning and words from one domain (in this case, the economic domain) to be used in the other (in this case, the moral domain)...Thousands of such metaphors contribute to our everyday modes of thought, (and) play an enormous role in characterizing our worldviews. (Lakoff, 2002, p. 63)

Reminiscent of Jungian amplification, Lakoff describes *radial conceptual categories* as the most common form of metaphorical structure. "The radial categories show how the coherent ideologies in each category fit together and what the relationships among them are." (Lakoff, 2002, p. 14) Radial categories are structured according to variations on a central model. As an example, he uses the category *mother* and explains that the term is characterized by four metaphorical sub-models: The birth model, the genetic model, the nurturance model, and the marriage model. (Lakoff, 2002, p. 8) All of the frames in the radial categories are couched in little metaphorical stories within stories. Presumably, if "mother" is the prototype and sub-model variants of the "mother metaphor" constitute radial categories or variations on the central model, discovering how the categories fit together and how they are related in a contextual field would result in discovery of meaning.

In-order to compensate for psychic dissonance caused by unconscious sources of collective pathology, a culture's media can be analyzed as *media dreams* according to the *Amplification Method* of dream analysis developed by Dr. Carl G. Jung as-well-as current neuro-linguistic research. Because *Psych*ecology video games (PEG) have all the fundamental dynamics and the same dream structure used by Jung to heal patients suffering from mental illness, PEGs become contextual dream analogs for the purpose of researching dream states and using research findings to address collective psychosis. Also, the *Amplification Method*—which is a system of metaphorical extension—is consistent with principles of research employed in neuro-linguistic research on cognitive framing; therefore, data from the most recent research in framing the cognitive unconscious can be correlated with the use of real-time images of neurobiological processes recorded with the use of imaging technology during game play.

Published in 1952, the George Orwell's book, *1984* seemed far into the future. As pointed out by Neil Postman (1985), the premise of Orwell's book is that it teaches us about the spiritual devastations of tyranny whereas the premise of Aldous Huxley's *Brave New World* teaches us that in the age of advanced technology, spiritual devastation comes from a culture's collective unconscious rather that its projected reality.

No one needs to be reminded that our world is now marred by many prison cultures whose structure Orwell described accurately in his parables. If one were to read both 1984 and Animal Farm, and then for good measure, Arthur Kunstler's Darkness at Noon, one would have a fairly precise blueprint of the machinery of thought-control as it currently operates in scores of countries and on millions of people. Of course, Orwell was not the first to teach us about the spiritual devastations of tyranny. (Postman, 1985, p. 155)

Postman goes on to comment on an important distinction as to premise between the future worldview of Orwell and of Aldous Huxley in *Brave New World*:

What Huxley teaches is that in the age of advanced technology, spiritual devastation is more likely to come from an enemy with a smiling face than from one whose countenance exudes suspicion and hate. In the Huxleyan prophesy, Big Brother does not watch us, by his choice. We watch him, by ours. There is no need for wardens or gates or Ministries of Truth. When a population becomes distracted by trivia, when cultural life is redefined as a perpetual round of entertainments, when serious public conversation becomes a form of baby-talk, when, in short, a people becomes an audience and their public business a vaudeville act, then a nation finds itself at risk; culture-death is a clear possibility.

Postman's exegesis of both these science fiction classics illustrates how dramatic-metaphorical structure and premise can be used to glean meaning from film as archetypal representation and prospective dream. One cannot view the current political vaudeville act without deep concern, so Postman (by way of Orwell and Huxley) succeeded in prophesizing the current crisis--the mediated archetypal representation—with a high degree of accuracy and current relevance. Though it is tempting, exploring political advertising is the subject for another time.

One rarely discovers science fiction that is not dark, so a good reason for using Jungian principles of dream analysis for exegesis is that—as mentioned above—their compensational dynamic provides a way to harmonize (neutralize) the fundamental duality perceived as good and evil.

Compensation is the psychological key to healing the Psyche (Soul) with archetypal representations. Projections perceived as evil can be adjusted with QF dynamics in-order to create coherence of frequency between conscious and unconscious dimensions of psyche. It may be stipulated that the fundamental premise of sci-fi has to do with the desolation of Earth, destroyed by humans and machines in one catastrophe or in combined catastrophic assaults such as environmental destruction built up over time. If one does find a positive projection, it is expressed as hope and comes at the very end of the narrative. If they are carefully observed and analyzed, films and stories always verbalize the premise. As quoted above, *lysis* (the answer) to the premise of *Total Recall* is that *the Heart holds the answers to critical problems.*

The premise of the Terminator Series comes at the very end of *Terminator Salvation* when the protagonist machine-organism sacrifices himself by allowing his strong human heart to be transplanted to the dying John Connor who is the key to future survival of humanity.

What Dreams May Come is all about heart, and exploration of the dream-as-afterlife. The climax comes when the protagonist of this sophisticated story becomes finally willing to sacrifice himself—relinquish control, "Sometimes when you lose, you win"—completely in the paradoxical effort to save his wife who, in despair, has committed suicide and is forever trapped in a closed loop of the cognitive framework. "Sometimes when you win, you lose."

Avatar is all about the heartfelt ethics of the coherent alien planet *Pandora* where all life forms are interconnected and threatened by humans from earth who intend to destroy Pandora by mining its *unobtanium*. Unobtanium may be understood as a rare element—unobtainable on earth—that has all the characteristics of the alchemical *Philosopher's Stone or the heart's coherence.* It symbolizes harmony among the natural-symbolic elements of earth, fire, air, and water. (Secret Teaching, p. 129) "A tiny particle of the philosopher's Stone, if cast upon the surface of water, will…immediately begin a process of recapitulating in miniature the history of the universe, for instantly the tincture—like the Spirits of

Elohim—moves upon the face of the waters." (Secret Teachings, p. 508) Building upon this brief indication—this brief allusion to QFT—it seems probable that—with insight as to meaning—Avatar has a vast scope for prospective healing potential.

The classic sci-fi film, Blade Runner, includes a-number of themes that have profound relevance for morality in the future, and the moral (premise) of the story again has to do with development of the coherent Heart. In the film's prologue, we see that cyborg-*Replicants*—a new life form—have achieved excellence superior to their human creators. Motivated by fear, those humans have enslaved Replicants, outlawed them, and limited their ability to develop emotions (Soul) by curtailing their lifespan.

The story's premise is evoked by the desire of the Replicants to have "more life" in-order to develop more heart. Again, it is the universal Hero's Journey with the quirk that humans create and then misuse a superior race of heroes. At the end of the story, in-spite of his short lifespan, Roy, the lead Replicant, has evolved to a sense of heart. He demonstrates Heart by sparing the protagonist Deckard's life (the Blade Runner). The parting speculation as to why he does it sums up the premise: "Maybe he had come to value all life." Such valuation is a Soul characteristic.

Dramatic nuances relating to life and love exist unquestionably in the framed structure of the story. In his parting monologue, Roy mentions Tannhäuser Gate which alludes to the *Bußlied* (Poem on Atonement); Tannhäuser became the subject of legend, first attested in 1430, and propagated in ballads from 1450. The legendary account makes Tannhäuser a knight and poet who found the Venusberg, the subterranean (unconscious) home of Venus, Goddess of Love. After a traumatic series of emotional pain and just before his death (symbolized by the flight of a dove), Roy says, "I've seen things…all those moments will be lost in time like tears in rain." This statement suggests the quantum collapse of QFT. (Brooks, 2016, p. 47)

So, what do people want? The archetypal representations above suggest that at unconscious levels they want to understand the meaning of life--relief from the trauma of chaotic living. Relief is almost universally understood as a spiritual state that exists in the form of a culture of Heart—a coherent culture of harmonized conscience.

DRAMATIC ARCHITECTURE AS COGNITIVE FRAMEWORK FOR SYMBOLIC RESONANCE AND MEANING

When we refer to *drama as a dynamic of the dream paradigm*, "narrative architecture" should be understood as a matrix or map of the cognitive unconscious. Projected as what Jung called archetypal representations, the structure of language mirrors the structure of archetypal subjective reality. Jung studied dream images as projections that divulge symbolic meaning about these subjective states, so the meaning imbedded in dreams mirrors the meaning of unconscious contextual life. In terms of software, such embedded meaning is similar to the imbedded meanings of steganography, but with an important difference. Steganographic meaning is programmed with pixels by humans, while meaning in archetypal representations is programmed by a "contextual individuality"—from dimensions of what we call the quantum unconscious where meaning is "entangled". *The cause of this dynamic is not yet known, but it must "live" in the dimensions of quantum field sentience.*

Theoretically, media projections or Media-dreams have the same purpose for collectives as dreams have for individuals. So, if the entangled source of dreams can heal an individual with meaningful insight, media-dreams can heal collectives according to the same dynamic. When applied to accessing the meaning of dreams, the media-dream becomes an analogue—based on the frequency articulations of Fourier transforms—to Jung's amplification method of dream interpretation. These analogs have "real" existence that is experienced multi-*functonally*--physically as when watching a film from the perspective of being seated on a roller-coaster; or when watching an emotive drama. Discovery of dream meaning according to the method of Jungian amplification is based upon a *narrative economy* that allows for the orderly management (harmonization) of the psychological resonances communicated by symbols (mostly images) in the dream.

What we are calling *psychological resonances* are frequencies computationally linked to qualitative states such as emotion and intuition. The superficial use of psychological resonances has been well established. The use of EEG, polygraphs, and voice stress detection technology is common, and the use of magnetic resonance technologies has revolutionized research during the last decades. Yes, magnetic resonances can be considered psychic in nature, and neurological researchers at the University of California at Berkley, hope soon to reconstruct the visual content of dreams with magnetic resonance devices. (Wired, 2008)

PEGs could tap the subtle resonances of the psychic intuition. (Gillin et al., n.d.; Bradley, 2006; Tomasino, 2007) In-order to accomplish this within the parameters of computational veridicality, models of psychological resonance comparable in complexity to existing models of color spectra could be calculated. In the world of game simulation, algorithms for surface features such as shadows and fluid dynamics are already advanced and available, and evolution in the use of imaging algorithms and eye-brain-image interactions are moving at warp speed. Most important, the nature of intuition is already being verified with Heartmath coherence technology. (Tomasino, 2007; Bradley, 2006)

Development of a narrative economy with which to access psychological resonances would necessarily incorporate such research as cited above. For now, even the *narrative architecture* used for literary exegesis provides a viable framework within which to discuss a psychological approach to spatial mapping of game images. MIT Professor, Henry Jenkins, begins to discuss the *immersive potential* of game images in his article on *Narrative Architecture (2005).* He says, "Environmental storytelling [contextuality] creates the preconditions for an immersive narrative experience in at least one of four ways."

That which preconditions a symbolically loaded narrative experience is *the psychic unconscious*, and it is because of the recursive nature of symbolic patterns that they provide access to unconscious states. In literature, immersion in this semi-conscious state is understood as a suspension of disbelief. Enactive states may easily be understood in terms of this recursive interaction. It is in this sense that I believe Jenkins uses the term e*nacted stories.* "The staging ground where narrative events are held together by broadly defined goals and conflicts." The quintessential narrative staging ground is the *Journey of the Hero*, where the broadly defined conflicts of *the road of trials* (experience) lead to the goal (discovery of the solution, *lysis*, to the hero's premise). The hero *avatar* (simulated protagonist) is a projection of the player, and movement across the map constitutes a sensory-motor relationship (of some kind) between proximal and distal phenomena. More likely, a "co-consciousness" (SENTIENCE) exists. (Dainton, 2003) A unity of *consciousness* prevails between the cognitive state of the player and the cognitive state symbolized by images on the monitor. The degree of player-immersion subsumes synchronic and diachronic aspects between the player and the avatar.

The premise of the hero character—to achieve psychological transcendence upon the completion of the return journey—becomes the premise of the game player. "All the world's a stage," and the game players are choreographers that improvise steps on a stage projected by some archetypal programmer that is, itself, mediated by the player. The game map becomes a "lucid" dream projection—the phenomenal appearance of the story with which a player accesses archetypal meaning according to rhetorical rules of metaphorical and narrative resonance. Such improvisation may *seem* to be open ended, but it operates within the recursive parameters of psyche. Yet, autonomous choice-making by the game-player—if programmed according to authentic Jungian principles—preserves the dynamic from cognitive manipulation.

The practical value of the law of correspondence is that it allows us to read maps. As a recursion of isometric resonance, whether a map takes the form of a seed, DNA, blueprint, road map, maritime map, star map, color solid, astrological chart, or projection of the images in a game story, the map is a practical navigational tool. (Schafer, Stephen, 2006) The utility of the map is based on an assumption that a valid correspondence exists between the *pixilated* resonances on the map and the unconscious patterns of the territory that the resonances afford.

As with DNA, all the parts correspond with aspects of the whole, and the whole exists within the part.[5] As maps, the enactive images of dreams and games constitute contextual-psychological maps of the individual player and the culture within which the persona plays the game. An individual is like the DNA map for a psychological *cultural persona.* Individuals can be understood as seeds, and like the DNA molecule, the signature is imbedded in every cell of the cultural persona. This fact is expressed as the *Law of Correspondence* that is a recursive principle of both Hermetic philosophy and modern science. Any science or philosophy based on a premise of wholeness must necessarily be apprehended with the principle of recursive correspondence. Moving across the map and winning the game can be understood in terms of a psychological atonement, synchronization, or healing.

What Jenkins calls *Evocative Spaces* can be understood in terms of psychic resonance. He observes that *spatial stories* can evoke pre-existing narrative associations because, "Increasingly, we inhabit a world of trans-media story-telling, one which depends less on each individual work being self-sufficient than on each work contributing to a larger narrative economy." This narrative economy is essentially environmental, and epigenetic research is contributing unprecedented information relative to the importance of the sentient environment. (Lipton, 2005, pp. 49-94) Perhaps evocative spaces would qualify as what Andrew Glassner calls *inciting incidents*. (Glassner, 2004, p. 66) Both ideas—evocative spaces and inciting instances—are consistent with a quantum energetic approach to the subject of energetic potentials. The existence of a resonating field of information in *potentia* may be like the inverse of various mathematical functions in which the intensity or the velocity at any point in a field may be calculated—as a scalar function so related to the vector that the vector is its gradient. In a media economy, evocative spaces, like dreams, have energy potential. Learning to deal with this potential is critical in an incipient media age.

It seems clear that n*arrative associations* are *metaphorical associations* that are based on experience in a loosely defined personal-cultural memory. In *the waking dream* (Grasse, 1996), human beings organize their reality, think, and make their choices based on semiotic spontaneously enactive evocative spaces. Jung describes evocative spaces in terms of semiotic resonance with archetypal potential. We tend to forget how psychological, qualitative, and subjective real life actually is.

What Jenkins discusses as *Embedded Narratives* are evocative spaces amplified to the narrative scope of resonant potential. He says, "The game world becomes a kind of information space, a memory palace." One might say that embedded narratives are intrinsic stories that may be communicated in the form of percept and affordance—latent stories that are amplified by semiotic resonances.

The subject of employing symbols to deepen emotion in stories (including film and games) has been addressed by many commentators (Freeman, 2004; Isbister, 2006), relative to interactive storytelling, and the fact that symbols alter and influence people at psychic levels has been well established (Dretzin, Goodman, Rushkoff, 2004). Using thirty-five experiments based on psychological/sociological research,

Cambridge researchers have proven that, at subconscious levels, people perceive mediated images as real (Reeves & Nass, 1996).

Emergent narratives, according to Jenkins are "not pre-structured or pre-programmed." Presumably, emergent narratives are of the type often referred to as open-ended. The language used by Jenkins to explain emergent narrative sounds entirely consistent with the associative processes that characterize a psychological reality or a dream. In a game, such associations are programmed at computational levels. However, the associative processes are open ended in the sense that particular arrangements are probabilistic.

As authors, individuals construct their personal mind's eye maps, project them like dreams and become the protagonists in their own stories. As Jenkins says, "They test and reformulate their mental maps of the narrative action and the story space. In games, players are forced to act upon those mental maps, to literally test them against the game world itself." Jenkins cites two types of emergent narrative, "the structured set of all causal events" and "the viewer's mental construction of the chronology of those events." This sounds a lot like Andrew Glassner's plot sequence and view sequence.

In fact, knowledge of narrative/symbolic mediation between conscious and subconscious states is employed in a multitude of venues. A plethora of modern social scientific disciplines employ narrative to explore the subconscious states of patients. Therapy often entails writing biographies. Abused children disclose subliminal information by telling stories projected on dolls. Rorschach tests employ ink-blots to elicit stories that cast light as to meaning residing in subliminal states. Suspension of disbelief and emersion in drama and story are known to trigger catharsis. Psychological research is replete with the use of narrative (Papadopolous, & Byng-Hall, 2007); and, now, narrative is being used in law enforcement (Needham & Amado, 2007), and forensics to access memory. Storytelling is a seminal pattern that generates multi-functional (thinking, sensing, feeling, and intuiting) referents. Mythic narrative has always been understood as an art form that resonates within all dimensions of human being.

As noted above, the narrative structures of dreams and drama are informed by what Aristotle called the *dramatic unities* of time, place, and action. Perhaps Aristotle was saying, simply, that good drama is realistic in a psychological way. Almost certainly, he understood the correspondences between a physical universe and a psychological universe. Such understanding would be useful in the design of *Psych*ecology Games as research instruments for the exploration of Psyche and the media age.

DRAMATIC ARCHITECTURE: LINGUISTIC TOOLS FOR EXEGESIS

If anything is "real," it is human dependence on linguistic structure. The idea that language anchors thought is ancient, and we would be well-advised to study the relationships between Chomsky's Generative language and Jung's language of dreams. Jung developed his system of dream analysis based on the empirical observation that dreams (a series of dreams) have the structure of another rhetorical mode—good Greek Drama. (Jacoby, 1973, p. 83) Metaphor and the sophisticated rhetorical mode we call drama provide the linguistic structure for cognitive reality and the necessary frame of reference for discovery of conscious meaning within the functional parameters of Jungian psyche. In order to understand how *Psych*ecology Games will be programmed in order to evoke correlates across many scales, a more detailed explanation of dramatic-metaphorical structure is necessary. Aristotle defined good Greek drama in the following terms:

1. Dramatic Unities (place, time, & action)
2. Cast of characters or *dramatis personae*
3. Exposition or statement of the dramatic problem. Here is presented the central content: the unconscious frames the question to which it will reply in the course of the dream.
4. *Peripety*: this is the 'backbone' of the dream; the plot is woven, the action moves toward a climax, transformation or catastrophe.
5. *Lysis*, the solution, the outcome of the dream, its meaningful conclusion and the disclosure of its compensatory message.

In addition to these Aristotelian characteristics of drama, are linguistic elements that are important in their applications to an interactive media dream—especially as dreams may be simulated with PEG. These include plot sequence, view sequence, the nested masks of personality, and character premise—all of which are discussed in Andrew Glassner's *Interactive Storytelling* and have been further developed in *Premise, The Key to Interactive Storytelling* (Schafer, 2008)

1. **Nested Masks:** The nested masks of personality correspond to Jungian functions and the fourfold mythic dimensions of the human being. The following definitions incorporate language that corresponds to Glassner's definitions (2004, p. 43-45):
    a. World mask: Archetypal representation projected into the world and portrayed as the image on the monitor
    b. Conscience: Structures of the cognitive unconscious—like Lakoff's family metaphors—that transform the absolute archetypes of the *True Self* into the form of values.
    c. Self-image: "A character's notion of himself shapes the conscience's values into desires."
    d. True Self: The Soul, unified self
2. **Premise:** What a character (player-avatar) learns, so it is directly connected to purpose and to meaningful insight. "The value of the premise is that it provides a test for the inclusion or exclusion of every element in the story. If some action, dialog, scene, chapter, or other element contributes to the premise, then it is a contribution to the story and should be retained." (Schafer, 2008, p. 69-71) Meaningful insight on the part of a player-dreamer is directly related to Jungian compensation and the achievement of psychic coherence. In literary terms, dénouement results in meaningful insight.
3. **Plot Sequence**: Beginning, middle, and end—the causal chain of events that make up the plot of a story. Plot sequence corresponds to archetypal pattern, and it is best expressed in the archetypal representation we know as *The Hero's Journey*. Often, plot sequence remains unconscious until meaningful insight is achieved relative to a player-dreamer's premise. In terms of David Boehm's holographic physics, plot sequence is the *implicate* order with which he explains the phenomenon of wave theory in physics. (Talbot, p.4648)
4. **View Sequence:** Beginning, middle, and end; but not necessarily in that order. This includes such literary devices as flashback and flash-forward which are useful in the simulation of psychological dynamics such as memory, intuition, and foresight. This is the perspective of the viewer-player-dreamer which allows for increased player interactivity. In literary terms, view sequence is like stream-of-consciousness. In the holographic physics of Boehm, it is something like the *explicate* order with which he explains the phenomenon of particle physics. In Jungian theory, it is the archetypal representation.

Within a unified framework of time, space (place), and action, the energy-information dynamics of dramatic architecture provide a multidimensional matrix with which to track and correlate neuro-linguistic patterns that behave according to fundamental QF dynamics. Discreet elements of such a matrix are subject to contextual choices that are infinitely variable, but the evolutionary axis upon which interactivity and cognitive feedback are woven is consciousness and unconsciousness. Within the framework of this psychic loom, degrees and nuances of the conscious-unconscious can be interrogated according to the two pillars of Jung's theoretical work: 1) the principle of psychic totality, and 2) the principle of psychic energism. (Jacoby, 1973, p. 2)

Wholeness of the personality is achieved when the main pairs of opposites are relatively differentiated, that is, when both parts of the total psyche, consciousness and the unconscious, are linked together in a living field-relationship. But the dynamic gradient, the flow of psychic life in this context, is not subject to manipulation or even suggestion, for the unconscious can never be made wholly conscious and always has the greater store of energy. The wholeness is always relative and gives us something to work on as long as we live. (Jacoby, 1973, p. 26) In other words, choice-making in PEGs is authentic.

As mentioned above, Jung's understanding that dreams have dramatic structure is consistent with the most recent research in neurobiology, molecular biology, and neurolinguistics. All confirm that unconscious dimensions of human cognition have a structure that is understandable as the language of narrative and metaphor. In the language of the brain— the very building blocks—words and thoughts are defined relative to narrative frames and conceptual metaphors. George Lakoff echo's Yung:

The very fact that we recognize these cultural narratives and frames means that they are instantiated physically in our brains. We are not born with them, but we start growing them soon, and as we acquire the deep narratives, our synapses change and become fixed. A large number of deep narratives can be activated together. We cannot understand other people without such cultural narratives. But more important, we cannot understand ourselves— who we are, who we have been, and where we want to go—without recognizing and seeing how we fit into cultural narratives. (Lakoff, 2008, p. 33-34)

Because dramatic-metaphorical framing is a fundamental dynamic shared between dreams and *Psyche*cology Games, PEG become analogous to dreams as to their potential to foster coherent states: i.e. meaningful insight, learning, healing, and individuation. According to the analog, a matrix of computational relationships can be developed and refined through experimentation. For example:

- *Psych*ecology Games = PEG programmed according to Jungian psychic parameters
- Archetypal psyche = game software (plot sequence)
- Archetypal representation = image on game monitor (view sequence)
- The dreamer = the gamer who is interactively immersed via the avatar in the kaleidoscopic play of symbols on the screen
- Dream image = Monitor image
- Media dream = visual-spatial-linguistic interfaces that highlight contextual incoherence
- Avatar = simulated gamer personified
- Gamer = Soul (paraclete)
- Contextual persona = gamer personality profile (Myers-Briggs)
- Contextual character premise = articulated according to Lakoffian metaphors and semantic frames
- Incoherence = dissonance between conscious and unconscious psychic states

- Compensation = Amplification (feedback) process for achieving degrees of coherence
- Interactivity = conscious-unconscious player response to recursive computationally projected symbol patterns
- Gamer choices = confluence of conscious-unconscious dynamics (percepts + affordances)
- *Peripety* (experiential resolution of dramatic problem) = gamer realization of premise (climax)
- Game win = realization of premise & achievement of coherent psychic state
- Healing insight on the part of the dreamer = winning insight (as to premise or meaning) on the part of the gamer
- Coherence = degrees of synchronicity across conscious-unconscious scales

In terms of Global Workspace Theory and its correlates in other systems, the analog could be developed along the following lines:

- The computational dimensions (software & game machine) of the game = the cognitive unconscious, Jung's personal & collective unconscious
- Audience (passive unconscious) and backstage (directive unconscious) in the Theatre Metaphor = Jungian collective unconscious, Jungian ego, Jungian introversion-extroversion, Myers-Briggs Feeling/Judging
- Backstage = cast of archetypal characters, Jungian pre-conscious, Jungian ego, Jungian introversion-extroversion, Glassner's Conscience Mask
- Audience = Jungian subjective world, Glassner's World Mask
- The symbolic projections on the lighted stage of the game-screen = consciousness, Jungian persona

Premise of Heart Coherence in Selected Sci-Fi

As asserted before, a fundamental premise in sci-fi is the importance of the human heart, which is true of nearly every mythic context as-well-as current research dealing with coherent states. As dream analogs, *Psych*ecology Games could be employed as research instruments for unprecedented exploration of contextual embodied (neurobiological) criteria related to normative-conceptual (value formation) levels of cognition at both personal and cultural levels. Our hypothesis is based on an assumption that parallel dynamics of cognition are operative in the conscious-unconscious interpretation of projected images in dreams, PEG, and the media-at-large. Such projected images take the form of geospatial architecture that is symbolically meaningful and computationally recursive with cognitive structures at the archetypal source. Such research would provide robust data—especially relevant to coherence—with which to frame sustainable educational and media policy in an incipient age of meaningful media.

Multi-Dimensionality in Sci-Fi Films Can Be Confusing

The sci-fi film genre includes a plethora of films like, *The Day the World Stood Still, War of the Worlds, The Time Machine,* and *The Invisible Man* to name a very few. Though forward-looking, these films must be considered from a pre-modern perspective on science, so they were more fiction than science. Additionally, a great deal of low-budget trash like *Killer tomatoes, The Blob,* and the plethora of monster movies express neither heart nor mind. However, even the trash-films are relevant to our subject because the proto-patterns for such trash also exist in the collective unconscious as reflections of trashy human patterns of sentience. Some of these trash films have become cults, and these cults can provide deeply significant analysis of contextual collectives. It matters not that cult advocates may be immersed in the sheer strangeness exhibited by cult sci-fi, their popularity signifies an evolutionary urge—no matter how unrefined the urge may be. As archetypal representations, they provide invaluable information relative to the problems being expressed by the planetary crisis and what constitutes "entertainment" to many.

Compensation in Reality-Based Films

Moreover, a plethora of films based on real-life people and events bring so-called facts into the realm of Media-dreams where Jungian healing dynamics can be observed in action and correlated directly to authentic fact. These films—real-life dreams--fall into several general and overlapping categories:

- Political conspiracy against populist values and superconductive morality

All The King's Men, All The President's Men, Grapes of Wrath, Treasure of the Sierra Madres, The Bourne series, Green Zone, Body of Lies, The Killing Fields, Platoon, JFK, Search for Red October, Clear & Present Danger, Matewan, Norma Rae, Invictus, Cry Freedom, The Siege, Twelve O'Clock High, Enemy of the State, Breaker Morant, Black Rain, American Gangster, Thirteen Days, et al.

- Whistleblowers

The Insider, Erin Brockavich, et al.

- The Journey of the Hero

Ghandi, Invictus, Cinderella Man, Hoosiers, Deliver us from evil, American Gangster, Seabiscuit, Cinderella Man, For Love of the Game, Green Zone, et al.

Real Life Drama as Archetypal Representation: The Insider

An excellent indicator of the validity of the above-detailed interpretive model using archetypal representations is that the same results can be attained with exegesis of media based on real-life. A prime example of the psychological compatibility of real-life experience and dramatic architecture is *The Insider* which is the true story of a whistleblower responsible for disclosing the illegal practices of the Tobacco Industry. The dramatic quality of Jeffry Wigand's story is unsurpassed and includes all of the narrative architecture of the Hero's Journey. It is a complex study because the story of the journalist who published the story, Bergman, constitutes a parallel Hero's Journey. Combined, the two stories mirror and reinforce one-another and the quandaries[6] of each hero provide a narrative counterpoint to their psychological dynamics.

Exegesis of The Insider

Reiteration of Nested Masks

1. True Self: (soul)--Spirit, whose symbol is fire
2. Conscience: (transforms impulses from true self to values)--Mind, whose symbol is air
3. Self-image: (A character's notion of self as an individual shapes the conscience's values into desires.)--Emotion, whose symbol is water
4. World Mask (Based on the above inner impulses, the result is action.)--Material existence, whose symbol is flesh or earth

Jeffry Wigand's Premise: Keeping One's Word: *"Truth & integrity are arbitrated by the standards of the True Self not by lesser standards."*

Levels of Integrity: Wigand translates the *value* of integrity into "keeping his word."

1. Keeping one's word at the level of the Self Image (as to family responsibility)
2. Keeping one's word at the level of Conscience (as to employer's non-disclosure contract that essentially requires him to lie—"Am I a company man or a scientist?")
3. Keeping one's word by taking action—Breaks silence, tells the truth, maintains integrity at the level of true self integrating his World Mask with his True Self

In-order to realize his premise, Wigand must relive the Hero Journey that is intrinsic to all human experience and all good stories. The hero's conflict is created by dissonance between the true self (soul) and its outer expression (world mask): Wigand must synchronize his true self with his world mask. He must sacrifice his more limited personal sense of honesty (family & job) to a higher more impersonal sense of honesty (citizenship or service to the common good). The process involves an expansion of spiritual identity: Spiritual integrity vs. more personal values.

In Wigand's case, to maintain his true self integrity, he must seem to go against his conscience—his word (non-disclosure) to his employer. Wigand already demonstrates true self integrity; his value (conscience) is honesty; his desire (self-image) is to keep his word as a father & a professional; his action (world mask) is to put his action where his true self is—to make the deposition and become a whistle blower. This choice causes him to sacrifice his more limited self (his word to employer) and his family (with which he is totally identified). From a spiritual perspective, this is necessary and logical because the job and family must exist within the culture. If he had not sacrificed for the general wellbeing, he and his family would eventually suffer because he committed a sin of omission.

1. *True self* = spirit = integrity = transcendence
2. *Conscience* = values = honesty = mind
3. *Self-image* = desire = emotion = keeping his word as a father and as a professional
4. *World mask* = action = marriage & non-disclosure agreement vs. deposition = material life

Wigand's Quandaries unfold as the plot sequence of the story.

1. **Conscience (Mind) vs. Self-Image (Desire):** To maintain his professional honesty, he must test his own values against his word to his employer. Going with his value seems to cause a breaking of his word to his family—his role as father and provider. The choice is to be dishonest in his job in order to be honest in his role as provider. First he chooses in favor of family, but finding this impossible, he chooses in favor of his values (honesty arising in the true self). This is a good example of the fact that mind and emotion can't really be separated; both are invested in his decision, but the primary battle in the instance of his job is rational. The primary battle in the instance of his family is emotional.
2. **Self-Image vs. True Self:** In order to maintain his sense of integrity to family, he must seem to go against his larger responsibility as a citizen. He chooses in favor of his true self, the honest citizen within the larger family of humanity that must be regulated by measure of social integrity.
3. **True Self (Soul) vs. Conscience (Mind):** To maintain the integrity of his values, he needs to break his word to his employer (non-disclosure). The quandary has to do with the context of honesty. To whom should he be honest, his employer, his family, or the public? His final test is to choose to be honest with society—the True Self.

Bergman's Premise

Keeping One's Word: *"Truth & integrity are arbitrated by the standards of the True Self not by lesser standards."*

Levels of Integrity: Bergman translates the *value* of integrity into "keeping his word."

1. Keeping one's word at the level of the Self Image (as to his sources)
2. Keeping one's word at the level of Conscience (as to employer's keeping information from the public that essentially requires him to lie—"Am I a company man or a journalist?")
3. Keeping one's word by taking action—Breaks silence, tells the truth, maintains integrity at the level of true self integrating his World Mask with his True Self

With slight variation, Bergman has the same hero premise as Wigand. *In Bergman's case, to maintain his true-self integrity, he must seem to go against his conscience—his word to his source (Wigand). Bergman must distinguish between his real profession and his employer. Like Wigand, his value is honesty, his desire (self-image) is to tell the truth both professionally and personally, and his action (world mask) is to become a whistle blower.*

Like Wigand, Bergman must ultimately choose in favor of transcending lesser masks in order to express the unified integrity of the true self with worldly action. He must express the soul through integrity of mind, emotion, and action. He must achieve at-one-ment by listening to the mythic call of his true self and demonstrating his integrity by a process of being tested through three basic initiations: Choosing the true self over the mind, over personal desire, and over the personal individuated self. By so doing, he sacrifices (the limitations) of his personal self in order to express the unlimited integrated true self. The closest that humans can approximate this transcendent state while in the physical body is by way of personal integrity—an integrity that is tested sequentially in the process of the psychological journey called the Hero Journey by Joseph Campbell.

Bergman's Quandaries

1. **Conscience (Mind) vs. Self-Image (Desire):** His first quandary requires that he do something he doesn't want to do. His rational conscience (sense of values) knows he should blow the whistle on his employer (CBS) rather than leaving Wigand out on a limb, but his personal desire is not to do it. Whistle blowing is too much like not protecting his sources—something that goes against his every personal desire and sense of honesty. In his story, conscience won the day.
2. **Self-Image vs. True Self:** The second quandary seems to go against all his virtues as a personality (his value of honesty, his desire to tell the truth) and sacrifice their lesser manifestations (protecting the lies of his employer) to the higher good represented by telling the real truth to the public. These are subtle choices that lead to understanding his true self.
3. **True Self (Soul) vs. Conscience (Mind):** Finally, the investigative journalism that went into the tobacco scandal was a product of the true self. If Bergman had resigned from CBS, he would have done so as a matter of conscience or personal mind. Instead, he chose in favor of the true self as expressed by his unwillingness to return to work with CB—instead he works for Frontline.

CONCLUSION

Summary of Jungian Dream Dynamics

A brief summary of the dynamics of dream analysis is necessary in order to analyze film-as-dream.

- Dreams have narrative structure—the Hero's Journey or what Jung called *Individuation* (human evolution = The Hero's Journey)
- Symbolic images are the key to unconscious meaning and the potential insight for healing
- Dreams exist in four interactive functional dimensions: Sensing (Physical-earth), Feeling (emotional-water), Thinking (reasoning-air), Intuiting (spirit-fire)
- Dreams have a purpose—insight for healing

The contemplating of film and sci-fi as archetypal representations provides a robust context not only for understanding the unconscious cognitive workings of Collective Personae, but for bringing them into a state of coherence using mediated biofeedback. Such biofeedback would take the form of dramatic content in all formats of the media including their mathematical and quantum-molecular scales. Developing dramatic structure with knowledge of its psychological impact would begin to transform cultural morality in the same way that advertising has changed U.S. (and the world's) cultural values from needs-based to wants-based consumerism. (Curtis, 2009)

This is not a form of censorship. Rather—if done correctly—biofeedback would provoke self-induced coherence. Simply referring to the Jungian dynamic of Archetypal Representation would provide the Collective Persona the tools of exegesis with which to choose its own future with distributive-contextual choices about media content.

REFERENCES

Alabdulgader, A., McCraty, R., Atkinson, M., Dobyns, Y., Vainoras, A., Ragulskis, M., & Stolc, V. (2018). Long-Term study of heart rate variability responses to changes in the solar and geomagnetic environment. *Scientific Reports.* Retrieved from https://www.heartmath.org/assets/uploads/2018/03/long-term-study-of-hrv-responses-to-changes-in-solar-geomagnetic-enviroment.pdf

Anonymous. (2014). The curious evolution of artificial life. *MIT Technology Review.* Retrieved 7/23/2017 from https://www.linkedin.com/grp/post/1039687-5900533809526378496

Anonymous. (2016a). Google Unveils Neural Network with "Superhuman" Ability to Determine the Location of Almost Any Image. *MIT Technology Research.* Retrieved 7/23/2017 from https://www.technologyreview.com/s/600889/google-unveils-neural-network-with-superhuman-ability-to-determine-the-location-of-almost/

Anonymous. (2016b). Neural networks learn to identify criminals by their faces. *MIT Technology Review.* Retrieved 7/23/2017 from https://www.technologyreview.com/s/602955/neural-network-learns-to-identify-criminals-by-their-faces/?utm_campaign=newsletters&utm_source=newsletter-weekly-computing&utm_medium=email&utm_content=20161

Anonymous. (2017). How we build machines that can think for themselves. *BBC.* Retrieved from http://www.bbc.com/future/story/20161114-how-we-built-machines-that-can-think-for-themselves

Anonymous. (2018). The machine vision challenge to better analyze satellite images of Earth. *MIT Technology Review.* Retrieved from https://www.technologyreview.com/s/611198/the-machine-vision-challenge-to-better-analyze-satellite-images-of-earth/

Attanasio, A. A. (2016). *Science Fiction.* Retrieved from http://dreadfuljoy.blogspot.com/2016/12/welive-in-world-not-of-science-but-of.html

Banafa, A. (2016). The Last Mile of loT. *Open Mind.* Retrieved 11/15/2016 from: https://www.bbvaopenmind.com/en/the-last-mile-of-iot-artificial-intelligence-ai/#.WCMHOfLODIA.linkedin

Bradley, R. T. (2006). The psychophysiology of entrepreneurial intuition: A quantum-holographic theory. *Proceedings of the third AGSE International Entrepreneurship Research Exchange.*

Brock, S. S. (2018). Gaming Cultural Atonement. Hershey, PA: IGI Global.

Brooks, R. A. (2016). Fields of color: The theory that escaped Einstein. Silver Springs, MD: Universal Printing, LLC.

Curtis, A. (2009). *Century of the self: Happiness machines for free.* British Broadcasting Corporation. Retrieved from http://freedocumentaries.org/film.php?id=140

Dainton, B. (2003). *Stream of Consciousness: Unity and continuity in Conscious Experience.* London: Routledge, International Library of Philosophy. Retrieved from http://psyche.cs.monash.edu.au/symposia/dainton/prec.html

Deep Neural Networks Learns to Judge Books by their Covers. (2016). *MIT Review.* Retrieved 11/15/2016 from https://www.technologyreview.com/s/602807/deep-neural-network-learns-to-judge-books-by-their-covers/?utm_campaign=newsletters&utm_source=newsletter-weekly-computing&utm_medium=email&utm_content=201611

Dretzin, R., Goodman, B., & Rushkoff, D. (Producers). (2004). *The Persuaders* [Video file]. Retrieved from https://www.pbs.org/wgbh/frontline/film/showspersuaders/

Equilibrium. (n.d.). Burbank, CA: Buena Vista Home Entertainment, Inc.

Fillmore, C. (1975). An alternative to checklist theories of meaning. In *Proceedings of the First Annual Meeting of the Berkeley Linguistic Society*. Berkeley, CA: Berkeley Linguistics Society.

Freeman, D. (2004). *Creating emotion in games*. Boston: New Riders Publishing. doi:10.1145/1027154.1027179

Garling, C., & Cully, A. (2015). Robots walk off Injuries: Robots can now keep moving after an injury without human intervention. *MIT Technology Review.* Retrieved from http://www.evolvingai.org/files/mit_tech_review_robot_adapts.pdf

Geril, P. (n.d.). *Smart dust is getting smarter.* EUROSIS—The European Multidisciplinary Society for Modeling and Simulation Theory. Retrieved 7/11/2017 from https://www.linkedin.com/groups/1039687/1039687-5922629195124473858

Gillin, L.M., Lapira, F., McCraty, R., Bradlley, R., Atkinson, M., Simpson, D., & Scicluna, P. (n.d.). *Before cognition: The active contribution of the heart/ANS to intuitive decision making as measured on repeat entrepreneurs in the Cambridge Technopol.* Australian Graduate School of Entrepreneurship.

Ginsberg, J. P., Berry, M. E., & Powell, D. A. (2010). Cardiac coherence and PTSD in combat veterans. *Alternative Therapies in Health and Medicine*, *16*(4), 52–60. PMID:20653296

Glassner, A. (2004). *Interactive storytelling*. Natick, MA: A.K. Peters.

Goffman, E. (1974). *Frame analysis; an essay on the organization of experience*. New York: Harper & Row.

Grasse, R. (1996). *The waking dream*. Wheaton, IL: The Theosophical Publishing House.

Hall, M. P. (2003). *The Secret Teachings of All Ages*. New York: Penguin.

How we build machines that can think for themselves. (2016). *BBC.* Retrieved 11/15/2016 from http://www.bbc.com/future/story/20161114-how-we-built-machines-that-can-think-for-themselves?ocid=ww.social.link.linkedin

Huxley, A. (2016). The perennial philosophy quotes. *Goodreads.* Retrieved 10/30/2016 from https://www.goodreads.com/work/quotes/889435-the-perennial-philosophy

Isbister, K. (2006). *Better game characters by design, a psychological approach*. San Francisco: Morgan Kaufmann Publishers—Elsevier.

Jacobi, J. (1973). *The psychology of CG Jung*. London: Yale University Press.

Jenkins, H. (2005). *Game design as narrative architecture*. Retrieved 11/7/2016 from http://www.electronicbookreview.com/thread/firstperson/lazzi-fair

Jung, C. G. (1933). *Modern man in search of a soul*. Harcourt Brace & Company.

Koriat, A., & Goldsmith, M. (1996). Memory metaphors and the real-life/ laboratory controversy: Correspondence versus storehouse conceptions of memory. USA. *Behavioral and Brain Sciences*, *19*(02), 167–228. doi:10.1017/S0140525X00042114

Kosov, A. A.,, Broderick, B. Y., Yorkin, C. S. (Producers) & Villeneuve, D. (Director). (2017). *Blade Runner 2049* [DVD]. USA: Alcon Media Group.

Lakoff, G. (2002). *Moral politics*. Chicago: The University of Chicago Press. doi:10.7208/chicago/9780226471006.001.0001

Lakoff, G. (2008). *The political mind*. Viking Penguin.

Lawrence, M., Davis, J., Dow, T., Godfry, W. (Producers) & Proyas, A. (Director). (2004). *I Robot* [DVD]. USA: Twentieth Century Fox.

Lewis, T. (2014). This computer chip can think like a human brain. *Live Science*. Retrieved from https://www.livescience.com/47240-ibm-computer-chip-simulates-brain.html

Lipton, B. (2005). The biology of belief. Mountain of Love/Elite Books.

Malin, S. R. C., & Strivatsava, B. J. (1979). Correlation between heart attacks and magnetic activity. *Nature*, *277*(5698), 646–648. doi:10.1038/277646a0 PMID:423962

Mann, M., & Brugge, P. J. (Producers) & Mann, M. (Director). (1999). *The Insider* [DVD]. USA: Touchstone Pictures.

McCarthy, J., Minsky, M. L., & Rochester, N. (1955). *Dartmouth summer research project on artificial intelligence*. Retrieved from BBC Future: http://www.bbc.com/future/story/20161114-how-we-built-machines-that-can-think-for-themselves?ocid=ww.social.link.linkedin

McCraty, R. (2004). The Energetic Heart: Bioelectromagnetic Communication Within and Between People. *Heartmath Institute*. Retrieved from https://www.heartmath.org/research/research-library/energetics/energetic-heart-bioelectromagnetic-communication-within-and-between-people/

McCraty, R. (2005). *Enhancing emotional, social, and academic learning with heart rhythm coherence feedback*. Boulder Creek, CA: Heartmath Research Center, Institute of Heartmath.

McCraty, R. (n.d.). *Coherence: Bridging personal, social, and global health*. Institute of Heartmath.

McCraty, R., & Childre, D. (2010). Coherence: bridging personal, social, and global health. Alternative Therapies, 16(4).

McCraty, R., & Schafer, S. (2016). *Exploring dimensions of the media dream: Functional context in collective personae*. Hershey, PA: IGI Global. doi:10.4018/978-1-4666-9891-8.ch001

Moritz, N. H., & Jaffe, T. (Producers) & Wiseman, L. (Director). (2012). *Total Recall* [DVD]. USA: Columbia Pictures Industries, Inc.

Moyers, B. (1985). *The Secret Government: Constitution in Crisis*. PBS.

National Institutes of Natural Sciences. (2016). World's fastest quantum simulator. *Institute for Molecular Science, National Institutes of Natural Sciences, Japan, Nature Communications*. Retrieved from PHYSORG: http://phys.org/news/2016-11-world-fastest-quantum-simulator-atomic.html

Needham, W. P., & Amado, C. A. (2007). Division of Natural Sciences. State University of New York at Purchase.

Papadopolous, R., & Byng-Hall, J. (Eds.). (2007). Narrative Based Medicine: A narrative approach to medicine. Academic Press.

Postman, N. (1985). *Amusing ourselves to death*. New York: Penguin Books.

Reeves, B., & Nass, C. (1996). The media equation. Stanford, CA: Center For the study of Language and Information & Cambridge University Press.

Robinson, D. G. (2007). Plant neurology: No brain, no gain? *Elsevier: Science Direct*. Retrieved from http://www.linv.org/images/about_pdf/Trends%202007%20Alpi.pdf

Schafer, S. (2008). The Psychecology Game I-V; Using Story-based Video Games to Capture Veridical Data on Subliminal States of Being. In *Toward a Science of Consciousness 2008*. Consciousness Center & the University of Arizona. Retrieved from https://bandura.sbs.arizona.edu/login/consciousness/pubreport.aspx?aid=2761

Schafer, S. (2011b). Articulating the paradigm shift: Serious games for psychological healing of the collective persona. In *Business Social Networking: Organizational, Managerial, and Technological Dimensions*. IGI Global. Retrieved from http://www.igi-global.com/book/handbook-research-serious-games-educational/5

Schafer, S. (2007). Premise: The key to interactive storytelling. *Gamasutra*. Retrieved from http://www.gamecareerguide.com/features/357/premise_the_key_to_interactive_.php?page=1

Schafer, S. (2006). Curriculum and the dream paradigm. *Gamasutra*. Retrieved from http://www.gmecareerguide.com/features/288/curriculum_and_the_dream_paradigm.php

Schafer, S., & Palamides, T. (2011c). Designing an Ethical structure for Social Influence Marketing. In *Business Social Networking: Organizational, Managerial, and Technological Dimensions*. IGI Global. www.igi-global.com

Schafer, S., & Yu, G. (2011a). Meaningful video games: drama-based video games as transformational experience. In M. M. Cruz-Cunha, V. H. Valvalho, & P. Tavares (Eds.), *Business, Technological, and Social Dimensions of Computer Games: Multidisciplinary Developments* (pp. 312–329). Hershey, PA: IGI Global. doi:10.4018/978-1-60960-567-4.ch019

Schafer, S. B. (2012e). Optimizing Cognitive Coherence, Learning, and Psychological Healing with Drama-based Games. In *Video Game Play and Consciousness*. Nova Science Publishers. Retrieved from http://www.facebook.com/insights/?sk=po_261980910540561#!/pages/Video-Game-Play-and-Consciousness/261980910540561?sk=wall

Schafer, S. B. (Ed.). (2016a). *Exploring the collective unconscious in a digital age. In Exploring the Collective Unconscious in a Digital Age*. Hershey, PA: IGI Global. doi:10.4018/978-1-4666-9891-8

Schafer, S. B. (2016b). *The dream analog: Psychecology video games. Exploring the Collective Unconscious in a Digital Age*. Hershey, PA: IGI Global. doi:10.4018/978-1-4666-9891-8.ch002

Schafer, S. B. (2016c). *The media-sphere as dream: Researching the contextual unconscious of collectives. In Exploring the Collective Unconscious in a Digital Age*. Hershey, PA: IGI Global. doi:10.4018/978-1-4666-9891-8

Schafer, S. B. (2016d). Fostering psychological coherence with ICTs. *International Journal of Reliable and Quality E-Healthcare*, *4*(4), 1-31. DOI doi:10.4018/IJRQEH.2015100101

Schafer, S. B. (2016e). *Employing dynamics of Jungian compensation to foster psychocological coherence with ICTs. International Journal of Privacy and Health Information Management*, 4(1).

Schafer, S. B. (2018a). Synergy with the World Soul: Establishing the humanitarian organism. International Journal of Civic Engagement and Social Change, 4(3). Doi:10.4018/IJCESC.20170701

Schafer, S. B. (2018b). *Generating the superconductive culture of conscience*. Dusseldorf, Germany: Lambert Academic Publishing. Retrieved from https://www.morebooks.de/store/gb/book/generating-a-superconductive-culture-of-conscience/isbn/978-620-2-07643-2

Simon, S., & Bain, B. (Producers) & Ward, V. (Director). (1998). *What Dreams May Come* [DVD]. USA: PolyGram Films.

Tomasino, D. (2007). The psychophysiological basis of creativity and intuition: Accessing the "zone" of entrepreneurship. *International Journal of Entrepreneurship and Small Business & Institute of Heartmath.*

Wilson, C., Kalogrides, L., Cameron, J., & Landau, J. (Peoducers) & Cameron, James (Director). (2010). *Avatar* [DVD]. United States: Fox.

Wiltshire, A. (2016). History of AI beating humans. *PCGamer.* Retrieved from http://www.pcgamer.com/artificial-intelligence-beating-humans/

Wu, X., & Zhang, X. (2016). *Neural network correctly identify criminals and noncriminals with an accuracy of 89.5 percent. MIT Technology Review*. Retrieved from https://www.technologyreview.com/s/602955/neural-network-learns-to-identify-criminals-by-their-faces/

ENDNOTES

1 The term "analog" comes from the fact that the computer solves by analogy by using physical quantities to represent numbers; i.e., physical quantities such as weight, temperature, or area are represented by voltage which is an electrical analog of the variable being analyzed. In PEG design, for example, Myers-Briggs numbers stand for personality-type characteristics.

2 Carl Jung calls dreams of the future "prospective dreams".

3 A cyborg short for "cybernetic organism" is a being with both organic and biomechatronic body parts. The term cyborg is not the same thing as bionic, biorobot or android; it applies to an organism that has restored function or enhanced abilities due to the integration of some artificial component or technology that relies on some sort of feedback. While cyborgs are commonly thought of as mammals, including humans, they might also conceivably be any kind of organism.

4 Kama Manas, वेद Veda (Sanskrit: from kāma, "desire" + manas, "mind") creates the relationship between our thoughts and our emotions within the design concepts of meaning and purpose. In humanity Manas is the human person, the reincarnating ego, immortal in essence, enduring in its higher aspects through the entire manvantara. When embodied, manas is dual, gravitating toward buddhi (the Soul) in its higher aspects and in its lower aspects toward kama (personality). Lower aspects are the intuitive mind, the animal, ratiocinative consciousness, and the lower mentality and passions of the personality.

5 Relative to the nature of the DNA molecule, this statement is accurate, but current research by Bruce Lipton, PhD (The Biology of Belief) and other epigeneticists argue even deeper molecular dynamics of sentience.

6 Quandaries are based on conflicts between levels of the nested masks, and they require choices that change the character and the direction of the plot.

Chapter 10
Heart–Brain Neurodynamics:
The Making of Emotions

Rollin McCraty
Institute of Heartmath, USA

ABSTRACT

As pervasive and vital as they are in human experience, emotions have long remained an enigma to science. This chapter explores recent scientific advances that clarify central controversies in the study of emotion, including the relationship between intellect and emotion and the historical debate on the source of emotional experience. Particular attention is given to the intriguing body of research illuminating the critical role of ascending input from the body to the brain in the generation and perception of emotions. This discussion culminates in the presentation of a new, systems-oriented model of emotion in which the brain functions as a complex pattern-matching system, continually processing input from both the external and internal environments. From this perspective, it is shown that the heart is a key component of the emotional system, thus providing a physiological basis for the long-acknowledged link between the heart and our emotional life.

INTRODUCTION

From the perspective presented in this paper, HeartMath interventions affect several aspects of the emotional process and result in what may be understood as the ethical restructuring of emotional patterns. Emotional restructuring techniques produce a positive emotion-driven shift in the heart's rhythmic patterns, and thus a change in the pattern of cardiac afferent input to the brain. The coupling of a more organized pattern of afferent input with an intentionally self-generated positive emotion reinforces the natural conditioning between the coherent physiological mode and the positive emotion. This subsequently strengthens the ability of a positive emotional shift to initiate a physiological shift towards increased coherence that facilitates the experience of a positive emotion that conforms to ethical coherence in the Psychecology.

DOI: 10.4018/978-1-5225-9065-1.ch010

HEART-BRAIN NEURODYNAMICS: THE MAKING OF EMOTIONS

The Mental and Emotional Systems

The relationship between mind and emotions has been deliberated at length throughout history, with most schools of thought drawing a boundary between them. Perception, appraisal, arousal, attention, memory, thinking, reasoning, and problem solving are often grouped together under the broader heading of cognition, or the mental system. The emotional system, on the other hand, encompasses feelings, which can span a range of intensity. The importance of gaining a deeper understanding of the emotional system has become increasingly recognized as an important scientific undertaking, as it has become clear that emotions underlie the majority of the stress we experience, influence our decisions, provide the motivation for our actions, and create the textures that determine our quality of life. In recent years, the concept of "emotional intelligence" has emerged, claiming that emotional maturity is as important as are mental abilities in both personal and professional spheres, and that emotional competencies often outweigh the cognitive in determining success (Goleman D., 1998).

The tendency to view emotions as operating separately and apart from rational or intellectual capacities dates back to the times of the ancient Greeks. Thus, historically, thinking and feeling—or intellect and emotion—have often been portrayed as opposing forces engaged in an incessant battle for control over the human psyche. Plato maintained that strong emotions made it impossible for him to think and described emotions as wild horses that had to be reined in by the intellect, while Christian theology has traditionally regarded many emotions as sins and temptations to be overcome by reason and willpower. Traditionally, the intellect was held in high regard, while emotions were considered "irrational" and received little recognition. However, a modern-day examination of emotions presents us with an entirely new perspective, providing a more comprehensive understanding of the emotional system and illuminating the critical roles that emotions play in human experience, performance, and rationality.

Most contemporary researchers agree that cognition and emotion are distinct functions mediated by separate but interconnecting neural systems. A number of research centers, rather than studying these systems in isolation, are attempting to understand the essential dynamic interactions that occur between them. From a neuroscience perspective, several intriguing forms of interaction have been discovered that link the cognitive centers with the emotional processing areas of the brain. For example, bidirectional neural connections that exist between the frontal cortex and the amygdala permit emotion-related input from the amygdala to modulate cortical activity and cognitive input from the cortex to modulate the amygdala's emotional information processing (Pribram KH, McGuinness D., 1975; LeDoux JE., 1994, pp. 216-223; LeDoux J., 1996).

Beyond these hard-wired neural connections, biochemical bridges also link key components of the mental and emotional systems. The cortex, for instance, has been found to contain a high density of receptors for many neuropeptides that are also heavily concentrated in the brain's subcortical areas, which are associated with emotional processing (Pert CB, Dreher HE, Ruff MR., 1998, pp. 116-149). Evidence suggests, moreover, that communication channels linking the mental and emotional systems are essential for the expression of our full range of mental capacities (Damasio AR., 1999).

In his book, *Descartes' Error*, neurologist Antonio Damasio presents evidence that patients with brain damage in the frontal lobes, a key site of integration of the cognitive and emotional systems within the brain, can no longer function effectively in the day-to-day world, even though their intellectual abilities are perfectly intact. Damasio presents a powerful argument supporting the seemingly counterintuitive position that input from the emotional system to our thought centers not only facilitates, but is actually indispensable to, the process of rational decision-making (Damasio AR., 1994).

Emotions influence nearly every type of cognitive activity in subtle yet crucial ways. Emotions can direct attention. This phenomenon is known as the "mood-congruity effect" (Bower GH., 1990). Thus, people in a given emotional state pay more attention to stimuli that are emotionally congruent with their current emotional state. Emotions also influence memory and learning, an effect known in neuroscience as "emotion state-dependent memory" (Bower GH., 1992, pp. 3-31). This is why information learned or obtained in a given emotional state may be more easily retrieved if the individual returns to an emotional state similar to the one that prevailed during the original learning. Emotions can also affect judgment, as well as the cognitive processing style employed during problem solving. This effect is readily demonstrable in the laboratory, as well as in everyday life (Clore GC., 1994, pp. 103-111).

While two-way communication between the cognitive and emotional systems is hard-wired into the brain, the actual number of neural connections going from the emotional processing areas to the cognitive centers is greater than the number going the other way (LeDoux J., 1996). This goes some way to explain the powerful influence of emotions on thought processes. It also provides insight into how emotional experience, in contrast to thought alone, can often be a powerful motivator of future attitudes and behavior, influencing moment-to-moment actions as well as both short-term and long-term performance. While emotions can easily dispel non-emotional events from conscious awareness, non-emotional forms of mental activity, such as thoughts, do not so easily displace emotions from the mental landscape. Likewise, experience reminds us that the most pervasive thoughts, least easily dismissed, are typically those fueled by the greatest intensity of emotion.

Interestingly, the seventeenth century philosopher René Descartes noted this same point over three hundred years ago. In commenting on the function of human emotion in his *Treatise on the Passions of the Soul*, Descartes wrote:

The utility of all passions consists alone in their fortifying and perpetuating in the soul thoughts, which it is good it should preserve, and which without that might easily be effaced from it. And again, all the harm which they can cause consists in the fact that they fortify and conserve these thoughts more than necessary, or that they fortify and conserve others on which it is not good to dwell. (Descartes R., 1649, art. 74)

Descartes' views on emotions were clearly more sophisticated than the simplistic notion that emotions are antagonists to rational thought. Descartes considered emotions a double-sided coin. They give substance and sustenance to what otherwise may have been ephemeral thoughts. As a result, they can work both for and against us. Descartes was really highlighting the contrast between the potential of effectively managed emotions and the harm caused by unmanaged emotions. Whereas effectively managed emotions work in synchrony with the mind to facilitate its activity, unmanaged emotions can be the source of mental chaos.

Mental and Emotional Coherence

To further refine Descartes' premise and express it within the context of the concepts discussed in this paper, we can say that when there is *coherence* within and between the mental and emotional systems, they interact constructively to expand awareness and permit optimal psychological and physiological functioning. Conversely, when the mental and emotional systems are out-of-phase, they lack synchronization and thus interact in a conflicting manner, thereby compromising performance. For example, people commonly tell themselves to "think positive" about a challenging task, yet emotionally they may still dread doing it. When our emotions are not aligned with getting the task accomplished we lack motivation and enthusiasm, which limits our access to creativity and insight, and thus impedes our overall performance. In other words, as many of us have likely experienced, positive thoughts or affirmations are often only superimposed on an underlying internal environment of emotional turmoil. In such cases, "positive thinking" is rarely able to produce an enduring shift in the negative feelings.

To better understand an experience such as this, it is important to realize that many common emotion regulation strategies operate on the assumption that all emotions follow thought, and thus by changing one's thoughts, one should be able to gain control over one's emotions. However, in the last decade, research in neuroscience has made it quite clear that emotional processes operate at a much higher speed than thoughts, and frequently bypass the mind's linear reasoning process entirely (LeDoux J., 1996). In other words, emotions do not always follow thought; in many cases, in fact, emotions occur independently of the cognitive system and can significantly bias or color the cognitive process and its output or decision (LeDoux JE., 1994, pp. 216-223; LeDoux J., 1996).

Since the mind and emotions affect a wide range of abilities and responses, mental and emotional coherence are of the utmost importance. Vision, listening ability, reaction times, mental clarity, problem solving, creativity, and performance in a wide range of tasks are all influenced by the degree of coherence of these two systems at any given moment. Because emotions exert such a powerful influence on cognitive processes, emotional incoherence often leads to mental incoherence. Furthermore, emotional incoherence is often the root cause of "mental" problems and stress. Mental health is maintained by emotional hygiene—emotional self-management—and mental problems, to a large extent, reflect a breakdown of emotional order or stability.

On the other hand, increasing stability in the emotional system can often bring the mind into a greater sense of peace and clarity as well. When the mental and emotional systems are in sync, we have greater access to our full range of potential and a greater ability to manifest our visions and goals, as the power of emotion is aligned with the mind's capacities. Even more intriguing, we can gain more conscious control over this process than previously believed through the application of tools and techniques designed to increase emotional stability. Empirical research on the outcomes of such techniques indicates that increased mental and emotional coherence, in turn, can lead to a higher degree of physiological coherence, manifested as increased efficiency and synchronization in the functioning of physiological systems (McCraty R, Childre D, 2002).

The positive emotion-focused coherence-building techniques developed by the Institute of HeartMath engage the heart as a point of entry into the psychophysiological networks that underlie emotional experience (McCraty R, Childre D., 2002; Childre D, Martin H., 1999; Childre D, Rozman D., 2002). One of the research focuses of our laboratory over the last decade has been the study of the patterns and rhythms generated in various physiological systems during the experience of different emotions. Through experimenting with numerous physiological measures, we have found that heart rate variability (heart rhythm) patterns are consistently the most dynamic and reflective of changes in one's emotional state. We have demonstrated that positive and negative emotions can be readily distinguished by distinct changes in heart rhythm patterns. Sustained positive emotions are associated with a noticeably coherent (*i.e.*, ordered, smooth, and sine wave-like) heart rhythm pattern, whereas negative emotions are characterized by a jagged, erratic pattern in the heart's rhythms. (Tiller WA, McCraty R, Atkinson M., 1996, pp.52-65) Moreover, further exploration led us to discover that unhealthy individuals could be greatly facilitated towards improved physical and emotional health through learning how to generate the coherent heart rhythm patterns displayed by healthy, high-functioning individuals.

An important implication of this work, in relation to the ideas developed in this monograph, is that the rhythmic patterns generated by the heart are not only *reflective* of emotions, but actually appear to play a key role in *influencing* moment-to-moment emotional perception and experience. In short, through its extensive interactions with the brain and body, the heart emerges as a critical component of the emotional system. Before developing this concept further, we place it in perspective by offering a brief historical review of the evolution of scientific thinking about emotions, leading up to a summary of current scientific understandings in this field.

The Source of Emotional Experience: An Evolving Model

Current scientific knowledge regarding the physiology of emotions has its roots in Galenic medicine. Galen's influence on scientific thinking persisted well into the 1800s, with the notion that thoughts ("spirits") circulate in the ventricles of the brain, and emotions circulate in the vascular system. Medical thinking at that time maintained that temperament was determined by four "humors" or secretions: sanguine, choleric, phlegmatic, and melancholic. Modern biomedical research has supplemented this simplistic model with a rich array of endocrine and exocrine hormones, which are invoked in any serious biological discussion of emotion. According to neuropsychologist Karl Pribram, who oversaw the brain research center at Stanford University for 30 years, the retreat from this perspective has been slow and guarded for two reasons: Old theories do not die easily, and there is an aspect of truth to this view. (Pribram KH., 1984) The "spirits" circulating in the ventricles have turned out to be neural electrical activity, and the "humors" flowing through the vascular system, endocrine secretions.

An arguably defining characteristic of emotions is that they involve greater activation of the autonomic nervous system and more conspicuous participation of the body than do mental states. This intimate relationship between emotions and physiology has been expressed for centuries in song, poetry, and prose. Even ordinary conversation pertaining to emotional experiences contains numerous physiological allusions. There is no question that emotions are accompanied by a vast array of physiological changes. This is why people so often tend to describe emotional experiences in physiological terms, such as "My heart was pounding," "My throat went dry," "My blood ran cold," "My skin crawled," "It was gut-wrenching," and "It took my breath away." That these figures of speech have become so engrained in everyday language attests to our experience of emotional states being intricately intertwined with, if not inseparable from, their bodily manifestations.

But is what is the ultimate *source* of emotions—the body or the brain? Do emotions originate as bodily sensations that are then perceived by the brain, or do they originate in the brain as a product of cognitive processes and only then trickle down into the body? This fundamental controversy has formed the core of a lively debate that has raged for over a century, yielding a fascinating and illuminating progression of ideas.

The James-Cannon Debate

In 1884, the debate over the source of emotional experience formally began with a proposal by psychologist-philosopher William James in his seminal article entitled "What is an Emotion" (James W., 1984, pp. 188-205)? James believed that emotional experience is not only accompanied by, but actually *arises from* organic changes that occur in the body in response to an arousing stimulus. These physiological signals (*e.g.*, racing heart, tight stomach, sweaty palms, tense muscles, and so on) are subsequently fed back to the brain, and only *then* felt consciously as a true emotion. James proposed that we can sense what is going on inside our body much the same as we can sense what is going on in the outside world. The awareness of the immediate sensory and motor reverberations that occur in response to a perception (*e.g.*, the pounding heart, the clenched jaw, etc.) is what makes that perception emotional. Thus, the feeling aspect of emotion is dictated by the physiology and not vice-versa. According to James:

Our natural way of thinking about...emotions is that the mental perception of some fact excites the mental affection called the emotion, and that this latter state of mind gives rise to the bodily expression. My thesis on the contrary is that the bodily changes follow directly the PERCEPTION *of the exciting fact, and that our feeling of the same changes as they occur* IS *the emotion. (James W., 1984, pp. 189-190)*

James maintained that the precise pattern of sensory feedback relayed from the body to the brain gives each emotion its unique quality. Thus, anger feels different from sadness or love because it has a characteristic physiological pattern or signature. James maintained that physiological responses contributing to emotion were "almost infinitely numerous and subtle," (James W., 1984, p. 191) reflecting the nuances of physiology and its emotional counterpart.

In fairness to James, it should be noted that his original premise—that the sensation of bodily changes is a *necessary condition* of emotion—was subsequently oversimplified by many of his contemporaries, as well as by many modern authors (Ellsworth PC., 1984, pp. 222-229). The oversimplification of James' views suggested that emotions are *nothing but* the sensation of bodily changes. In fact, when using the term "perception" in his writings, James did acknowledge the role of interpretation or cognitive appraisal of the exciting stimulus in the initiation of emotional experience. However, he argued that the emotional "feeling" was not a primary feeling directly aroused by appraisal, but rather a secondary feeling indirectly aroused by the organic changes that occurred following the appraisal.

James' perspective was called into question in the 1920s by the prominent experimental physiologist Walter Cannon (Cannon WB., 1927, pp. 106-124). Cannon believed that the essential mechanisms of emotion occurred within the brain alone and that bodily responses and afferent input to the brain were not needed to fully experience emotions. He argued, in brief, that bodily feedback, especially from the viscera, was both too slow and not sufficiently differentiated to explain the dynamic range and variety of emotional expression. Though Cannon felt that bodily sensations could not account for differences between emotions, he believed that they nevertheless played an important role in giving emotions their characteristic sense of intensity and urgency.

To support his views, Cannon demonstrated that artificially induced visceral responses alone do not produce emotions and that animals still show "emotional behavior" when feedback from the viscera is surgically eliminated. Of course, here Cannon was forced to rely solely on behavioral evidence to define the parameters of emotion in his animal subjects. In place of the visceral theory, Cannon proposed a brain (thalamic) theory of emotions. He suggested that emotional expression results from the operation of hypothalamic structures, while emotional feeling results from stimulation of the dorsal thalamus. This theory was based on the observation that emotion-like behavior could be elicited in decorticated and decerebrated animals, but not when thalamic structures were ablated as well. Further, a variety of expressive and bodily responses were obtained when the thalamus was electrically stimulated (Marshall LH, Magoun HW., 1998).

In Cannon's view, the thalamus and hypothalamus discharged simultaneously to the body to produce physiological responses and to the cortex to produce emotional experiences. In measuring the amount of time it took for electrical stimulation of the hypothalamus to produce visceral changes, Cannon concluded that these bodily responses were too slow to be the cause of emotions. He saw them rather as the effect, since his measurements suggested that we would already be feeling the emotion by the time these responses occur.

Much of Cannon's experimental research centered on autonomic nervous system (ANS) responses that occur in states of hunger or intense emotion. (Cannon WB., 1929) His research led him to propose the concept of an emergency reaction—the "fight-or-flight response"—to describe a specific physiological response that accompanies any state in which physical energy must be expended. The sympathetic division of the ANS, which he believed to act in a uniform way regardless of how or why it was activated, mediated this response. Cannon held that the visceral changes accompanying emotion were part of this nonspecific arousal, and thus that all emotions had the same ANS signature.

Cannon's arguments won over the weight of scientific opinion of the day, and his view consequently spawned a search for emotional mechanisms in the brain. Others such as Lindsley and Papez built upon Cannon's theory by mapping out additional subcortical and limbic structures and communication pathways involved in the brain's emotion-regulating networks (Lindsley DB., 1951, pp. 473-516; Papez JW., 1937, pp. 725-743).

Experimental evidence demonstrated the existence in the hypothalamic region of an energy-conserving or *trophotropic* process working primarily through the parasympathetic branch of the ANS, and a mobilizing or *ergotrophic* system working through the sympathetic branch. (Hess WR., 1954) It was assumed that the hypothalamus and dorsal thalamus were at the apex of the hierarchy of control of visceral and autonomic functions and were the key to understanding emotional processes.

Neuropsychologist Karl Lashley was the first to criticize this assumption. He pointed out several flaws in the theory by using lesion studies showing that emotional disturbances (on which the Cannon theory was based) could also be observed following lesions elsewhere, such as in the afferent paths in the nervous system or between the forebrain and thalamic structures. (Lashley KS., 1960) He also noted that neither the James nor Cannon theories could account for the dissociation between outward emotional expression and inner feelings, which is a common clinical and experimental observation.

The Limbic Theory

An important breakthrough came in 1937 when James Papez, a professor of neuroanatomy at Cornell University, described a circuit between centers in the brain and suggested that it might constitute the neural substrate for emotion, thus introducing the idea of a circuit or system rather than a single center. (Papez JW., 1937, pp. 725-743) He suggested that blockage of information flow at any point along this circuit would cause disorders of emotions. Now known as the *Papez circuit*, this model described the flow of information from the hippocampal formation to the thalamus, then to the cingulate gyrus, and back again to the hippocampal formation.

This was later elaborated on by Paul MacLean, chief of the laboratory for brain evolution and behavior at the National Institute of Mental Health. In the 1950s, MacLean introduced the concept of the "limbic system" to denote the interacting regions of the brain involved in emotional processing (MacLean PD., 1949, pp. 338-353; MacLean PD., 1952, pp. 407-418). In addition to the areas of the Papez circuit, MacLean included regions such as the amygdala, septum, and prefrontal cortex in the limbic system. Later, he also originated the *triune brain* model, which delineated three functional brain systems that he believed developed successively in response to evolutionary needs (MacLean PD., 1970, pp. 336-349; MacLean PD., 1990). Although MacLean's theory has had little impact on neurobiology, it has become popular in the lay press and with psychotherapists. However, it should be noted that extensive work in comparative neurobiology unequivocally contradicts the evolutionary aspects of his theory (Butler AB, Hodos W., 1996).

MacLean believed that emotional experience could be most accurately described as a response to the *composite* of stimuli the brain receives from the external environment, as a result of ongoing perceptions of the outside world, *and* internal sensations or feedback transmitted to the brain from bodily organs and systems. The limbic system came to be viewed as the receiving station or site for the association and correlation of these varied stimuli, being strategically located to correlate every form of *internal* and *external* perception. MacLean also emphasized the importance of memory and provided data showing that the limbic cortex exceeds the neocortex in the turnover of protein, a measure of the demand for new RNA in memory formation (Flanigan S, Gabrieli ER, MacLean PD., 1957, pp. 588-594).

Here at last was the seat of emotion—the visceral brain. Karl Pribram summed it up with the following:

The persuasive power of this suggestion is great: Galen, James..., [and] Cannon...are all saved; visceral [bodily] processes are the basis of emotion; and an identifiable part of the brain is responsible for emotional control and experience because of its selective relations with viscera...The path from the "emotions in the vascular system" to "emotions in the forebrain" had finally been completed, and each step along the way freed us from preconceptions popularly current when the step was taken. (Pribram KH., 1984, p. 16)

Despite its popularity, there are problems with the limbic theory of emotions and it falls heir to the same criticisms leveled against Cannon. The idea of a specific center (*i.e.*, the thalamus) as a privileged site for emotional experience did not hold up; and the same problem arises with relations between the limbic structures and bodily input, and for that matter, the limbic system itself and emotions (Pribram KH., 1984). For example, it was found that emotional changes can be observed to accompany lesions in parts of the brain other than limbic areas. Further, ablation and stimulation of limbic structures influence problem solving and other cognitive behaviors in selective ways that cannot be attributed to changes in emotion. In fact, obvious and specific "memory" defects follow limbic lesions, while changes in emotions cannot be found (Marshall LH, Magoun HW., 1998). Obviously, the Papez-MacLean theory, like its predecessors, presented only part of the picture.

With the development of newer techniques for electrical brain stimulation, Pribram and others showed that the so-called "limbic" brain regions were under the surveillance and control of the neocortex. Brain structures such as the hippocampus, amygdala, cingulate cortex, septum, thalamus, hypothalamus, and prefrontal cortex came to be viewed as interpreting experience in terms of feelings rather than "intellectualized" representations. It now appears that the whole brain as well as the ascending input from the body, both neurological and hormonal, are necessary in the full experience of emotion.

Memory

An important aspect of emotional experience is memory. The first associations of memory with specific parts of the limbic system appear to have been made in 1900 by the Russian neurologist-anatomist Vladimir Mikhailovich Bekhterev when he observed memory deficits in a patient with hippocampal degeneration (Marshall LH, Magoun HW., 1998). The story of the search for memory is far beyond the scope of this monograph; however; the work of Canadian psychologist Donald Hebb has special relevance to this paper's theme. In 1949, Hebb predicted a form of synaptic plasticity based on temporal activity, which was verified decades later with the discovery of long-term potentiation (Seung HS., 2000, p. 1166).

Hebb believed that synaptic connections were the material basis of mental associations; however, he went well beyond the naïve connectionism theories of that time period in two important respects. First, he argued that an association could not be localized to a single synapse. Instead, neurons were grouped in "cell assemblies," and an association was distributed over their synaptic connections. Secondly, Hebb rejected the concept that input-response behaviors could be explained by simple reflex arcs connecting sensory neurons to motor or output neurons. He believed that sensory stimulation could initiate patterns of neural activity that were maintained by circulation in synaptic feedback loops. This reverberatory activity made it possible for a response to follow a delay that was characteristic of thought.

In essence, Hebb argued for a dual-trace mechanism of memory. Reverberatory neural activity was the trace for short-term memory, and synaptic connections were the trace for long-term memory. He hypothesized the conversion of short-term memory into long-term memory by the stabilization of reverberatory activity patterns. Once such an activity pattern was stored, in a redistribution or change in the strength of synaptic connections, it could be recalled repeatedly by an excitation from sensory neurons or from other reverberatory activity patterns occurring in other cell assemblies that provide inputs. In the past fifty years, several aspects of Hebb's theory have been confirmed, while the technology needed to prove or disprove other aspects does not yet exist.

In the 1970s, new insights into the question of what happens in the brain during the time interval between stimulus and response were made possible with the discovery of long-term potentiation. This and the first neural network models of delay activity provided a candidate for Hebb's "reverberatory" activity. For example, it has been demonstrated that certain prefrontal cortex neurons remain active during delays of many seconds and encode information about the preceding stimulus or the impending response. Changes in distribution and strength of synapses have been confirmed, and this aspect of his theory is not in doubt. What remains unknown is whether the delay between stimulus and response is truly due to a reverberatory type of activity, and if so, if the reverberatory activity is stabilized by long-term potentiation. Also, Hebb's concept of only two memory traces may be incorrect, as it is now known that synaptic plasticity involves many processes operating on different time scales (Abbott LF, Nelson SB., 2000, pp. 1178-1183).

Current Perspectives on the Nature of Emotion

Most theorists now agree that emotion involves, at the most fundamental level, the registration and interpretation of a stimulus based on memory processes in addition to information from physiological responses and subjective feeling states. In more recent years, attempts have been made to determine the "correct" sequence of these components in the generation of emotional experience. However, when interpretation, subjective feeling, and bodily responses are all considered as *processes*, rather than discrete events or simple input-output relations, the source of a large part of the controversy dissolves (Ellsworth PC., 1994, pp.222-229).

We find that it is indeed possible to have emotional processing in specific brain areas simultaneously with input from the body to the brain, each building on the other to contribute to the dynamic process of emotion. Recent elucidation of the numerous afferent pathways through which the body transmits signals to the brain and the interaction of this information with higher-level brain processes provides strong support for this perspective. Elmer Green, Menninger Clinic physician and pioneer of the biofeedback approach to treatment of disease, offered an astute summation of this highly debated topic: "Every change in the physiological state is accompanied by an appropriate change in the mental emotional state, conscious or unconscious, and conversely, every change in the mental emotional state, conscious or unconscious, is accompanied by an appropriate change in the physiological state" (Green E, quoted in CB Pert, 1997, p. 137).

The remaining element of the controversy, namely the specificity of physiological responses, must now take into account new data revealing that communication between the body and the brain is much more sophisticated and complex than previously imagined. The generation of such data has been made possible, in part, due to the development of more sophisticated recording techniques and instrumentation that more clearly capture the subtleties and complexities of communication between different bodily systems and between the body and brain. In addition, technological advances have enabled us to achieve finer measurements of neuroendocrine and immune activity, thereby offering a wider view into the array of physiological responses at the cellular level that accompany different emotional states.

Before introducing a new model of emotion that synthesizes and further develops many of the perspectives discussed here thus far, a brief review of the role played by activity in both the efferent and afferent pathways of the nervous systems in emotional experience is relevant.

Specificity of Autonomic Responses

Let's return to Cannon's assumption that all emotions are associated with the same basic state of non-specific arousal or activation of the ANS. In the 1960s, Stanley Schachter and Jerome Singer, social psychologists at Columbia University, embraced this view by suggesting that a cognitive interpretation of a basically undifferentiated state of physiological arousal within the social or environmental context of the arousing stimulus was the missing factor in determining the specificity of emotion. (Schacter, Stanley & Singer, Jerome, ND) Schachter and Singer's model, called the *two-factor theory*, proposed that emotions are produced by both feedback from the body and the cognitive appraisal of what caused those responses. In other words, we label the response according to what we think is causing the response. This theory had a profound influence on the thinking on the subject of emotion at the time. However, in the last thirty years the tide has turned, as increased evidence has emerged to indicate that autonomic responses in different emotional states are much more complex than previously assumed, and certainly far from uniform.

In contrast to the thinking in Cannon's day, which attributed emotional arousal to sympathetic nervous system activation alone, we now understand that simultaneous and complex changes in the patterns of efferent activity in both the sympathetic and parasympathetic branches of the ANS are involved in the experience of different emotions. The sensations produced in any given emotional state depend on the extent to which sympathetic effects are balanced by parasympathetic influences; thus sympathetic/parasympathetic balance has become an important measure in psychophysiological research.

Many emotional states are associated with complex patterns of sympathetic/parasympathetic activity in different tissues. For example, in states of aggression and resentment, increased sympathetic discharges occur in the vascular system while parasympathetic discharges predominate in the gastrointestinal tract. Conversely, increased sympathetic activity occurs in both the cardiovascular and gastrointestinal systems in states of fear. Further, autonomic responses vary both quantitatively and qualitatively with the degree of emotional intensity (Gellhorn E, Loofbourrow GN., 1963).

A number of experiments conducted in the 1950s provided evidence that different emotions could be differentiated psychophysiologically (Malmo RB, Shagass C, Davis FH., 1950, pp. 362-376; Ax AF., 1953, pp. 433-442; Schachter J., 1957, pp. 17-29; Graham DT, Stern JA, Winokur G., 1958, pp. 446-457). These findings have been confirmed recently (Levenson RW, Ekman P, Friesen WV., 1990, 363-384; Ekman P, Levenson RW, Friesen WV., 1983, pp. 1208-1210; Hubert W, de Jong-Meyer RD., 1991, pp. 73-93; Collet C, Vernet-Maury E, Delhomme G, Dittmar A., 1997, pp. 45-57; Sinha R, Lovallo WR, Parsons OA., 1992, pp. 422-435). For example, in an experiment by Ekman and colleagues (Ekman P, Levenson RW, Friesen WV., 1983, pp. 1208-1210) at the University of California in San Francisco, subjects experienced different emotional states (happiness, surprise, disgust, sadness, fear, and anger) both by reliving past emotional experiences and by constructing facial prototypes of emotion, muscle by muscle, according to instruction. Specific differences in the autonomic parameters of heart rate, finger temperature, and skin resistance were found among the six different emotions measured. The response patterns differed not only between positive and negative emotions, but also among the negative emotions of disgust, sadness, anger, and fear. These differences were consistent across profession, age, gender, and culture (Ekman P, Levenson RW, Friesen WV., 1983, pp. 1208-1210). While this and other research provided convincing evidence of autonomic variation among different emotional states, the variation measured was often small and present in only some of the physiological parameters, or experienced by only a subset of subjects.

A more recent study measuring multiple autonomic parameters showed that six basic emotions (happiness, surprise, anger, fear, sadness, and disgust) could be fully differentiated on the basis of electro-dermal variables (skin resistance, skin conductance, and skin potential), thermos-vascular variables (skin blood flow and skin temperature), and a respiratory variable (instantaneous respiratory frequency) Collet C, Vernet-Maury E, Delhomme G, Dittmar A., 1997, pp. 45-57). These results clearly support the concept of emotion-specific ANS activity, which can be demonstrated with the aid of careful experimental procedures providing that a sufficient number of autonomic variables are considered.

Individual differences in patterns of autonomic discharge during emotional states have also been identified and associated with personality characteristics. For instance, individuals who have been characterized as “impulsive” personality types display rhythmic bouts of palmar sweat secretion and increases in heart rate even at rest, while in others, little change occurs in these physiological parameters under similar circumstances (Gellhorn E, Loofbourrow GN., 1963).

The Importance of Afferent Input

In addition to understanding how complex patterns of efferent autonomic activity correlate to differing emotions, many scientists are beginning to understand the critical role played by the afferent neural signals that flow from the body to the brain. Afferent feedback from bodily organs has been shown to affect overall brain activity and to exert a measurable influence on cognitive, perceptual, and emotional processes.

Physiology textbooks are replete with diagrams that illustrate nervous system pathways from the brain to autonomically innervated organs. However, many of these illustrations do not complete the communication circuit. They frequently omit the extensive systems of visceral afferent fibers, which carry messages from receptors in the body to the brain. The nerve pathways connecting most organ systems to the brain are, in fact, composed of as many afferent fibers as there are efferent connections; (Leek BF. 1972, pp. 113-160) while in some visceral nerves, such as the abdominal vagus, up to 90 percent of the fibers are afferen (Andrews PLR., 1986, pp. 65-86). Remarkably, we now know that the heart sends more neural traffic to the brain than the brain sends to the heart. While afferent pathways were identified during the early years of autonomic research, their study was not emphasized. However, research conducted primarily in the 1950s through the 1970s began to illuminate the importance of afferent input from the thoracic, abdominal, and neck cavities back to the brain—and the effects of this input on brain activity and emotional experience.

One of the earliest contributors to our understanding of the importance of afferent neural traffic was the German internist, Ludwig van Müller. He was particularly interested in the perception of sensory stimuli arising from internal organs and their role in the regulation of different bodily states and sensations (Neundörfer B, Hilz MJ. Ludwig Robert Müller [1870-1962] pp. 1-5). He pointed out, in 1906, that emotions influence heart rate, and conversely, that heart rate influences emotions. For example, he observed that cardiac palpitations can induce emotions (Müller LR., 1906, pp. 14-16).

Early neurophysiological evidence of the influence of afferent input on brain activity dates back to 1929. Tournade and Malméjac, followed by Koch two years later, showed that stimulation of the carotid sinus nerve (contributing afferent fibers which enter the brain stem), or an increase in pressure in the carotid sinus itself, produced a decrease in muscle tone in anaesthetized animals (Heymans C, Neil E., 1958). Koch also demonstrated that by sharply increasing the pressure within the carotid sinus he could inhibit motor activity and induce prolonged sleep. These results were confirmed in later investigations, which showed that distention of the carotid sinus produced marked changes in cortical electrical activity, from low-voltage fast to high-voltage slow waves (characteristic of sleep), and inhibited activity of the pyramidal nerve cells in the motor cortex, which control muscle movement (Heymans C, Neil E., 1958).

In the 1950s, French and Italian neurophysiologists performed a variety of experiments investigating the effects of changes in heart rate and blood pressure on brain activity. Changes in heart rate and blood pressure are detected by receptors in the heart, the aortic arch, and the carotid sinus. Information from these receptors is transmitted to the brain stem via the vagal and glossopharyngeal nerves (Lacey JI, Lacey BC., 1970, pp. 205-227). In one study, Bonvallet and Allen demonstrated that elimination of the glossopharyngeal and vagal input to the brain resulted in a prolongation of cortical activation and skeletal muscle activity (1963, pp. 969-988). Then in 1974, French researchers Gahery and Vigier, working with cats, found that stimulating the vagus nerve reduced the electrical response in the cuneate nucleus of the brain to about half its normal rate (Gahery Y, Vigier D., 1974, pp. 241-246).

Since that time, extensive experimental data have been gathered documenting the role played by afferent input in modulating such varied processes as pain perception, (Randich A, Gebhart GF., 1992, pp. 77-99) hormone production, (Drinkhill MJ, Mary DA., 1989, pp. 299-313) electrocortical activity, and cognitive functions (Rau H, Pauli P, Brody S, Elbert T., 1993, pp. 322-325; Sandman CA, Walker BB, Berka C., In: Cacioppo JT, Petty RE, eds., 1982, pp. 189-222; van der Molen MW, Somsen RJM, Orlebeke JF. In: Ackles PK, Jennings JR, Coles MGH, eds., 1985, pp. 1-88). Animal studies have now demonstrated that a variety of brain regions are involved in the processing of visceral afferent information, including the hypothalamic and thalamic nuclei, amygdala, hippocampus, cerebellum, somatosensory cortex, prefrontal cortex, and insula (Schandry R, Montoya P., 1996, pp. 75-85; Montoya P., 1994). Thus, it has become clear that the influence of cardiovascular afferent signals on the brain is far more pervasive than previously considered.

Uncovering Conversations Between the Heart and Brain

Among the first modern psychophysiological researchers to systematically examine the "conversations" between the heart and brain were John and Beatrice Lacey (Rosenfeld SA., 1978). During 20 years of research throughout the 1960s and 1970s, they observed that afferent input from the heart and cardiovascular system could significantly affect *perception and behavior*. Their research produced a body of behavioral and neurophysiological evidence suggesting that sensory-motor integration could be modified by cardiovascular activity (Lacey JI, Lacey BC., 1970; Lacey JI., In: Rubinstein E, Parloff M, eds., 1959:, pp. 160-208; Lacey JI, Kagan J, Lacey BC, Moss HA., 1963, pp. 161-196; Lacey JI., In: Appley MH, Trumbull R, eds., 1967, pp. 14-42; Lacey BC, Lacey JI., In: Obrist PA, Black AH, Brener J, DiCara LV, eds., 1974, pp. 538-564).

The Laceys' observations directly challenged the "arousal" or "activation" theory proposed by Cannon. In essence, Cannon believed that all of the physiological indicators underlying emotion—heart rate, blood pressure sweating, pupil dilation, narrowing of certain blood vessels, and so on—moved predictably *in concert* with the brain's response to a given stimulus. Thus, Cannon had suggested that when we are aroused, the sympathetic nervous system mobilizes us to fight or flee. In contrast, in quieter moments, the parasympathetic nervous system relaxes our inner systems. Presumably, autonomic responses all increased together when we were aroused and decreased in unison when we were at rest, and the brain was entirely in control of both these processes.

The Laceys observed that all physiological responses did not always move together, and in essence, the heart seemed to behave as if it had a mind of its own. They noticed that this view of activation as a single dimension only partially matched actual physiological behavior. As their research evolved, they found that the heart, in particular, seemed to have its own peculiar logic that frequently diverged from the direction of other ANS responses. In laboratory studies of reaction time and operant responses, the Laceys observed that, in response to certain stimuli, all autonomic variables recorded did not exhibit the expected response pattern typical of arousal. At times, for example, heart rate decelerated and blood pressure *decreased*, while simultaneously recorded parameters such as skin conductance, respiration rate, and pupillary dilation all increased as expected. The Laceys called this phenomenon "directional fractionation" and noted that it appeared to be dependent upon the nature of the stimulus and the type of mental processing involved (Lacey JI., In: Rubinstein E, Parloff M, eds., 1959:, pp. 160-208).

The Laceys found that tasks requiring mental concentration or attention to *internal* stimuli (*e.g.*, mental arithmetic, reverse spelling, or making up sentences) produced an acceleration in heart rate and an increase in skin conductance. In contrast, tasks requiring attention to the *external* environment (*e.g.*, detecting colors and patterns or empathizing with a dramatic recitation) produced a marked deceleration in heart rate, although skin conductance still increased.

They also showed that patterns of physiological responses were affected as much by the *context* of a specific task and its requirements as by emotional stimuli. Thus, heart rate, for example, tends to decrease, even in the presence of a distressing emotional context, when subjects are attending visually or auditorially to events in their external environment; on the other hand, heart rate accelerates when subjects mentally recall and think about the very same unpleasant emotional material (Lacey, JI, 1959; Lacey JI, Kagan J, Lacey BC, Moss HA, 1963, pp. 161-196; Lacey JI, 1967, pp. 14-42). Subsequent research also revealed an intriguing link between the heart rate response (but not other autonomic responses) to different environmental stimuli and an individual's cognitive style, or attitude toward the external environment (Israel NR., 1968; Israel NR., 1969, pp. 499-509).

The selectivity of the heart's response indicated that *it was not merely mechanically responding to an arousal signal from the brain.* Even more intriguing, in simple reaction time experiments, which required attention to external cues, an *anticipatory* deceleration in heart rate was observed during the preparatory interval, and subjects' reaction times were faster during periods when their heart rate was slowing (Lacey JI, Lacey BC., 1970). This led the Laceys to propose that cardiovascular afferent feedback to the higher brain centers plays a role in *facilitating either the intake or rejection of environmental stimuli*, in accordance with the nature of the mental processing required for a given task (Lacey BC, Lacey JI., 1974). In brief, such a mechanism would permit us effectively to "tune out" potentially disruptive external environmental events when performing tasks requiring internal cognitive elaboration, and, conversely, to focus in on external inputs when our activities demanded close attention to our environment.

To support this hypothesis, the Laceys and others found evidence that in humans under normal physiological conditions, brain activity varies in relation to cardiovascular events (Lacey JI, Lacey BC., 1970; Koriath JJ, Lindholm E., 1986, pp. 183-195). Thus, increased heart rate and the resulting increased afferent discharge inhibits (desynchronizes) cortical activity. Conversely, decreased heart rate occurring prior to sensory intake promotes cortical facilitation and processing by reducing brain inhibition. (Lacey BC, Lacey JI., 1974) In their reaction time experiments, the Laceys discovered that the greater the cardiac deceleration, the greater the cortical activation, and the greater the behavioral efficiency (*i.e.*, the faster the speed of response). In other words, *afferent input to the brain from the heart can either inhibit or facilitate the brain's activity, which, in turn, can affect perception and motor activity.*

Evoked Potential Studies

A useful technique for the study of how and where information flows through the brain is evoked potential analysis. Evoked potentials (also sometimes referred to as event-related potentials) are obtained using signal averaging, a procedure for separating a known repetitive signal from other signals. Evoked potential analysis can be used to study the flow of information through many different pathways in the brain. Common applications of the technique are to study the visual, auditory, and somatosensory systems. In the case of the visual system, for example, the flow of information through the nervous system produced in response to a series of light flashes or a changing visual stimulus of any kind can be traced through the different visual pathways as it is processed. In this case, the resulting waveforms are called visual evoked potentials.

It is also possible to examine the flow of afferent input through the brain from many other sensory systems, such as the auditory and tactile systems, or to assess how a change in afferent signals generated by one system affects the processing of information in another system. For example, the effects of cardiac afferent input on sensory perception have been studied by looking at how these signals affect processing in the visual system. It has been shown that the processing of visual information is significantly changed as heart rate and carotid pressure change. These findings provide confirmation of the

Laceys' earlier behavioral evidence that cardiovascular activity influences sensory intake (Walker BB, Sandman CA., 1982, 520-527).

While these data indirectly support the view that cardiovascular afferent information interacts with higher central nervous functions, experiments by the German researcher Rainer Schandry and others have provided more direct psychophysiological evidence for this perspective. Their work has demonstrated that cardiovascular events like heartbeats are detectable as a signal in the EEG and evoke cortical responses analogous to "classical" sensory event-related potentials (Schandry R, Montoya P., 1996, pp. 75-85; Schandry R, Sparrer B, Weitkunat R., 1986, pp. 261-275; Montoya P, Schandry R, Muller A., 1993, pp. 163-172). When the heart's afferent signals are being studied, the ECG R-wave is used as the timing source for the signal averaging and the resulting waveforms are called heartbeat evoked potentials (HBEPs).

These experiments have shown that the processing of afferent input from the cardiovascular system is accompanied by specific electrical activity in the brain. This processing of cardiovascular afferent information is most pronounced at the frontocortical areas, a brain region known to be particularly involved in the processing of visceral afferent information. Recent findings have demonstrated that the HBEP is significantly diminished in diabetic patients with autonomic neuropathy, and reduced amplitude of the HBEP is significantly correlated with reduced awareness of body sensations (Schandry R, Leopold C., 1997, pp. 249-250). In other words, when the communication of afferent signals from the heart to the brain is compromised, there is less awareness of feeling sensations in the body.

Furthermore, psychological factors, such as motivation, attention to cardiac sensations, and general perceptual sensitivity, have been found to alter HBEPs in the brain in a manner analogous to the cortical processing of external stimuli (Schandry R, Montoya P., 1996; Montoya P, Schandry R, Muller A., 1993). These findings confirm our own data demonstrating that focusing attention in the area of the heart and generating a positive emotion alters HBEPs, thus indicating an modulation of cortical processing. Taken together, these data suggest that perception and processing of information arising from bodily processes is comparable to perception and processing of external events, and the effects of both sources of input on perceptual and emotional experience must be considered.

In summary, evidence now clearly demonstrates that afferent signals from the heart significantly influence cortical activity. Specifically, we now know that afferent messages from the cardiovascular system are not only relayed to the brain stem to exert homeostatic effects on cardiovascular regulation, but also have separate effects on aspects of higher perceptual activity and mental processing. Furthermore, as discussed next, there are now data from both animals and humans to support the premise that central *emotional processing* is also altered by afferent input from the heart.

Afferent Input Influences Emotional Processing: The Role of the Amygdala

The influence of cardiovascular afferent input to the brain on emotional processes is highlighted by recent evidence suggesting that psychological aspects of panic disorder are often created by unrecognized paroxysmal supraventricular tachycardia (PSVT), a sudden-onset atrial arrhythmia. According to one study, DSM-IV criteria for panic disorder were fulfilled in more than two-thirds of patients with these sudden-onset arrhythmias. In those patients in whom PSVT was unrecognized at initial evaluation, symptoms were attributed to panic, anxiety, or stress in 54 percent of the cases. In the majority of cases, once the arrhythmia was recognized and treated, the panic disorder disappeared (Lessmeier TJ, Gamperling D, Johnson-Liddon V, Fromm BS, Steinman RT, Meissner MD, Lehmann MH., 1997, pp. 537-543).

Interestingly, this study confirmed the observations of pioneer ANS researcher Müller, who reported the induction of emotions by cardiac palpitations over 90 years earlier. (Müller LR., 1906) Likewise, our research has also shown that changing the pattern of afferent information generated by the cardiovascular system can significantly influence perception and emotional experience (McCraty R, Childre D., 2002; McCraty R, Barrios-Choplin B, Rozman D, Atkinson M, Watkins AD. 1998, pp. 151-170).

The amygdala has been the subject of intense scrutiny in recent years. This brain center plays a key role in emotional memory, emotional processing, and dreaming. (Aggleton JP, ed., 1992) Several studies have investigated the effects of cardiovascular afferent input on the amygdaloid complex (*i.e.*, the amygdala and associated nuclei). For example, in cats, spontaneous neural activity in the central nucleus of the amygdala has been shown to be synchronized to the cardiac cycle and to be modulated by afferent input from the aortic depressor and carotid sinus nerves (Zhang JX, Harper RM, Frysinger RC., 1986, pp. 193-207).

Similarly, data from humans undergoing surgery for epilepsy demonstrated that cells within the amygdaloid complex specifically responded to information from the cardiac cycle (Frysinger RC, Harper RM., 1990, pp. 162-171). Pribram, who did much of the original mapping of the functions of the amygdaloid complex, found it has extensive projections to both the brain stem autonomic nuclei and the higher cognitive centers, and is thus uniquely placed to coordinate affective, behavioral, immunological, and neuroendocrine responses to environmental stimuli (Pribram KH., 1984; Pribram KH, Bagshaw MH., 1953, pp. 347-375).

The observed interaction of afferent cardiac input with this brain region supports the view that visceral information not only influences emotional processing and emotional experience, but can also influence hormonal and immune responses (McCraty R, et al., 1998).

Taken together with the demonstrated role of the amygdala in the regulation of viscero-autonomic activity and the resultant effects on familiarization, considered below, a new view of emotional processing and regulation emerges.

The Role of Familiarization in Emotional Processing

To further unfold our understanding of the emotional system and the heart's role in emotion, we now review the model of emotion first developed by Karl Pribram (Pribram KH, Melges FT., 1969, pp. 316-341).

Simply said, in Pribram's model a memory, or stable pattern of activity, is formed and maintained in the neural architecture of the brain as we gain experience both in internal self-regulation and in interacting with the external environment. These stable patterns are updated and modified as we encounter new experiences and learn how a certain action usually leads to specific result. All ongoing or current sensory input to the brain, from both the internal and external sensory systems, is compared to these stable patterns. When a mismatch between current input and a stable pattern occurs, novelty is sensed.

These stable patterns create a set of "expectancies" against which breathing, eating, drinking, sleeping, alerting, sexual, and other behaviors are evaluated. The stable neurological pattern acts as a set point against which an input is matched, and therefore determines what is familiar and what is novel, and perhaps exciting.

The set point, based on previous experience, becomes a reference point for evaluating current and future experience, and is biased or adjusted according to ongoing experience. To maintain stability as we encounter life's events, we must make adjustments that return us to the "familiar" set point. These adjustments require us to take an "action"—which can be either an outward action (*i.e.*, control of some kind over the external environment) or an internal adjustment (*i.e.*, self-control of our inner environment). Since our psychophysiological systems are designed to maintain stability and resist change, returning to familiar set points gives us a sense and feeling of security, while remaining in unfamiliar territory causes unrest. Interestingly, this is true even if the established set point is one of chaos and confusion.

Attention

No conscious awareness of anything, including our emotions, is possible until it has captured our attention. Sensory neurons in our eyes, ears, nose, and body are in continuous action, day and night, whether we are awake or asleep. The brain receives a steady stream of information about all the events the sense organs are capable of detecting. It would be bewildering if we were continuously aware of all the incoming information. In fact, we completely ignore most of the information arriving at the brain most of the time. Yet any input is capable of shifting and dominating our attention. In order for this process to function, there must be mechanisms and processes that direct *selective* attention. The attention mechanisms must continuously scan the available information and assign priority, usually based on biological importance. Large, sudden, novel occurrences typically have the ability to grab our attention. Emotions also have the ability to capture and focus attention, and attention is involved in the management of our emotional state.

In 1890, William James described attention thus:

Everyone knows what attention is. It is taking possession of the mind, in clear and vivid form, of one out of what seems several simultaneously possible objects or trains of thought. Focalization, concentration of consciousness, are of its essence. It implies withdrawal from some things in order to deal effectively with others, and is a condition which has a real opposite in the confused, dazed, scatterbrained state… (James W., 1890, pp. 403-404)

Many laboratories around the world have investigated the brain structures involved in awareness and attention. Generally there have been two approaches to attention research: (1) recording physiological or behavioral responses against a background of regular, repeating sensory events and (2) pairing of the outcome of the response to sensory events.

When a new stimulus is presented to the brain, a change in activity in the central and autonomic nervous systems is produced. If the response is short-lived (1–3 seconds), it is called *arousal* or an *orienting reflex*. If, however, the stimulus or event is recurrent, the brain rapidly adapts and we *habituate*. For example, people who live in a noisy city adapt to the ambient noise and eventually become unaware of it. However, when they take a trip to the quiet countryside, the lack of noise seems strange and noticeable. Thus, any change in the stimulus will cause the reappearance of the arousal response, or the orienting reflex. The arousal reaction therefore reflects a *mismatch* between the new information and the familiar representation stored in the brain. A change in brain potentials can be measured during the arousal response to a novel stimulus, and is called *mismatch negativity*. (Näätänen R., 1992)The observed changes in the nervous system can be separated into a *phasic* component, which habituates quickly, and a long-lasting *tonic* component, which habituates more slowly (Sharpless S, Jasper H., 1956, pp. 655-680).

James, and more recently Pribram and McGuiness, also distinguished two types of attention. Pribram and McGuiness called these involuntary and voluntary. *Involuntary* or *primary attention*, as James called it, is provoked by certain classes of stimuli that are novel, salient, or intense, which impinge upon our awareness regardless of ongoing activity. *Voluntary attention*, on the other hand, describes the process whereby the individual voluntarily determines the contents of his/her own awareness and the duration of focus. In the Pribram and McGuinness model, the distinction between involuntary and voluntary attention identifies two aspects of attentional control: one regulates *arousal* resulting from a mismatch in sensory input; the other controls the preparatory *activation* of potential responses. In addition, there is a third aspect of attention that serves to coordinate involuntary arousal and voluntary activation, and this aspect of attention requires effort (Pribram KH, McGuinness D., 1975).

Pattern-Matching and the Maintenance of Stability

In their book titled *Plans and the Structure of Behavior,* Miller, Galanter, and Pribram propose that in order for an organism to maintain continued stability, it must be able to maintain a match between its current experience or "reality" and its neural and hormonal set points or "*programs*" (Miller GA, Galanter EH, Pribram KH., 1960). These programs consist of hierarchies of nested neural feedback loops that determine what is familiar. Incongruities or differences in the input (new experiences) arouse or activate us depending upon the degree of mismatch, and, in most cases, determine what action is needed to reestablish stability. When the differences (mismatch) are of sufficient magnitude, there is a temporary discontinuity; importantly, *it is this discontinuity or mismatch—effectively a departure from the familiar—that gives rise to the experience of emotion.* In this context, it is interesting to note that the word "emotion" derives from the Latin *emovere*, which means "to move out or away from."

Pribram, in his book *Languages of the Brain,* carries the theory further. When the input to the brain does not match the existing program, an adjustment must be made in an attempt to achieve control and return to stability. One way to reestablish control is by taking an outward action. We are motivated to eat if we feel hungry, run away or fight if threatened, do something to draw attention to ourselves if feeling ignored, etc. Alternatively, we can reestablish stability and gain control by making an internal adjustment (without any overt action). For example, a confrontation at work may lead to feelings of anger, which can prompt inappropriate behavior (*i.e.*, outward actions such as yelling, hitting, etc.). However, through internal adjustments, we can self-manage our feelings in order to inhibit these responses, reestablish stability, and maintain our job. Thus, stabilization is achieved through external action on the environment or through internal self-control. These processes are referred to, respectively, as motivational control and emotional control (Pribram KH., 1967, pp. 830-838). Ultimately, when we achieve stability through our efforts, the results are feelings of satisfaction and gratification. By contrast, when there is a failure to achieve stability or control, feelings such as anxiety, panic, annoyance, apprehension, hopelessness, or depression result.

Pribram and many others have conducted numerous experiments providing evidence that these sorts of internal adjustments, although commonplace, represent a complex interplay between peripheral and central processes. For example, the afferent input systems and even their receptors are modulated by the central nervous system, which alters information processing in the sensory input channel (Pribram KH., 1971). In other words, the higher brain centers can inhibit or "gate" the information flowing into the brain. There are many examples of how we can control input channels. Where we focus our attention has a powerful effect on modulating inputs and thus on determining what gets processed at higher levels. In a noisy room filled with many conversations, we have the ability to tune out the noise and focus on a single conversation of interest. In a like manner, we can modulate pain from a stubbed toe or headache or desensitize ourselves to sensations like tickling.

Arousal

There is ample support that arousal, measured as EEG desynchronization, occurs in response to novel or unfamiliar input, and that arousal is one of the elements of emotional experience. In classical models of arousal theory, the amount of neural and/or hormonal activity generated in response to a given stimulus or event determines whether the experience leads to familiarization or disruption. Arousal theory states that a correlation exists between the amount of a specific hormone or amount of neural excitation and the amount of emotional arousal.

However, this is only part of the story. Arousal can at times be associated with an increase the *amount* of neural activity, but arousal can also occur without any increase in neural activity. In the latter case what does change, instead, is the *pattern* of activity in the nervous system (for example, variations in the time intervals between sequential firings of a neuron or group of neurons, or in which efferent pathways are active). Therefore, the amount of neural activity does not always necessarily indicate the level of arousal (Pribram KH, Melges FT., 1969). This is an important realization, as it shifts the focus from thinking in terms of amount of activity alone to understanding the importance of the pattern of activity.

This is also related to the observation that differing emotions are reflected in the patterns of the heart rhythm. For example, during an emotional state shift, the *pattern* of beat-to-beat heart rate variability can shift dramatically, while the *amount* of variability remains exactly the same. This is not to imply that changes in the amount of neural activity or amount of heart rate variability are not also important sources of information that contribute to ongoing emotional experience. However, in the broader context of the model presented here, such variations can also be considered as changes in pattern relative to a familiar baseline or set-point.

The Role of the Heart

Monitoring the alterations in the rates, rhythms, and patterns of afferent traffic is a key function of the cortical and emotional systems in the brain. Pribram was well aware of the influence of afferent input from the heart and other organ systems in determining the set points, or what becomes the familiar pattern, as far back as 1969, when he wrote:

Visceral feedback constitutes, by the nature of its receptor anatomy and diffuse afferent organization, a major source of input to this biasing mechanism; it is an input which can do much to determine set-point. In addition, cardiovascular and autonomic events are repetitiously redundant in the history of the organism. They vary recurrently, leading to stable habituations; this is in contrast to external changes which vary from occasion to occasion. Habituation to visceral and autonomic activity makes up, therefore, a large share...of the stable base-line from which the organism's reactions can take off (Pribram KH, Melges FT., 1969, p.322).

These set points establish a background against which blood pressure, hormonal balance, and all regularly recurring behaviors are initiated and maintained. For example, when we sense a mismatch between our actual heart rate and the habituated heart rate, we generate a feeling (*e.g.*, excitement or anxiety if heart rate is accelerated). The specific feeling experienced may reflect the nature of the mismatch. Importantly, a mismatch may be registered not only due to changes in heart rate but also due to changes in the pattern of the afferent traffic.

Although input originating from many different bodily organs and systems is involved in the processes that ultimately determine emotional experience, it has become clear that the heart plays a particularly important role. The heart is the primary and most consistent source of dynamic rhythmic patterns in the body. Furthermore, the afferent networks connecting the heart and cardiovascular system with the brain are far more extensive than the afferent systems associated with other major organs.

Additionally, the heart is particularly sensitive and responsive to changes in a number of other psychophysiological systems. For example, heart rhythm patterns are continually and rapidly modulated by changes in the activity of either branch of the ANS, and the heart's extensive intrinsic network of sensory neurons also enables it to detect and respond to variations in hormonal rhythms and patterns (Armour JA., 1994, pp. 219-244).

In addition to functioning as a sophisticated information processing and encoding center, (Armour JA, 2003) the heart is also an endocrine gland that produces and secretes hormones and neurotransmitters (Cantin M, Genest J., 1986, pp. 76-81; Mukoyama M, Nakao K, Hosoda K, Suga S, Saito Y, Ogawa Y, Shirakami G, Jougasaki M, Obata K, Yasue H, et al., 1991, pp. 1402-1412; Gutkowska J, Jankowski M, Mukaddam-Daher S, McCann SM., 2000, pp. 625-633; Huang M-H, Friend DS, Sunday ME, Singh K, Haley K, Austen KF, Kelly RA, Smith TW., 1996, pp. 1298-1303).

Thus, with each beat, the heart not only pumps blood, but also continually transmits dynamic patterns of neurological, hormonal, pressure, and electromagnetic information to the brain and throughout the body (McCraty R, Atkinson M., 2003). Therefore, the multiple inputs from the heart and cardiovascular system to the brain are a major contributor in establishing the dynamics of the baseline pattern or set point against which the "now" (current input) is compared.

The repeating rhythmic patterns generated by the heart, whether they are ordered or disordered, become familiar to the brain. At the brain stem level, these patterns are compared to set points that control blood pressure, affect respiration rate, and gate the flow of activity in the descending branches of the autonomic system. From there, these signals cascade up to a number of subcortical centers, such as the thalamus, hypothalamus, and amygdala, which are involved in the processing of emotion.

With the understanding that the emotional system operates essentially as a pattern recognition system, the finding that a significant proportion of people diagnosed with panic disorder actually have an unrecognized atrial arrhythmia is easily understandable. When a sudden-onset arrhythmia occurs, there is a large and sudden change in the pattern of afferent signals arriving at the amygdala and hippocampus, resulting in a significant mismatch between the current input and the familiar, stable pattern. The system is unable to achieve stability through an outward action or through an internal adjustment; the mismatch therefore captures attention and gives rise to feelings of fear and anxiety, which build to panic. In cases where the arrhythmia is constant or occurs more frequently, the system adapts or habituates—in other words, the new input pattern becomes familiar.

On the other hand, a change in the pattern of afferent cardiovascular input that accompanies a more coherent or ordered heart rhythm, such as those that occur with certain breathing techniques or the use of HeartMath positive emotion-focused tools, results in a "pattern match" associated with security and positive emotional experience. These coherent rhythms are familiar to a "healthy" system as they have occurred spontaneously many times during sleep and positive emotional states. However, in many individuals, a coherent pattern is rare and relatively unfamiliar to the brain. In this case, with the practice of self-generating coherent rhythms, they become the familiar baseline pattern and that which the system attempts to maintain.

Emotional Instability

When the neural systems that maintain the baseline reference patterns are in an unstable state (due to stress, anxiety, chemical stimulants, etc.), sensory input from either internal or external sources that would ordinarily be processed smoothly can be perceived as a mismatch and give rise to an uncomfortable feeling. Thus, patterns of neural activity in the brain can effectively predispose the individual towards either stability or instability. The reference patterns can be temporarily destabilized by large, sudden changes in the pattern of afferent activity, such as those that occur in the example of a sudden-onset arrhythmia or during an emotionally charged situation. If a reference pattern is destabilized, a mismatch can be perceived even in the absence of novel input. This explains why we can have an upsetting interaction with our spouse, and even though things may have been smoothed over and the event consciously forgotten, we could subsequently be set off by what we perceive as a funny look from a coworker upon arriving at the office. Physiologically, the instability is still in our system. Under normal circumstances, the look would have gone unnoticed. Likewise, had we been able to stabilize our neural systems by clearing the emotional residue on the way to work, the look from the coworker would not have thrown us off.

In addition to processes that monitor the input and controls for maintaining stability (pattern matching) in the here-and-now, there are also matching processes that appraise the degree of congruity or incongruity between the past and the now and between the now and the projected future. Furthermore, these prospective appraisals can be divided into optimistic and pessimistic (Pribram KH., 1970, pp. 41-53). If the appraisal does not result in a projected ability to return to stability, feelings of fear and anxiety can result. This appraisal could be due to past experience of similar situations or a lack of experience in the projected future situation. However, as we encounter novel situations and learn that we are able to maintain stability, we can apply that experience to similar future situations without fear.

Pribram states that when a homeostatic system becomes stabilized and a new pattern has become familiar, new sensitivities develop and different strategies and programs are added to handle the acquired sensitivities (Pribram K., 1963, pp. 113-159). In essence, we mature. Encountering novel situations or obstacles requires that we develop new strategies: we either take an external action to gain control or self-manage our internal systems. Once we learn how to handle the new challenge effectively and maintain stability, the strategy (complex pattern) for dealing with the challenge becomes familiar and part of our repertoire. Through this process, we increase our internal self-control and management of emotions as well as our ability to effectively deal with external situations.

The baseline patterns maintained in the neural architecture are modified by other sources of neural and hormonal input that affect the "bias" or sensitivity of the system. Because the neural systems involved in comparing the incoming sensory information are made up of short, fine fibers with many branches, they are especially sensitive to hormonal influences. Thus, the system is readily affected by changes in the patterns of hormonal input associated with different psychophysiological states. In this way hormones provide important influences on the brain processes involved in the experience of emotion.

The Making of Emotions: A Converging View

In summary, we can see earlier theories of emotion, coupled with current research, converging into a more complete and comprehensive view of emotions. Endocrine research significantly advanced the previous view of emotions as "humors." The visceral theory acknowledged an arousal mechanism that provides feelings of interest, novelty, and familiarity, as well as more painful disruptions of stable states.

James emphasized the communication of bodily responses to the brain. Cannon's thalamic theory contributed by offering evidence of the thalamus as a prime locus for processing emotional information from the body's chemical homeostatic systems. Papez and MacLean introduced the idea of emotional circuits and systems instead of a single center and added the possibility of a memory component to the emotional system. With Pribram's cortical control of afferent input and monitoring of a departure from stable, familiar patterns, it becomes clear that *both the brain and the entire body are involved in the full experience and expression of emotions.*

With this understanding in mind, we can view the experience of emotion as emerging from an intricate array of interactions occurring within a complex system. Broadly speaking, its main components include the brain and nervous system, the hormonal system, and body. Although there are numerous sources of bodily input to the brain, the heart is given particular relevance in the emotional system due to its unique degree of afferent input and its consistent generation of dynamic rhythmic patterns that are closely coupled with changes in emotional state. From a generalized perspective, one of the ways an emotion is generated is through the comparison of information received from the external sensory systems, (*e.g.*, sights, sounds, and smells) against preexisting memories. This processing occurs at unconscious levels, unless attention is captured, and results in changes in the patterns of descending autonomic activity flowing to the body. This leads to a wide variety of specific changes in biochemical outputs and biophysical states, such as alterations in patterns of muscle tension (especially in the face), adrenal secretions, vascular resistance, cardiac output, and heart rhythms. These alterations, in turn, result in changes in the afferent inputs from the body back to the brain, which are then compared to a set of preexisting reference patterns. This ascending bodily input is crucial to the felt experience of an emotion, and may or may not reinforce the cognitive level appraisal and labeling of the feeling. The process continues as the system makes external and internal adjustments in order to maintain stability, and, depending upon the outcome, can further color and add textures to the emotional experience. Of course, this is only one example, as the process can also be initiated by changes in the internal systems alone as well as through many combinations of the internal and external sensory systems' interactions with the reference patterns and memories.

Within the context of the model of emotion developed here, we can also gain new insight into the mechanisms underlying the efficacy of the HeartMath emotional restructuring techniques, which produce a positive emotion-driven shift in the heart's rhythmic patterns, and thus a change in the pattern of cardiac afferent input to the brain. The coupling of a more organized pattern of afferent input with an intentionally self-generated positive emotion reinforces the natural conditioning between the coherent physiological mode and the positive emotion. This subsequently strengthens the ability of a positive emotional shift to initiate a physiological shift towards increased coherence, and a physiological shift to facilitate the experience of a positive emotion.

From the perspective presented in this paper, HeartMath interventions affect several aspects of the emotional process. First, by reducing nervous system chaos, they stabilize the neural systems that maintain the baseline or reference patterns against which incoming information is compared. They also modify the baseline patterns by reinforcing the coherent psychophysiological patterns associated with positive emotions and allowing these patterns to become familiar, thus effectively establishing a new baseline or norm. Once this new reference pattern established, the system then automatically strives to maintain this state.

With practice of these techniques, as the neural architecture comes to recognize the patterns associated with coherent heart rhythms as familiar, it becomes progressively easier to intentionally generate coherent rhythms and their psychophysiological benefits, even during experiences of stress or challenge. Moreover, we have demonstrated that as people continue to practice intentionally self-generating states of psychophysiological coherence using heart-based techniques, they also begin to demonstrate a greater frequency of *spontaneous* heart rhythm coherence, without conscious use of the interventions. These data support the concept the techniques facilitate an actual repatterning process at the level of the neural architecture, which can be objectively assessed using electrophysiological measures.

In sum, consistent use of heart-based positive emotion-focused techniques reinforces existing neural pathways that the brain uses to control its input (self-manage) and facilitates the establishment of new control pathways, thus improving our ability to self-manage our emotions and regulate our physiological state. Experientially, the occurrence of a system-wide repatterning process with consistent use of the HeartMath interventions is supported by reports from thousands of individuals who have noted enduring improvements in many aspects of health, well-being, and performance, increased emotional stability and new capabilities for dealing with stress and challenges. In a very real sense, we become the architects of our own neural landscape.

ACKNOWLEDGMENT

I would like to express my appreciation to Dr. Karl Pribram for his careful review of this monograph and his insightful input on its content.

HeartMath is a registered trademark of the Institute of HeartMath.

REFERENCES

Abbott, L. F., & Nelson, S. B. (2000). Synaptic plasticity: Taming the beast. *Nature Neuroscience*, *3*(S11), 1178–1183. doi:10.1038/81453 PMID:11127835

Aggleton, J. P. (Ed.). (1992). *The Amygdala: Neurobiological Aspects of Emotion, Memory and Mental Dysfunction*. New York: Wiley-Liss.

Andrews, P. L. R. (1986). Vagal afferent innervation of the gastrointestinal tract. *Progress in Brain Research*, *67*, 65–86. doi:10.1016/S0079-6123(08)62757-0 PMID:3823483

Armour, J. A. (1994). Peripheral autonomic neuronal interactions in cardiac regulation. In J. A. Armour & J. L. Ardell (Eds.), *Neurocardiology* (pp. 219–244). New York: Oxford University Press.

Armour, J.A. (2003). *Neurocardiology—Anatomical and functional principles.* Boulder Creek, CA: HeartMath Research Center, Institute of HeartMath, Publication No. 03-011.

Ax, A. F. (1953). The physiological differentiation between fear and anger in humans. *Psychosomatic Medicine*, *15*(5), 433–442. doi:10.1097/00006842-195309000-00007 PMID:13100546

Bonvallet, M., & Allen, M. B. Jr. (1963). Prolonged spontaneous and evoked reticular activation following discrete bulbar lesions. *Electroencephalography and Clinical Neurophysiology, 15*(6), 969–988. doi:10.1016/0013-4694(63)90141-X PMID:14132294

Bower, G. H. (1990). Mood-congruity of social judgements. In J. Forgas (Ed.), *Emotions and Social Judgements*. Oxford, UK: Pergamon.

Bower, G. H. (1992). How might emotions affect learning? In S. A. Christianson (Ed.), *The Handbook of Emotion and Memory: Research and Theory* (pp. 3–31). Hillsdale, NJ: Lawrence Erlbaum.

Butler, A. B., & Hodos, W. (1996). *Comparative Vertebrate Neuroanatomy: Evolution and Adaptation*. New York: Wiley-Liss.

(1959). C. Psychophysiological approaches to the evaluation of psychotherapeutic process and outcome. InRubinstein, E., & Parloff, M. (Eds.), *Research in Psychotherapy* (pp. 160–208). Washington, DC: American Psychological Association. doi:10.1037/10036-010

Cannon, W. B. (1927). The James-Lange theory of emotion: A critical examination and an alternative theory. *The American Journal of Psychology, 39*(1/4), 106–124. doi:10.2307/1415404

Cannon, W. B. (1929). *Bodily Changes in Pain, Hunger, Fear and Rage: An Account of Recent Researches into the Function of Emotional Excitement* (2nd ed.). New York: D. Appleton & Company. doi:10.1097/00007611-192909000-00037

Cantin, M., & Genest, J. (1986). The heart as an endocrine gland. *Scientific American, 254*(2), 76–81. doi:10.1038cientificamerican0286-76 PMID:2935933

Childre, D., & Martin, H. (1999). *The HeartMath Solution*. San Francisco: HarperSanFrancisco.

Childre, D., & Rozman, D. (2002). *Overcoming Emotional Chaos: Eliminate Anxiety, Lift Depression and Create Security in Your Life*. San Diego: Jodere Group.

Clore, G. C. (1994). Why emotions are felt. In P. Ekman & R. J. Davidson (Eds.), *The Nature of Emotion* (pp. 103–111). New York: Oxford University Press.

Collet, C., Vernet-Maury, E., Delhomme, G., & Dittmar, A. (1997). Autonomic nervous system response patterns specificity to basic emotions. *Journal of the Autonomic Nervous System, 62*(1-2), 45–57. doi:10.1016/S0165-1838(96)00108-7 PMID:9021649

Damasio, A. R. (1994). *Descartes' Error: Emotion, Reason and the Human Brain*. New York: G.P. Putnam's Sons.

Damasio, A. R. (1999). *The Feeling of What Happens*. Orlando, FL: Harcourt.

Descartes, R. (1649). Treatise on the Passions of the Soul (E. Haldane & G. Ross, Trans.). In *The Philosophical Works of Descartes*. Cambridge, UK: Cambridge University Press.

Drinkhill, M. J., & Mary, D. A. (1989). The effect of stimulation of the atrial receptors on plasma cortisol level in the dog. *The Journal of Physiology, 413*(1), 299–313. doi:10.1113/jphysiol.1989.sp017655 PMID:2600852

Ekman, P., Levenson, R. W., & Friesen, W. V. (1983). Autonomic nervous system activity distinguishes among emotions. *Science, 221*(4616), 1208–1210. doi:10.1126cience.6612338 PMID:6612338

Ellsworth, P. C. (1994). William James and emotion: Is a century of fame worth a century of misunderstanding? *Psychological Review, 101*(2), 222–229. doi:10.1037/0033-295X.101.2.222 PMID:8022957

Flanigan, S., Gabrieli, E. R., & MacLean, P. D. (1957). Cerebral changes revealed by radioautography with S35-labeled l-methionine. *Archives of Neurology and Psychiatry, 77*(6), 588–594. doi:10.1001/archneurpsyc.1957.02330360046003 PMID:13423931

Frysinger, R. C., & Harper, R. M. (1990). Cardiac and respiratory correlations with unit discharge in epileptic human temporal lobe. *Epilepsia, 31*(2), 162–171. doi:10.1111/j.1528-1167.1990.tb06301.x PMID:2318169

Gahery, Y., & Vigier, D. (1974). Inhibitory effects in the cuneate nucleus produced by vago-aortic afferent fibers. *Brain Research, 75*(2), 241–246. doi:10.1016/0006-8993(74)90744-6 PMID:4366859

Gellhorn, E., & Loofbourrow, G. N. (1963). *Emotions and Emotional Disorders: A Neurophysiogical Study*. New York: Harper & Row.

Goleman, D. (1998). *Working with Emotional Intelligence*. New York: Bantam Books.

Graham, D. T., Stern, J. A., & Winokur, G. (1958). Experimental investigation of the specificity of attitude hypothesis in psychosomatic disease. *Psychosomatic Medicine, 20*(6), 446–457. doi:10.1097/00006842-195811000-00002 PMID:13614604

Gutkowska, J., Jankowski, M., Mukaddam-Daher, S., & McCann, S. M. (2000). Oxytocin is a cardiovascular hormone. *Brazilian Journal of Medical and Biological Research, 33*(6), 625–633. doi:10.1590/S0100-879X2000000600003 PMID:10829090

Hess, W. R. (1954). *Diencephalon: Autonomic and Extrapyramidal Functions*. New York: Grune & Stratton.

Heymans, C., & Neil, E. (1958). *Reflexogenic Areas of the Cardiovascular System*. Boston: Little, Brown.

Huang, M.-H., Friend, D. S., Sunday, M. E., Singh, K., Haley, K., Austen, K. F., ... Smith, T. W. (1996). An intrinsic adrenergic system in mammalian heart. *The Journal of Clinical Investigation, 98*(6), 1298–1303. doi:10.1172/JCI118916 PMID:8823294

Hubert, W., & de Jong-Meyer, R. D. (1991). Psychophysiological response patterns to positive and negative film stimuli. *Biological Psychology, 31*(1), 73–93. doi:10.1016/0301-0511(90)90079-C PMID:2021681

Israel, N. R. (1969). Leveling-sharpening and anticipatory cardiac response. *Psychosomatic Medicine, 31*(6), 499–509. doi:10.1097/00006842-196911000-00003 PMID:5361745

Israel, N. R. (1968). *Cognitive control and pattern of autonomic response*. Paper presented at meeting of the Eastern Psychological Association, Washington, DC.

James, W. (1884). What is an emotion? *Mind, 9*(34), 188–205. doi:10.1093/mind/os-IX.34.188

James, W. (1890). *Principles of Psychology*. New York: Holt.

Kaada, B. R., Pribram, K. H., & Epstein, J. A. (1949). Respiratory and vascular responses in monkeys from temporal pole, insula, orbital surface and cingulate gyrus. *Journal of Neurophysiology*, *12*(5), 347–356. doi:10.1152/jn.1949.12.5.347 PMID:18137711

Koriath, J. J., & Lindholm, E. (1986). Cardiac-related cortical inhibition during a fixed foreperiod reaction time task. *International Journal of Psychophysiology*, *4*(3), 183–195. doi:10.1016/0167-8760(86)90014-0 PMID:3793561

Lacey, B. C., & Lacey, J. I. (1974). Studies of heart rate and other bodily processes in sensorimotor behavior. In P. A. Obrist, A. H. Black, J. Brener, & L. V. DiCara (Eds.), *Cardiovascular Psychophysiology: Current Issues in Response Mechanisms, Biofeedback, and Methodology* (pp. 538–564). Chicago: Aldine.

Lacey, J. I. (1967). Somatic response patterning and stress: Some revisions of activation theory. In M. H. Appley & R. Trumbull (Eds.), *Psychological Stress: Issues in Research* (pp. 14–42). New York: Appleton-Century-Crofts.

Lacey, J. I., Kagan, J., Lacey, B. C., & Moss, H. A. (1963). The visceral level: Situational determinants and behavioral correlates of autonomic response patterns. In P. H. Knapp (Ed.), *Expression of the Emotions in Man* (pp. 161–196). New York: International Universities Press.

Lacey, J. I., & Lacey, B. C. (1970). Some autonomic-central nervous system interrelationships. In P. Black (Ed.), *Physiological Correlates of Emotion* (pp. 205–227). New York: Academic Press.

Lashley, K. S. (1960). The thalamus and emotion. In F. A. Beach, D. O. Hebb, C. T. Morgan, & H. S. Nissen (Eds.), *The Neurospsychology of Lashley*. New York: McGraw-Hill.

LeDoux, J. (1996). *The Emotional Brain: The Mysterious Underpinnings of Emotional Life*. New York: Simon and Schuster.

LeDoux, J. E. (1994). Cognitive-emotional interactions in the brain. In P. Ekman & R. J. Davidson (Eds.), *The Nature of Emotion: Fundamental Questions* (pp. 216–223). New York: Oxford University Press.

Leek, B. F. (1972). Abdominal visceral receptors. In E. Neil (Ed.), Handbook of Sensory Physiology (Vol. 3, pp. 113–160). Heidelberg, Germany: Springer.

Lessmeier, T. J., Gamperling, D., Johnson-Liddon, V., Fromm, B. S., Steinman, R. T., Meissner, M. D., & Lehmann, M. H. (1997). Unrecognized paroxysmal supraventricular tachycardia: Potential for misdiagnosis as panic disorder. *Archives of Internal Medicine*, *157*(5), 537–543. doi:10.1001/archinte.1997.00440260085013 PMID:9066458

Levenson, R. W., Ekman, P., & Friesen, W. V. (1990). Voluntary facial action generates emotion-specific autonomic nervous system activity. *Psychophysiology*, *27*(4), 363–384. doi:10.1111/j.1469-8986.1990.tb02330.x PMID:2236440

Lindsley, D. B. (1951). Emotion. In S. S. Stevens (Ed.), *Handbook of Experimental Psychology* (pp. 473–516). New York: Wiley.

MacLean, P. D. (1949). Psychosomatic disease and the "visceral brain": Recent developments bearing on the Papez theory of emotion. *Psychosomatic Medicine*, *11*(6), 338–353. doi:10.1097/00006842-194911000-00003 PMID:15410445

MacLean, P. D. (1952). Some psychiatric implications of physiological studies on frontotemporal portion of limbic system (visceral brain). *Electroencephalography and Clinical Neurophysiology*, *4*(4), 407–418. doi:10.1016/0013-4694(52)90073-4 PMID:12998590

MacLean, P. D. (1970). The triune brain, emotion and scientific bias. In F. O. Schmitt (Ed.), *The Neurosciences: Second Study Program* (pp. 336–349). New York: Rockefeller University Press.

MacLean, P. D. (1990). *The Triune Brain in Evolution: Role in Paleocerebral Functions*. New York: Plenum.

Malmo, R. B., Shagass, C., & Davis, F. H. (1950). Symptom specificity and bodily reactions during psychiatric interview. *Psychosomatic Medicine*, *12*(6), 362–376. doi:10.1097/00006842-195011000-00006 PMID:14797915

Marshall, L. H., & Magoun, H. W. (1998). *Discoveries in the Human Brain*. Totowa, NJ: Humana Press. doi:10.1007/978-1-4757-4997-7

McCraty, R., & Atkinson, M. (2003). Psychophysiological coherence. Boulder Creek, CA: HeartMath Research Center, Institute of HeartMath, Publication 03-016.

McCraty, R., Barrios-Choplin, B., Rozman, D., Atkinson, M., & Watkins, A. D. (1998). The impact of a new emotional self-management program on stress, emotions, heart rate variability, DHEA and cortisol. *Integrative Physiological and Behavioral Science*, *33*(2), 151–170. doi:10.1007/BF02688660 PMID:9737736

McCraty, R., & Childre, D. (2002). The appreciative heart: The psychophysiology of positive emotions and optimal functioning. Boulder Creek, CA: HeartMath Research Center, Institute of HeartMath, Publication No. 02-026.

Miller, G. A., Galanter, E. H., & Pribram, K. H. (1960). *Plans and the Structure of Behavior*. New York: Henry Holt & Co. doi:10.1037/10039-000

Montoya P. (1994). *Herzwahrnehmung und hirnelektrische Aktivität. Eine Analyse der topographischen Verteilung von herzschlag-synchron evozierten Potentialen (HEP)*. Frankfurt/Main: Peter Lang.

Montoya, P., Schandry, R., & Muller, A. (1993). Heartbeat evoked potentials (HEP): Topography and influence of cardiac awareness and focus of attention. *Electroencephalography and Clinical Neurophysiology*, *88*(3), 163–172. doi:10.1016/0168-5597(93)90001-6 PMID:7684965

Mukoyama, M., Nakao, K., Hosoda, K., Suga, S., Saito, Y., Ogawa, Y., ... Yasue, H. (1991). Brain natriuretic peptide as a novel cardiac hormone in humans. Evidence for an exquisite dual natriuretic peptide system, atrial natriuretic peptide and brain natriuretic peptide. *The Journal of Clinical Investigation*, *4*(4), 1402–1412. doi:10.1172/JCI115146 PMID:1849149

Müller, L. R. (1906). Ueber die Beziehung von seelischen Empfindungen zu Herzstörungen. *Munchener Medizinische Wochenschrift*, *53*, 14–16.

Näätänen, R. (1992). *Attention and Brain Function*. Hillsdale, NJ: Lawrence Erlbaum Associates.

Neundörfer, B., & Hilz, M. J. (1998). Ludwig Robert Müller (1870-1962) – A pioneer of autonomic nervous system research. *Clinical Autonomic Research*, *8*(1), 1–5. doi:10.1007/BF02267597 PMID:9532414

Papez, J. W. (1937). A proposed mechanism of emotion. *Archives of Neurology and Psychiatry*, *38*(4), 725–743. doi:10.1001/archneurpsyc.1937.02260220069003

Pert, C. B., Dreher, H. E., & Ruff, M. R. (1998). The psychosomatic network: Foundations of mind-body medicine. *Alternative Therapies in Health and Medicine*, *4*(4), 30–41. PMID:9656499

Pribram, K. (1963). Reinforcement revisited: A structural view. In *Nebraska Symposium on Motivation*. Lincoln, NE: University of Nebraska Press.

Pribram, K. H. (1967). The new neurology and the biology of emotion: A structural approach. *The American Psychologist*, *22*(10), 830–838. doi:10.1037/h0025031 PMID:4383680

Pribram, K. H. (1970). Feelings as monitors. In M. B. Arnold (Ed.), *Feelings and Emotions* (pp. 41–53). New York: Academic Press. doi:10.1016/B978-0-12-063550-4.50010-3

Pribram, K. H. (1971). *Languages of the Brain*. New York: Brandon House.

Pribram, K. H. (1984). Emotions: A neurobehavioral analysis. In K. R. Scherer & P. Ekman (Eds.), *Approaches to Emotion*. Hillsdale, NJ: Erlbaum.

Pribram, K. H., & Bagshaw, M. H. (1953). Further analysis of the temporal lobe syndrome utilizing fronto-temporal ablations. *The Journal of Comparative Neurology*, *99*(2), 347–375. doi:10.1002/cne.900990208 PMID:13109029

Pribram, K. H., & McGuinness, D. (1975). Arousal, activation, and effort in the control of attention. *Psychological Review*, *82*(2), 116–149. doi:10.1037/h0076780 PMID:1096213

Pribram, K. H., & Melges, F. T. (1969). Psychophysiological basis of emotion. In P. J. Vinken & G. W. Bruyn (Eds.), *Handbook of Clinical Neurology* (pp. 316–341). Amsterdam: North-Holland Publishing Company.

Randich, A., & Gebhart, G. F. (1992). Vagal afferent modulation of nociception. *Brain Research. Brain Research Reviews*, *17*(2), 77–99. doi:10.1016/0165-0173(92)90009-B PMID:1327371

Rau, H., Pauli, P., Brody, S., Elbert, T., & Birbaumer, N. (1993). Baroreceptor stimulation alters cortical activity. *Psychophysiology*, *30*(3), 322–325. doi:10.1111/j.1469-8986.1993.tb03359.x PMID:8497561

Rosenfeld, S. A. (1978). *Conversations between heart and brain*. Rockville, MD: National Institute of Mental Health.

Sandman, C. A., Walker, B. B., & Berka, C. (1982). Influence of afferent cardiovascular feedback on behavior and the cortical evoked potential. In J. T. Cacioppo & R. E. Petty (Eds.), *Perspectives in Cardiovascular Psychophysiology* (pp. 189–222). New York: The Guilford Press.

Schandry, R., & Leopold, C. (1997). The severity of diabetic autonomic neuropathy is reflected in the heartbeat evoked brain potential [abst.]. *Clinical Autonomic Research*, *7*(5), 249–250.

Schandry, R., & Montoya, P. (1996). Event-related brain potentials and the processing of cardiac activity. *Biological Psychology*, *42*(1-2), 75–85. doi:10.1016/0301-0511(95)05147-3 PMID:8770371

Schandry, R., Sparrer, B., & Weitkunat, R. (1986). From the heart to the brain: A study of heartbeat contingent scalp potentials. *The International Journal of Neuroscience*, *30*(4), 261–275. doi:10.3109/00207458608985677 PMID:3793380

Seung, H. S. (2000). Half a century of Hebb. *Nature Neuroscience*, *3*(S11), 1166. doi:10.1038/81430 PMID:11127829

Sharpless, S., & Jasper, H. (1956). Habituation of the arousal reaction. *Brain*, *79*(4), 655–680. doi:10.1093/brain/79.4.655 PMID:13396069

Sinha, R., Lovallo, W. R., & Parsons, O. A. (1992). Cardiovascular differentiation of emotions. *Psychosomatic Medicine*, *54*(4), 422–435. doi:10.1097/00006842-199207000-00005 PMID:1502284

Tiller, W. A., McCraty, R., & Atkinson, M. (1996). Cardiac coherence: A new, noninvasive measure of autonomic nervous system order. *Alternative Therapies in Health and Medicine*, *2*(1), 52–65. PMID:8795873

van der Molen, M. W., Somsen, R. J. M., & Orlebeke, J. F. (1985). The rhythm of the heart beat in information processing. In Advances in Psychophysiology (Vol. 1). London: JAI Press.

Walker, B. B., & Sandman, C. A. (1982). Visual evoked potentials change as heart rate and carotid pressure change. *Psychophysiology*, *19*(5), 520–527. doi:10.1111/j.1469-8986.1982.tb02579.x PMID:7134357

Zhang, J. X., Harper, R. M., & Frysinger, R. C. (1986). Respiratory modulation of neuronal discharge in the central nucleus of the amygdala during sleep and waking states. *Experimental Neurology*, *91*(1), 193–207. doi:10.1016/0014-4886(86)90037-3 PMID:3940875

Chapter 11
Liquid Humanitarianism

Syed Ismyl Mahmood Rizvi
https://orcid.org/0000-0003-4268-6509
Patna University, India

ABSTRACT

Evolution is fluid, and it adapts to the shapes of its containers. This chapter contemplates the idea of liquid humanitarianism in the process of adapting to its cultural containers. True humanitarianism is evolving from the earlier stage defined by the term "humanism," which has focused primarily on human self-interest rather than geocentric interest. Collective cognitive transcendence from a humanist state to a humanitarian state constitutes initiation into an infinite cosmic reality that is based on the emerging quantum perspective of science.

INTRODUCTION

Before the term *humanitas* came into use, it was an unconscious concept intrinsic to the human species—a survival mechanism practiced among humans long before the advent of ancient cultures and civilizations. The term describes a practical state of human welfare that was critical to human survival. Beginning within families and gradually extending throughout ancient cultures and civilizations, the concept of humanitas may be loosely equated with a general evolutionary trend in the expansion of consciousness to ever-larger civilized contexts—family, clan, neighborhood, town, city, state, and nation state. The essential characteristic of humanitas is empathy—without which the current global civilization would not have been possible. The development of systems of transportation, trade and commerce, education, value structures, language, ethics, and even military defense have always been based on existential aspects of empathetic humanitas. So, it may be stipulated that modern humanism is based on empathetic cooperation. However, from a psychological point of view, the concept of humanitarianism goes beyond humanism to incorporate super-human characteristics and scales of sentience.

From the psychological point of view, this is a longstanding but largely misinterpreted ontological concept that is directly correlated with modern holistic principles such as holography and quantum field theory (QFT). For example, a fundamental principle of alchemy is synthesized in the axiom, "As it is above, so it is below". The alchemical metaphor of turning base metal into gold expresses the same

DOI: 10.4018/978-1-5225-9065-1.ch011

principle as the axiom. In the metaphor, humanitarian sentience is the gold. The alchemical ontology is that "everything is in everything". The ancient Greek philosopher, Anaxagoras (fifth century B. C. E.) maintained that the original state of the cosmos was a mixture of all its ingredients (Anaxagoras, 2015). He propounded a physical theory of "everything in everything," and claimed that *nous* (intellect or mind) was the motive cause of cosmos. In his own time, his philosophy was largely misunderstood because criticism derived from contemporary theories that did not have the same scope. From the modern QFT perspective, his philosophy becomes very practical for understanding the evolution of humanitarian sentience.

Shared empathy is based on life-experience that demonstrates the similarities among human individuals. The term "mutuality" refers to the experience of recognizing the humanity of the not-self—that other who shares a physical body, consciousness, emotions, drives, and desires that are familiar to us through our self-knowledge. This mutual empathy evolves into a cognitive set characterized by a common denominator that links human beings. During early stages of development, mutual empathy is mostly generated by defensiveness—the need for security, health, sympathy and freedom from threat, illness, rejection and subjugation. Gradually, mutuality is extended to protection of posterity, enjoyment of mutual respect, fairness and appreciation of others, and finally to the quest for self-actualization. (Aloni, Nimrod, 2007)

According to Theory of Mind, human empathy is the capacity to put oneself into someone else's shoes, to imagine their thoughts and feelings (Baron-Cohen, S., 1995; Baron-Cohen, S., et al, 1985). Empathy verifies the belief of Greek philosophers that man by nature--a social being and that the state of nature is a social state-of-being. In the state of nature, natural right and natural law prevails, but this natural state is one of cooperation as clearly indicated by current environmental research which has displaced the Darwinian view of nature as "Red in tooth and claw". The distinction is important for understanding the difference between humanism and humanitarianism. The human view of nature radically effects human behavior and value structures. The consequences of human belief that the natural environment is inferior to humanitas result in "domination", whereas the humanitarian view of nature is "protective". Leo Strauss (1953) observed that humans should act in accordance with the principles of nature in order to function in harmony with society and with their inner social being.

Only in the nineteenth century was the concept of the Latin term *humanitas* incorporated into English as Humanism. The etymology of the term assists our understanding of the fluidity of meaning attached to words. Earlier the word *humanitas* was used by Cicero to describe the shaping of an orator after acquiring learning from *bonae litterae* ("good letter", i.e., classical literature especially poetry), which would cultivate his sense of understanding for an active life in public service. Cicero's injunction to cultivate the humanities was revived by Petrarch (Francesco Petrarca), who is often considered to be the father of humanism (Wilkins, E. H., 1961). Indeed, historians debate the meaning and origin of the term 'humanism'. A recent investigator has concentrated on the phrase *studia humanitatis*, 'cultural pursuits', which he finds attested in two passages of Cicero's speeches, *Pro Murena 61* and *Pro Caelio 24*, and then in many Italian writers from Coluccio Salutati in 1369 to the middle of the fifteenth century (Petersen, E., 1988). The writings of Italian humanists were "consequently explicitly meant to prepare the pupil for political activity and were rich in political connotations and allusions (Piccolomini, A., 2002). According to Machiavellian Intelligence—the capacity to politically engage within a social group—is likely to have played a part in hominid evolution. Social memory—the capacity to remember encounters and perceive dominance hierarchies—is another crucial element of social life (Dunbar, R., 2003). Erasmus, by contrast, argues somewhat tortuously that *humanitatis*—in *humanitatis literas*—comes from the ancients' conviction that (Yoran, H., 2010) "success in study depended basically on a relationship of good will between student and master" (Erasmus, D., Puer 338; LB 1: 511A). According to Northrop Frye: "The humanist was typically a scholar and critic, rather than poet" (Frye, N., 1971). Nevertheless, as Eugenio Garin believes, by their revival of antiquity humanist educators--through their philological, critical, historical understanding of the ancients

Enabled their pupils to know the great, exemplary figures of antiquity and hence to know themselves. This could not have occurred in the Middle Ages because antiquity was not then studied historically for its own sake but subjectively and uncritically, and so medieval classical studies – in so far as they existed – could not lead to the development of the whole man. (Black, R., 2001, p. 15)

Black (2001, p. 15) emphasizes the importance of Garin's view: "Objective self-knowledge is developed only through knowledge of others; to know himself, an individual must be able to take someone else's perspective". Garin says further that:

Acquisition of historical consciousness and critical consciousness, of awareness of self and others, of an understanding of the fullness of the human world and its development […] The revived study of the ancients, rediscovered as such, came to signify the discovery of a sense of human colloquium and collaboration, the initiation to the world of men. Educating youth in the classics truly thus helped to provide the beginning of an awareness of the human community in its development and its unity. (Garin, E., 1957, p. 103)

Carl Jung's definition of the collective unconscious Psyche:

The doctor in me refuses point blank to consider the life of a people as something that does not conform to psychological law. A people, in the doctor's eyes, presents only a somewhat more complex picture of psychic life than the individual. Moreover, taking it the other way round, has not a poet spoken of the 'nations' of his soul? And quite correctly, as it seems to me, for in one of its aspects the psyche is not individual, but is derived from the nation, from collectivity, or from humanity even. In some way or other we are part of an all-embracing psychic life, of a single 'greatest' man, to quote Swedenborg. And so we can draw a parallel: just as in me, a single human being, the darkness calls forth the helpful light, so does it also in the psychic life of a people." (Jung 1933, p. 209-210)

George Lakoff—a well-known neurolinguist-- has confirmed this Jungian neurobiological dynamic by affirming that without the existence of subconscious cultural story archetypes, we cannot know ourselves. "We live our narratives. The lived story is at the center of modern personality theory," and "the theory of neural computation…shows how our brains not only permit this but favor it…The roles of narrative that you understand yourself as fitting give meaning to your life, including the emotional color that is inherent in narrative structure (Lakofff, George 2008, p. 33). Therefore, humanitarians are limited only by the capacity of human containers to adapt empathetically to the fluidity of events.

BACKGROUND

By the fourteenth-century the term humanitas was well known to certain scholars and the *studia humanitatis* were firmly enshrined in the university curriculum. The term *umanista* was used in fifteenth-century Italian academic jargon to describe a teacher or student of classical literature and the arts associated with it, including that of rhetoric. The English equivalent 'humanist' makes its appearance in the late sixteenth century with a very similar meaning (Kraye, J., 1996).

During the Middle Ages, Christian Humanism spread across Europe, but the humanist revolution during the European Renaissance altered the Christian cultural hegemony. The god-centered view of the universe changed to the man-centered perspective. In the Renaissance, Humanism devoted itself to the revival of classical culture: the life, thought, language, and literature of ancient Greece and Rome (Henderson, G., & Brown, C., 1997), and the humane values that may be derived from them. The salient feature in Renaissance Humanism that has permanent value can be applied today is its insistence on freedom from religious control of knowledge; its immense intellectual vitality; its ideal of the well-rounded personality; and above all, its stress on enjoying to the full our life in this world (Lamont, C., 1997).

The classical notions of humanism and of super-human natural law and universal principles voiced by Greek, Roman, European and other philosophers seized the imagination of those who heard them. Those notions responded—and still respond—to deeply felt human wishes for a community of peace, order and tranquility. They had a powerful idealistic impact in a chaotic world (Smyser, W., 2003, p. 17). The philosophers of the Middle Ages tried to bring reason into this chaotic world. Often working and writing in monasteries, they expanded on the original Greek and Roman concepts of natural law and natural right. Saint Thomas Aquinas, writing during the thirteenth century, declared natural law to be universally valid, universally obligatory and linked to the divine. He endorsed St. Augustine's belief that the natural law was "written in the hearts of men" and could not be blotted out (Aquinas, T., 1992). But the philosophy of Aquinas, like that of his predecessors Aristotle, Plato, St. Augustine, Marcus Tullius Cicero, Meng-tzu, etc., did not prevail. Even as he wrote of a natural law that transcended human law, the world of a universal church collapsed. Secular lords, and especially kings, began to gather greater power. They claimed a more immediate authority than the Pope's. In the process of unmasking the pretensions of super-human authority claimed by the church, secular authorities not only rejected universal natural law but imposed their own separate laws. The validity of super-human authority was further besmirched as they fought each other for land, wealth and power (Smyser, W., 2003, p. 18). Violent wars raged across Europe as the continent faced both religious and political upheaval and as the new nation-states fought for position (Smyser, 2003, p. 21). The Peace of Westphalia, which ended the Thirty Years' War in a series of agreements signed in the German cities of Münster and Osnabrück in 1648, formally put an end to the Pope's and the Holy Roman Emperor's universal claims. It gave all rulers, whether kings or lords, full authority within their domains. It put an end to any hopes for a common universal order, whether of divine law or natural law. Secular rulers could decide the laws and even the religions that would govern their subjects (Smyser, 2003, pp. 21-22). The new secular doctrine has come down to the present as the principle of sovereignty. Under that principle, each national government has total authority within its own borders. It can do as it pleases within those borders. Any attack from any other state against or across those borders is an act of war (Sommer, T., 1998). The principle of absolute sovereignty has its origins in the nationalist doctrines laid down by the kings of England, France and Spain over 500 years ago (Smyser, 2003, p. 1).

The incipient modern humanitarian movement arose after national abuses of sovereignty became even more extreme than usual during the middle of the nineteenth century (Smyser, 2003, p. 1). Today, humanitarian agencies are evolving in their mandate to serve humankind and the planet. They are doing whatever they can to help those who need help. Before 1850, although its origins went back through centuries of humanist and religious philosophy, organized humanitarian sensibilities were nearly non-existent. Due to the wars of the eighteenth and early nineteenth centuries, organized humanitarian efforts such as the Red Cross began to emerge. Their purpose was mainly to care for soldiers and prisoners of war. After 1945, humanitarians suddenly had more to do, as Western leaders helped the world to recover from war and supported wider humanitarian action, peacekeeping and refugee care. (Smyser, 2003, p. 259).

The growth of humanitarian sensibilities increased as the evils of slavery and of colonization were partially renounced. Due to the destructive forces of the two world wars behind us, all societies became more sensitized and informed about the need for a just and stable world. After the main tenets of basic human rights had been established in 1948 (Abdi, A., & Shultz, L., 2008, p. 1), the international human rights regime emerged, with the founding of the United Nations and the adoption of the Universal Declaration of Human Rights 1948 (Helfer, L. R., & Graeme, W. A., 2011, p. xi).

Much remains to be done. Still, millions of people have been and are being persecuted, tortured, and killed based on who they are, or which groups they were born into. The killing fields of Cambodia, the genocide in Rwanda, the bloodstained hillsides of Bosnia-Herzegovina, the tragedy in Darfur, Sudan, and yes, the HIV/AIDS pandemic in sub-Saharan Africa (and potentially forthcoming in India and China), among a myriad of other failures, are symptomatic of a critical need for expanded humanitarian efforts. Ironically, the global human rights agenda seems to have betrayed many of the world's citizens (Abdi, A., & Shultz, L., 2008, p. 2). Such betrayal was not the doing of humanitarians, but of those who misused humanitarian principles for political purposes. As John R. Wallach asserts, the concept of human rights became more humanist than humanitarian:

Human rights talk rose to such prominence because their ideological fluidity and ambiguity allowed them to become a 'tool of the powerful.' On the one hand, they represent stasis, constraining political actors and institutions within a universal and international code, and on the other hand they represent a powerful moral charter to pursue social change. In the 1990s, the definition of rights shifted from the former to the latter, thus furnishing states with an 'ethics of power' that permitted them to reshape domestic policies, as well as to refashion foreign policy and intervene militarily in regions of political instability. While it is true that human rights came to coincide with the national self-interest of powerful states, in so doing, national self-interest was itself transformed. This was especially the case in a Europe pursuing greater economic and political integration, where seeking intergovernmental solutions to political conflicts became an ingrained way of conducting international affairs. (Wilson, R., 2005, p. 6)

Nevertheless, modern humanism—perhaps due to its over-emphasis on the hubris of human priority—seems to be devolving into a post-apocalyptic dystopia. Humankind is now perceived as the centre of the universe, but the virtues of this ascendancy are being questioned and debated by existential humanists. As Jean-Paul Sartre writes in *Existentialism is a Humanism*: "Man first of all exists, encounters himself, surges up in the world—and defines himself afterwards" (Sartre, J., 1946). Sartre confirms that: "Existentialism will never consider man as an end, because man is constantly in the making. And we have no right to believe that humanity is something we could worship, in the manner of Auguste Comte.

The cult of humanity leads ultimately to an insular Comteian humanism and – this needs to be said – to Fascism. We do not want that type of humanism" (Sartre, 1946).

As we know, Carl Jung proposed to distinguish between people on the basis of the degree of expression of each of the four basic mental functions (sensation, intuition, thinking, and feeling) in the context of their extraversion and introversion (Drozdovski, A., 2015, p. 162). According to Jung, what constitutes upward mobility for humanity must be psychologically defined in terms of both thinking and feeling functions (Jacoby, J., 1973, p. 12). Value intensity is a measure of contextual (personal) importance, and whether someone feels something pleasant or unpleasant is a measure of whether something is perceived as "good" or "bad". The latter is an emotional measure which, according to Jung, is a dynamic and fluid aspect in the healing process.

Most important, the therapeutic aspect of this "upward mobility" of humanity, this fluid evolutionary process implied by Sartre has to do with making intuitive (spiritual) choices; i.e., from the perspective of the general welfare a person can choose to act in positive or negative ways. Here it is also clear that since man can choose to be either cruel or good, he is, in fact, neither of these things essentially (Akhter, R., 2014, p. 86). It may be stipulated that, because of this choice-making dynamic, humans are constantly remaking themselves and determining their own collective future. For humans, this evolutionary plan seems to entail transcendence from a state of humanist *hubris* to a state of empathetic humanitarianism. Humanitarians must be engaged in a fluid state of decision-making that always looks to the future.

FLUID FRAMEWORK FOR ETHICS IN HUMANITARIAN DECISION-MAKING

Philosophical contemplation regarding the ethics of individual versus collective welfare are longstanding. Nevertheless, a consensus exists relative to the characteristics of fluid ethical behavior in complex social systems, and these characteristics are essentially humanitarian.

Aristotle defined ethics in terms of behavior consistent with the common good. However, in the emergent reality of physics, cognitive psychology, and the media, the human scientific perception of reality has been drastically complicated. Reality is no longer solid and stable. It has all the fluid dynamics of dreams. Ironically, the modern scientific reality has become metaphysical and—as many prominent physicists have pointed out—the principles of myth and science have converged. This integrated reality is characterized by non-locality (spooky action at a distance), human choices generated primarily from unconscious patterns projected as symbols (archetypal representations that appear to be real but are projections from the personal & collective unconscious), and the rapidly emergent mediated reality that is dreamlike (according to Jungian principles, dynamics, and purposes of dreams). In other words, determinations of what is ethical and humanitarian in a social context must be reconsidered, and a humanitarian ethics based on fluid ILLUSION must be created.

According to most modern socio-political assumptions, ethical behavior consists on a continuum of balance between the individual good and the collective wellbeing. Most importantly, in the scientific-metaphysical reality the idea of the "common good" must be extended and understood to apply to all environmental life forms. Disagreements in perceptions of values such as when a fetus becomes a human, whether experimentation should be done on animals, whether plant genetics should be modified (GMO), or whether unconscious human emotions should be manipulated for profit (advertising and neuromarketing) have raised many ethical issues. Ironically, because the common good has been defined in insular ways, arguments for the common good have often resulted in outrageously unethical behavior.

Given this dilemma, a refinement of definitions seems to be in order, and our definitions must be framed in the ethical context of liquid ethical paradigms—the fluid dynamics and interference patterns of the paradigm shift.

In these vague and unknown contexts of ethical social function, the fundamental values of myth must be reconsidered. Modern students of myth have observed universal parallels among the world's myths, and these parallels may be correlated with Carl Jung's principles of dream analysis. Joseph Campbell has observed that "myths are psychology". In other words, myths are dreamlike and they can be analyzed like dreams.

Jung's principles for analysis of dreams seem to provide a practical context and framework for contemplation of humanitarian social ethics—especially social media—during the emergent reality-shift. Jung's principles include "individuation" which is the point of complete "self" realization that is learned through experience. Another principle is "harmony", and harmony can be understood as Heart-coherent energy patterns (as in a concert). Other important principles include "empathy", "the dream has a purpose," and that purpose is "projected as images." By "projection", Jung meant that images are something like mirrors of subconscious patterns that act as a language that tells individuals and collectives how they might behave in-order do to learn and evolve.

Redefining patterns of the unconscious in terms of humanitarian patterns of ethics is the challenge. In order to optimize this exercise, we should redefine the sentience of a holographic media-sphere in terms of the structure and purpose of dreams. The objectives of our dreamlike experience should be characterized by empathy and harmony—which can now be measured as coherence. Fundamental mythic value contexts of social harmony become comprehensively practical within the framework of compassionate behavior. When authentic empathy and compassion are the arbiters, unethical behavior is minimized. Such compassion seems to thrive when people are informed, and the universal mythic pattern of the Hero's Journey provides a humanitarian training ground. Humanitarianism has to do with practicing heroic choices.

Liquid Humanitarianism

The social world is increasingly recognized as being fluid, fluxing and changing (Schultze, U., 2000). According to systems design, organizations must be both stable and responsive to the environment (Styhre, A., 2008, p. 13). Jung defined this design principle in terms of "Psychic contextuality". Social-sentient reality is fluid and changing rather than stable and fixed. It is heterogeneous and composite rather than unified and homogeneous (Styhre, 2008, p. 219). As a consequence, the distinction between humanism and humanitarianism is permeable and fluid: perhaps, then, the important distinction between humanism and humanitarianism has to do with the degree and flow of disciplined holistic choices made by cooperating humanitarians.

Making collective heroic choices is a challenge. One must ask, what is the greatest threat to humanity? The answer might be an uncontrolled viral disease, a massive asteroid strike, nuclear war, or any natural calamity capable of obliterating humanity. In point-of fact, it may be stipulated that inhumanity is the greatest threat to humanity and the planet. "We don't need any atom bomb. The uprooting of man has already taken place" (Heidegger, M., 1993).

Reactively, inhumanity is opposing all humanitarian progress in a self-serving effort to maintain its illusory humanistic supremacy and to destroy the seeds of the future that threaten its delusions. This opposition is now blatantly evident in the daily news. Opposing this destructive force constitutes the global human challenge, but the challenge is complicated by the fact that humans—both individual and collective—are, by nature, schizophrenic—because they contain both the rotten fruit of the past and the seeds of the future. To understand this paradox, the human being must be understood as a conflicting mixture of human and inhuman impulses that are constantly at war. These conflicting components of the human organism are like Siamese twins who cannot be cut away from each other. Though this fluid battle has many fronts, many defeats, and many victories, the human imperative is to establish coherent planetary sentience—to create single-minded cooperation on a global humanitarian scale of sentience. In other words, the cure for inhumanity is humanity itself. Science tells us that the healing process begins at sub-atomic, molecular scales of universal sentience.

These conflicting energies of human psyche have been comprehensively researched by Carl Jung who provided a way to neutralize the inhuman and to maximize the human with a process of dream analysis for the purpose of "compensating" for unhealthy cognitive patterns with coherent patterns which he called the overall process "individuation", and this process of "Individuation" constitutes a method for creating coherence (harmonious balance) between fluid conflicting cognitive energies. However, the conflict extends beyond the human physical nature to the effects that human thoughts and feelings have on the sentient universe. For example, in a sentient universe, human thoughts and feelings appear to have correspondences on other scales of sentience such as the atomic and biological (molecular, plant, animal) and super-human levels that exist as pervasive subjective force fields such as intuition, inspiration, compassion, hope, and love.

Selecting humanitarian aid projects requires complex fluid decisions about where to go (location), what to do (activities), when to start (initiation) or not (rejection), and when to extend or end activities (prolongation and termination). These decisions need to be made in contexts which can be characterized as 'complex humanitarian crises', referring to the fact that these crises are both complex in cause and consequence (Duffield, M., & Prendergast, J., 1994; Harriss, J., 1995; Albala-Bertrand, J.M., 2000).

Most humanitarian crises develop from a series of interrelated causes, such as intra-societal tensions, a colonial heritage, environmental degradation, economic decline, or an unequal distribution of power and economic resources (Field, J., 1993; Duffield, M., & Prendergast, J., 1994). These causes result in conflict, hunger, destruction of political, social and economic infrastructures, and population movements within or crossing the national borders (Heyse, L., 2006). In such a complex context, the decision-making process about humanitarian aid interventions is constrained by many obstacles (Cuny, F., 1983; Smillie, I., 1995).

Access is the first obstacle to conquer, obstructing humanitarian aid organizations to independently decide about their project locations. Humanitarian aid organizations must often deal with the absence of stable political structures that require ongoing negotiations relative to their access. Of course, it is a well-known fact that humanitarian aid is often diverted by, the local ruling elite for its own benefit by allowing the organization to provide aid to some groups while not to others (Heyse, L., 2006). Humanitarian crises vary widely in nature and setting, but they will always be characterized by: violence, direct exposure to violence, witnessing violence, loss of family members, displacement, food scarcity, increased exposure to communicable diseases, reduced access to health services, plus a range of other socioeconomic setbacks (Hill, K., 2004).

All of these problems are complex, but Paula Reed Lynch of the U.S. Department of State has listed four mantras for working in complex humanitarian emergencies. First, use local resources. Second, use global resources. Third, learn from previous research and program experiences. And fourth, you cannot easily change directions when a program has already begun. For example, if a feeding program has begun by giving food to men but discovers that it is better to give the food to women, it may be difficult to change the established routine. Therefore, research must be built into a program from the beginning, and programs must build on knowledge gained from research in previous emergencies. (Reed, T., 2002)

Human agency always entails complex permutations of flexible response semi-predictable outcomes to which organizational routines and operations must be adjusted. Human agency thus serves as an intermediary role between organizational operating policies and ideals. In any case, agency must be fluid. (Bourdieu, P., 1990, 1977; de Certeau, M., 1984). According to Alasdair Macintyre social "practice" denotes "coherent and complex forms of socially established cooperative human activity" (Macintyre, A., 1981, p. 187). In other words, social practice is cooperative not competitive and it should be prospective—looking to future improvement. Humanitarian resolution of the current crisis depends upon the collective capacity to develop a futurist collective perspective. This future-orientation of human agency is rare, and developing this future perspective requires close ongoing scrutiny and a high order of creativity in-order to make discriminating choices. The future perspective requires the ability to see through an iron curtain of archaic rational-materialistic thoughtforms and defensive self-serving cognitive patterns.

The longstanding fabrications of the human world have generated processes of dehumanization and empowered domains of the inhuman (Tester, K., 1995). Currently the trajectory of dehumanizing processes is very powerful, and this trajectory must be changed. However, changing the trajectory of inhumanity resides in perfecting the quality of humanitarian sentience. Ongoing research on the dynamics of unconscious human sentience has provided new tools for understanding human sentience and for empowering its positive aspects. These tools are based on understanding the sentient states of quantum agency that can be generated within the dramatic structure of a coherent media.

Going Beyond Conformity to Humanistic Limitations

Why can't most people think beyond established religious beliefs and cultural practices? Why are human brains scrambled with stored information based upon what other people have believed and the perspectives of others instead of trusting their own perceptions? Why have humans become so blind? Do we still qualify as human beings or have we become something less and created the worst age of humanitarian crises? Why have some been more effected by inertia than others? The answers to these question and many more reside in understanding universal sentience and using this humanitarian concept to shape the world around us.

It is clear that humanism is characterized more by conformity than by evolution. The primary dynamic of humanistic thinking is a kind of conformity that is moribund. The pejorative dynamics of "blending in" "following the crowd", and even "going with the flow" are actually manifestations of the herd mentality that has little to do with making the innovative choices required of humanitarians. However, the stasis of herd mentality is gradually evolving into a greater appreciation of individual responsibilities within a collective. Moreover, the collective now includes all life and sentience, planetary and cosmic. The complexity of quantum humanitarian flow can no longer be ignored. The reason underlying this complex flow of sentience among people is the infinity of differences that result from value systems, race, religion, ideology, gender, attitude, diet, etc. Problem-solving among future humanitarians must be addressed within the context of a human agency that is based verisimilitude with authentic quantum-cognitive fluidity.

Our decisions are a mix of positive and negative—generated from past experiences or from the future condensed into a fluid momentary present. In a word, because our sentient processes are fluid, perhaps they could be adjusted according to the mathematics of fluid dynamics. We often deviate from our personal decisions and judgements, but we are held together within a *social force field*. The dynamics of such force fields may be seen in the seemingly superconductive events of the times: unification of the working class as the Union movement throughout the world, mobilization of the poor, and *The Communist Manifesto* by Karl Marx brought a new perspective to class distribution. Sometimes, single acts can trigger spontaneous collective flow. Mohammed Al Bouazizi's self immolation spurred a people's revolution in Tunisia. Ghiath Matar, or little Gandhi, became a symbol of peaceful resistance of the Syrian revolution. The renaming of an Abu Dhabi mosque in honour of Mary, Mother of Jesus, as 'Mariam Umm Eisa' mosque, signifies to promote religious tolerance among Christians and Muslims. The mosque is located next to Catholic, Anglican and Coptic Orthodox churches (Crompton, P., 2017).

Humanitarian Sentience

The evolution of a new humanity characterized by humanitarian sensibilities is flowing toward the future like a river. "No man ever steps into the same river twice, for it's not the same river and he's not the same man." "One cannot step twice into the same river, nor can one grasp any mortal substance in a stable condition, but it scatters and again gathers; it forms and dissolves, and approaches and departs." (Heraclitus, LI, D. 91).

Heraclitus shared an important understanding among Greek philosophers—a consensus about the fluid nature of existence. Greek natural philosophy from Anaximander to Aristotle and beyond shared this understanding of what amounts to a quantum-dynamic ontology. (Kahn, C. H., 1979, p. 166). It is precisely this notion of pervasive physical change that the Eleatics rejected as false (Melissus fr. 8.3), while the physicists took it for granted as an obvious truth of experience (Diogenes fr. 2). Plato has this in mind when he speaks of a doctrine of continual change and becoming that is held by Protagoras and Heraclitus and Empedocles "and by all the wise men except Parmenides" (Theaetetus 152E) and which he finds in the Heraclitean image of the river, where "everything moves on and nothing stands still". (Cratylus 402 A8) Plato's interpretation of opposing forces--cold-warm and moist-dry—is another way of expressing the thesis of radical flux. (Kahn, C. H., 1979)

The nature/nurture debate is longstanding, and both sides of the argument have been extensively studied. The debate as to the extent to which the environment or genetics influence social behaviors remains unresolved. As Bryan Turner stresses: "For the individual and the group, the body is simultaneously an environment (part of nature) and a medium of the self (part of culture)" (Turner, B., 1984). However, the issue extends far beyond the physical body to all of the classical human functional bodies; Mind, Emotion, Perception, and Intuition.

In today's scientific environment—just as in the theological environment—humanitarian sentience must be contemplated from the perspective of quantum integration of these functions. For example, abundant research has verified the obvious—that thoughts, emotions, perceptions, and intuitions influence and are influenced by the environment. All of the functions are integrated, and the reason the physical body is studied as the source of the other functions is that flesh is convenient for research. But what if the individual body is not the source of all ideas, thought, feeling, and moral sensibility? Research already verifies the influence of digestive bacteria on the dietary choices made by humans (Kuspa, Adam, MD,

2018; Sheikh, Knuval, 2017; Collins, Sonya, ND). It appears that any further contemplation of environmental influences on social function must be extended to quantum-sentient levels.

Perhaps common sense based on experience—resulting in spontaneous superconductivity--would provide a more accurate "flow" of compensational data upon which to act. Universal sentience is based on the quantum electrodynamic concept of pervasive integration. Applied to issues of social mechanism the implication that thought adheres to electro-magnetic principles seems obvious. Images offer a key for researching thought as electro-magnetic--Mind has invisible magnetic effects expressed as the interplay of past experiences, pictures stored as memories, and fantasies. It is probable that the receptor brain attracts and repels information according to contextual situations. As Niccolò Machiavelli says in his main work *The Prince* "Men and women acted as they had to act within such a world, not as they might choose to act in a perfect world. An immoral world justified immoral means" (Smyser, 2003, p. 20). This constitutes the present state of reductive affairs.

According to both Jung and Lakoff, group behavior is based on the same dynamics as personal behavior (Jung 1933, p. 202; Lakoff, 2008, pp. 33-34). Understood as intuition, (McCraty, Rollin & Schafer, Stephen, 2016; McCraty, Rollin & Childre, Doc, 2010; Bradley, Raymond Trevor, 2006) the collective brain transmits signals according to the group behavior like antenna of ants transmit signals to communicate with each other. This signal could include 'prediction error' that partially explains behavior that is both positive and negative (a prediction error arises from the difference between expected and actual outcomes). This signal could be conformist--a true adoption of what other people think.

Group-think is a factor that characterizes much social research. There are a number of writers who have suggested that whether it be the contemporary age, the second modernity (Ulrich Beck) or the late modern (Anthony Giddens), hypermodern (John Armitage), super-modern (Marc Augé) or postmodern society, is an age characterized of transient social relations and fluid and fluxing institutions and coalitions (Bauman, 2000; Urry, J., 2000). These social characteristics signify the "liquidity" of evolution beyond conformity to humanist principles.

CONCLUSION

The metaphors of 'liquidity' and 'fluidity' attempt to grasp the current state of society and humanity. According to Philip Ball, "liquid water is essential for the kind of delicate chemistry that makes life possible. Liquid water is a universal solvent, a mediator of life's chemical reactions, and it has a structure unlike that of any other liquid" (Ball, P., 2001). Every cell in a human body is 'pushed and pulled' effectively by some invisible living-sentient forces largely unknown to us. Collectively, the entire human body is made of 'liquid cells' which are easily stimulated to act according to the prescribed situations. Human relationships, human nature, human fabrication and human behavior are changeable and constantly influenced by others in the environment. Therefore, more emphasis should be placed on research of collective and universal "humanitarian" sentience.

REFERENCES

Abdi, A. A., & Shultz, L. (2008). *Educating for human rights and global citizenship*. Albany, NY: State University of New York Press.

Akhter, R. (2014). *Existentialism and its relevance to the contemporary system of education in India: Existentialism and present educational scenario*. Hamburg, Germany: Anchor Academic Publishing.

Albala-Bertrand, J. M. (2000). Complex emergencies versus natural disasters: An analytical comparison of causes and effects. *Oxford Development Studies*, *28*(2), 187–204. doi:10.1080/713688308

Aloni, N. (2007). *Enhancing humanity: The philosophical foundations of humanistic education*. Dordrecht, The Netherlands: Springer.

Anaxagoras. (2015). Anaxagoras of Clazomenae. In *Stanford Encyclopedia of Philosophy*. Retrieved from http://plato.stanford.eddu/entries/anaxagoras.

Annette, B. S., Oernbo, E. K., Stoica, A., Gerkau, N. J., Barbuskaite, D., Tritsaris, K., ... MacAulay, N. (2018). Cotransporter-mediated water transport underlying cerebrospinal fluid formation. *Nature Communications*, *9*(1). PMID:29867199

Aquinas, T. (1986). The order of learning the sciences (A. Maurer, Trans.). In *The division and methods of the sciences* (pp. 99–102). Toronto: Pontifical Institute of Mediaeval Studies.

Aquinas, T. (1992). Summa theologica, question 94, on the natural law. In M. L. Morgan (Ed.), Classics of moral and political theory (pp. 482–483). Indianapolis, IN: Hackett.

Aristotle. (1947). Analytica Posteriora (G. R. G. Mure, Trans.). In McKeonR. (Ed.), *Introduction to Aristotle* (pp. 9–109). New York: Modern Library.

Aristotle. (2004). *Nicomachean Ethics (C. Rowe, Trans.)*. Oxford, UK: Oxford University Press.

Ascham, R. (1970). Report of the affair and state of Germany. In W. A. Wright (Ed.), *English Works* (pp. 123–169). Cambridge, UK: Cambridge University Press.

Augustine. (1998). *The city of God against the pagans* (R. W. Dyson, Trans. & Ed.). Cambridge, UK: Cambridge University Press.

Ball, P. (2001). *Life's matrix: A biography of water*. Berkeley, CA: University of California Press. Retrieved from https://www.nasa.gov/vision/universe/solarsystem/Water:_Molecule_of_Life.html

Baron-Cohen, S. (1995). *Mindblindness: An essay on autism and theory of mind*. Boston, MA: MIT Press/Bradford Book. doi:10.7551/mitpress/4635.001.0001

Baron-Cohen, S., Leslie, A. M., & Frith, U. (1985). Does the autistic child have a 'theory of mind'? *Cognition*, *21*(1), 37–46. doi:10.1016/0010-0277(85)90022-8 PMID:2934210

Baudrillard, J. (1998). When Bataille attacked the metaphysical principles of economy. In F. Bottig & S. Wilson (Eds.), *Bataille: A critical reader*. Oxford, UK: Blackwell.

Bauman, Z. (2000). *Liquid modernity*. Cambridge, UK: Polity Press.

Black, R. (2001). *Humanism and education in Medieval and Renaissance Italy: Tradition and innovation in Latin schools from the twelfth to the fifteenth century*. New York: Cambridge University Press. doi:10.1017/CBO9780511496684

Bourdieu, P. (1977). *Outline of a theory of practice*. Cambridge, UK: Cambridge University Press. doi:10.1017/CBO9780511812507

Bourdieu, P. (1990). *The logic of practice*. Cambridge, UK: Polity Press.

Bourdieu, P. (2004). *The science of science and reflexivity*. Chicago, IL: The University Of Chicago Press.

Bradley, R. T. (2006). *The psychophysiology of entrepreneurial intuition*. Auckland, New Zealand: Institute for Whole Social Science.

Cara, C. (2009). Creating a coherent culture: Building a culture that enables innovation and change. *International Journal of Knowledge, Culture and Change Management, 9*(4), 15-34.

Certeau, M. de. (1984). Practices of everyday life. Berkeley, CA: University of California Press.

Collins, S. (2014). What Is Your Gut Telling You? *WEBMED*. Retrieved from https://www.webmd.com/digestive-disorders/news/20140820/your-gut-bacteria#1

Crompton, P. (2017). *Mary, Jesus' Mother' is new name for UAE mosque: Mohammad Bin Zayed orders name change to 'consolidate bonds of humanity between followers of different religions'*. Retrieved from https://gulfnews.com/news/uae/culture/mary-jesus-mother-is-new-name-for-uae-mosque-1.2043544

Cuny, F. (1983). *Disasters and Development*. New York: Oxford University Press.

Drozdovski, A. K. (2015). The connection between typological complexes of properties of the nervous system, temperaments, and personality types in the professions and sports. *Open Access Journal of Sports Medicine, 6*, 161–172. doi:10.2147/OAJSM.S75612 PMID:26056499

Duffield, M., & Prendergast, J. (1994). *Without troops and tanks: The emergency humanitarian desk and the cross border operation into Eritrea and Tigray*. Lawrenceville, NJ: The Red Sea Press.

Dunbar, R. I. M. (2003). The social brain: Mind, language, and society in evolutionary perspective. *Annual Review of Anthropology, 32*(1), 163–181. doi:10.1146/annurev.anthro.32.061002.093158

Durham, W. H. (1991). *Coevolution: Genes, culture, and human diversity*. Stanford, CA: Stanford University Press.

Erasmus, D. (1965b). Christian Humanism and the Reformation (J. C. Olin, Ed. & Trans.). New York: Harper and Row.

Erasmus, D. (1965a). *The colloquies of Erasmus (C. R. Thompson, Trans.)*. Chicago: University of Chicago Press.

Erasmus, D. (1985a). De pueris instituendis / De recta pronuntiatione. In *Collected Works of Erasmus* (Vol. 4). Toronto: University of Toronto Press.

Erasmus, D. (1985b). A Declamation on the Subject of Early Liberal Education for Children. Beert C. Verstraete (Trans.). In J. K. Sowards (Ed.), *Collected Works of Erasmus* (Vol. 26, pp. 291–346). Toronto: University of Toronto Press.

Erasmus, D. (1985c). On Good Manners for Boys. Brian McGregor (Trans.). In J. K. Sowards (Ed.), *Collected Works of Erasmus* (Vol. 25, pp. 269–289). Toronto: University of Toronto Press.

Field, J. O. (Ed.). (1993). *The challenge of famine: Recent experiences, lessons learned*. West Hartford: Kumarian Press.

Frye, N. (1971). *The critical path: An essay on the social context of literary criticism*. Bloomington: Indiana University Press.

Garin, E. (Ed.). (1957). L'educazione in Europa (1400–1600). Rome: Bari.

Garin, E. (1965). *Italian humanism (P. Munz, Trans.)*. Oxford, UK: Blackwell.

Gay, P. (1995). *The Enlightenment: The rise of modern paganism*. New York: W. W. Norton & Company. (Original work published 1966)

Grafton, A. (2004). *Bring Out Your Dead: The Past as Revelation*. Cambridge, MA: Harvard University Press.

Harriss, J. (Ed.). (1995). *The politics of humanitarian intervention*. London: Pinter Publishers.

Heidegger, M. (1978). Basic Writings (D. F. Krell, Ed.). London: Routledge & Kegan Paul.

Heidegger, M. (1993). "Only a God can save us": Der Spiegel's Interview with Martin Heidegger (1966). In R. Wolin (Ed.), *The Heidegger controversy: A critical reader* (pp. 105–106). Cambridge, MA: The MIT Press.

Helfer, L. R., & Austin, G. W. (2011). *Human rights and intellectual property mapping the global interface*. New York: Cambridge University Press.

Henderson, G. E., & Brown, C. (1997). *Glossary of Literary Theory*. Toronto: University of Toronto Press.

Henrich, J., & McElreath, R. (2003). The evolution of cultural evolution. *Evolutionary Anthropology: Issues. News Review (Melbourne)*, *12*(3), 123–135.

Heyse, L. (2006). *Choosing the lesser evil: Understanding decision making in humanitarian aid NGOs*. Hampshire, UK: Ashgate Publishing Limited.

Hill, K. (2004). *War, humanitarian crises, population, displacement, and fertility: A review of evidence*. Washington, DC: The National Academies Press.

Huizinga, J. (1984). *Erasmus and the Age of Reformation*. Princeton, NJ: Princeton University Press. (Original work published 1924) doi:10.1515/9781400858071

Jung, C. G. (1933). *Modern man in search of a Soul*. New York: Harcourt Brace & Co.

Kahn, C. H. (1960). Anaximander and the Origins of Greek Cosmology. New York: Columbia University Press.

Kraye, J. (1996). *The Cambridge companion to Renaissance humanism.* Cambridge University Press. doi:10.1017/CCOL0521430380.016

Kristeller, P. O. (1979). Renaissance thought and its sources (M. Mooney, Ed.). New York: Columbia University Press.

Kuspa, A. (2018). *Lectins help social amoeba establish their own microbiome.* Baylor College of Medicine. Retrieved from https://phys.org/news/2018-07-lectins-social-amoeba-microbiome.html

Kwan, V. S. Y. (2008). Missing links in social cognition: The continuum from nonhuman agents to dehumanized humans. *Social Cognition, 26*(2), 125–128. doi:10.1521oco.2008.26.2.125

Lakoff, G. (2008). *The political mind.* New York: Penguin Publishing Group.

Laurie, T. (2015). Becoming-animal is a trap for humans: Deleuze and Guattari in Madagascar. In H. Stark & J. Roffe (Eds.), *Deleuze and the Non-Human* (pp. 142–162). Hampshire, UK: Palgrave Macmillan. doi:10.1057/9781137453693_9

Luhmann, N. (1995). *Social systems.* Stanford, CA: Stanford University Press.

Machiavelli, N. (2008). *The Prince.* New York: Random House Modern Library.

Macintyre, A. (1981). *After virtue.* London: Duckworth.

McCraty, R., & Childre, D. (2010). Coherence: Bridging personal, social, and global health. Alternative Therapies, 16(4).

McCraty, R., & Schafer, S. (2016). *Exploring dimensions of the media dream: Functional context in collective personae.* Hershey, PA: IGI Global. doi:10.4018/978-1-4666-9891-8.ch001

McNiell, Hardy, W. (1981). Discrepancies among the social sciences. *Conspectus of History, 1*(7), 37–38.

Morawski, S. (1994). The hopeless game of Flânerie. In K. Tester (Ed.), *The Flâneur* (p. 181). London: Routledge.

Parsons, T. (1951/1991). *The social system.* London: Routledge.

Petersen, E. (1988). The communication of the dead: notes on the *studia humanitatis* and the nature of humanist philology. In A. C. Dionisotti, A. Grafton, & J. Kraye (Eds.), *The Uses of Greek and Latin: historical essays* (pp. 57–69). London: Warburg Institute and University of London.

Piccolomini, A. S. (2002). The education of boys (C. W. Kallendorf, Trans. & Ed.). In *Humanist Educational Treatises* (p. 43). Cambridge, MA: Harvard University Press.

Reed, H. (2002). *Research ethics in complex humanitarian emergencies: Summary of a workshop.* Washington, DC: National Academy Press.

Reed, T. J. (2015). *Light in Germany: scenes from an unknown Enlightenment.* University of Chicago. doi:10.7208/chicago/9780226205243.001.0001

Robinson, G. E., Fernald, R. D., & Clayton, D. F. (2008). Genes and social behavior. *Science, 322*(5903), 896–899. doi:10.1126cience.1159277 PMID:18988841

Sartre, J.-P. (1946). *Existentialism is a humanism.* Yale University Press.

Schafer, S. B. (2016). *Employing dynamics of Jungian compensation to foster psychocological coherence with ICTs. International Journal of Privacy and Health Information Management*, 4.

Schultze, U. (2000). A confessional account of an ethnography about knowledge work. *Management Information Systems Quarterly*, *24*(1), 3–41. doi:10.2307/3250978

Sheikh, K. (2017). How Gut Bacteria Tell Their Hosts What to Eat: By suppressing or increasing cravings, microbes help the brain decide what foods the body "needs". *Scientific American: Neuroscience.* Retrieved from https://www.scientificamerican.com/article/how-gut-bacteria-tell-their-hosts-what-to-eat/

Smillie, I. (1995). *The Alms Bazaar: Altruism under fire, non-profit organizations and international development.* London: Intermediate Technology Publications. doi:10.3362/9781780446127

Smith, M. P. (2001). *Transnational urbanism: Locating globalization.* Malden, MA: Blackwell.

Smyser, W. R. (2003). *The humanitarian conscience: Caring for others in the age of terror.* Hampshire, UK: Palgrave Macmillan.

Sommer, T. (1998). Teure Frucht des wüsten Krieges. *Die Zeit (Hamburg)*, *22*, 94.

Strauss, L. (1953). *Natural Right and History.* Chicago: University of Chicago Press.

Stuart Hughes, H. (1952). *Consciousness and society: The reorientation of european social thought 1890 - 1930.* New York: Random House, Vintage Books.

Styhre, A. (2008). *Science-based innovation from modest witnessing to pipeline thinking.* Hampshire, UK: Palgrave Macmillan.

Tester, K. (1995). *The inhuman condition.* London: Routledge.

Turner, B. (1984). *The body and society.* Oxford, UK: Basil Blackwell.

Universal Declaration of Human Rights. G.A. Res. 217A, art. 27, U.N. GAOR, 3d Sess., 1st plen. mtg., U.N. Doc. A/810 (Dec. 12, 1948).

Urry, J. (2000). *Sociology beyond societies: Mobilities for the twenty-first century.* London: Routledge.

Vergerio, P. P. (2002). The character and studies befitting a free-born youth (C. W. Kallendorf, Trans. & Ed.). In *Humanist Educational Treatises* (p. 43). Cambridge, MA: Harvard University Press.

Wallach, J. R. (2005). Human rights as an ethics of power. In R. A. Wilson (Ed.), *Human rights in the 'war on terror'.* New York: Cambridge University Press. doi:10.1017/CBO9780511511288.006

Weissbrodt, D. (2001). *Selected International Human Rights Instruments* (3rd ed.). Cincinnati, OH: Anderson Publishing Co.

Wilkins, E. H. (1961). Life of Petrarch. University of Chicago Press.

Williams, B. (1985). *Ethics and the Limits of Philosophy.* Cambridge, UK: Cambridge University Press.

Wilson, R. A. (Ed.). (1997). *Human Rights, Culture & Context: Anthropological Perspectives*. London: Pluto Press.

Wilson, R. A. (2005). *Human rights in the 'war on terror*. New York: Cambridge University Press. doi:10.1017/CBO9780511511288

Wilson, R. A., & Mitchell, J. P. (Eds.). (2003). *Human Rights in Global Perspective: Anthropological Studies of Rights, Claims and Entitlements*. London: Routledge.

Yoran, H. (2002). The humanist critique of metaphysics and the foundation of the political order. *Utopian Studies*, *13*, 1–19.

Yoran, H. (2007). Florentine civic humanism and the emergence of modern ideology. *History and Theory*, *46*(3), 326–344. doi:10.1111/j.1468-2303.2007.00413.x

Yoran, H. (2010). *Between utopia and dystopia: Erasmus, Thomas More, and the humanist republic of letters*. Plymouth, UK: Lexington Books.

Chapter 12
Gaming Cultural Atonement:
Healing Collective Personae With Mediated Biofeedback

Stephen Brock Schafer
Pacific Rim Enterprises, USA

ABSTRACT

According to theological precedents and science—just as sub-human sentient scales exist—so must super-human scales of sentience exist in a quantum field structured as fractal geometry. If the human organism is multi-functionally "conscious," so is the quantum field in which it is embedded. Just as differences exist among the frequencies of light, differences exist in harmonic scales of the universal frequency spectrum. The loosely defined term "consciousness" is too limited to express the cosmic range of "sentient" access to frequencies. Though humans have unique choice-making agency, theological-philosophical and experiential precedents which place limits on its coherence. With their emphasis on coherent choice-making, PSYCHecology video games (PEGs) could be designed as "dreamlike" analogs to harmonize functional agency. According to Jungian principles, these dream analogs could be employed to reinforce coherent humanitarian sentience with mediated biofeedback for the purpose of healing the collective unconscious.

INTRODUCTION

The primary issue at hand is whether collective humanity will take responsibility for the choices they make before they fulfill the current suicidal course of their agency. Only if humans—as planetary stewards—choose to take responsibility for all sub-human sentient lives, will they fulfill their evolutionary responsibility. However, achieving cooperative humanly responsible choices will require a prodigious struggle. For this reason, this chapter is addressed primarily to those having "humanitarian" values. This is because a consensus already exists among humanitarians as to the benign morality that should govern planetary life. It may be asserted that this humanitarian organism nearly constitutes a critical mass of responsible tendencies. Within a quantum field of mediated energy and information, today the mystical significance of prayer and the deep esoteric significance of moral laws can be understood from

DOI: 10.4018/978-1-5225-9065-1.ch012

a scientific perspective on such realities as the sentient impacts of light, sound, thought, and feeling at a distance. This hypothesis of practical respect for universal sentience can then be applied to research on artificial intelligence, robotics, the media, and the karmic resonances of these force fields over time.

Most important, according to both mystic and scientific ontology, maintenance of direct contact between humanitarians and super-human dimensions of sentience could now become stipulated fact. Given the existing critical mass of humanitarian sentient qualities, mediated biofeedback projected from the collective unconscious into the mediated "contextual dreamscape" could be employed as a tool for harmonizing the irascible tendencies of the human collective and synchronizing with higher-order "intelligence". The purpose of maintaining coherent states of being between the human organism and the super-human Hierarchy of sentient bandwidths is to better understand collective human responsibility. Of course, this will not be easy, nor will it happen overnight.

However, as many instances of things "going viral" on the Internet demonstrate, the time is ripe for humanitarian sentience to use the superconductive character of the global media to alter chaotic patterns of the collective unconscious and to replace them with harmonious patterns. Managing complete agreement among infinitely different value patterns is not necessary, nor is the creation of global bureaucracies. The simple amplification of a good idea through social media can have a salutary but ephemeral impact on the healing energies of the geosphere, (McCraty, Atkinson, Stolc, Alabdulgader, Vainoras & Ragulskis, 2017) but a more comprehensive effort by global humanitarian organizations working cooperatively could have emphatic and long-term effect (Schafer, 2018; Schafer, 2017a; Schafer, 2017b; Schafer, 2012a; Schafer, 2012b).

In an emergent world of crisis and "wonders" where blatant lies become real and the destiny of nations can be determined by covert psychological hacking of the media, it is not idealistic to suggest that synergy with the human soul can be achieved with the use of a coherent media. This question has been further argued and defined in the above publications. Making sense of the strategy requires some knowledge of physics (quantum field theory), psychology (neurobiology, neurolingusitics, cognitive framing), mathematics (Fourier transforms and fractal geometry), myth (the universal Hero's Journey), coherence, media science, linguistics, and principles of Jungian dream analysis in order to synchronize specific elements into the practical strategy for gaming cultural coherence.

Hypothetically, propagation of fields according to quantum field theory (QFT) can be correlated with cognitive neurobiological dynamics at many scales in-order to better understand and establish the scientific basis for universal "sentience." Robust research on quantum neurobiology (genetic & cognitive framing dynamics), cell membrane research (signal transduction) sentience in plants and animals, the sentience of water, autonomous robots, sentience in artificial intelligence and robots, and the super-human sentience of planets (astrology) and other super-human fields is both ancient and current. Yet the primary significance of sentience is being ignored—not as-a consequence of scientific discipline but as a result of prejudice.

In the incipient age of mediated illusion characterized by the internet, artificial intelligence, robotics, political hacking, the neuro-marketing psychology of the unconscious, and the proven integrity of species, it would be foolhardy not to expand the concept of "consciousness" and to explore the dynamics of sentience. Such exploration would lead to a New Order of research on universal morality within the context of sentient Life—especially super-human life.

The subject of sentience is vast, and many scientific and mystical perspectives contribute to the hypothesis. The irony of modern science is that according to the materialistic "Spirit of the Age" (Jung, 1933, pp. 173-195), since the seventeenth century (Newton's theory of gravity in 1666), science has become increasingly subjective and susceptible to an ontological paradigm shift that closely approximates prior mystical worldviews based on Divinity, Logos, or God. Richard Feynman's title, *The God Particle* (referring to the Higgs boson), was certainly not intended as a treatise on theology, but the title is more suggestive than he suspected. The title (chosen by the publisher) prophecies the emergent field integrity of science and myth rather than a quantum mechanical (QM) and fundamentally atheistic reality.

Masterfully enumerated by Rodney A. Brooks in *Fields of Color: The Theory that escaped Einstein*, the process of discovery leading to QFT has been long and very complex; It has included a brilliant sequence of scientific discoveries ranging from Newton's gravity to Einstein's gravity through electromagnetic fields, force and matter fields, self-fields, Special Relativity, General Relativity, and renormalization, J. Schwinger's final solution.

The paradoxical quality of QFT's discrete pervasiveness is that which connects science with the mystical and—if carefully observed—accords with human perceptual experience in dreams. The difference between quantum field (QF) theory and quantum mechanics (QM) is that—though fields often behave like particles—they are "fields" that are paradoxically discreet and pervasive whereas QM is "a theory in which reality is made of particles, described by a wave function that gives the probability of finding the particle at a given point." (Brooks, p. 93).

Within a quantum-cognitive framework, the apparent implausibility of discreet pervasiveness must be addressed. The entire edifice of quantum science appears to be implausible, so researching QFT from the perspective of Jungian dream dynamics makes more sense than researching it from the orthodox perspective of reality based on atomic particles. In the QF, the alchemical axiom, "As it is below, so it is above." takes on new significance that may also be titled discreet pervasiveness or diversity in unity. The axiom can also be paraphrased as "each part contains the whole," so each microcosm holds the potential of the macrocosm. To materialistic scientists, the idea of the "whole contained by every part" has been an enigma until recently, but now this potential may be scientifically stipulated as the objective of human evolution. To mention only a few of the accepted scientific realities, the parts of the DNA molecule contain the whole. This idea accords with the idea of the alchemical homunculi. Quantum fields could be understood in terms of the alchemical alkahest (universal solvent) or the quintessence (active principle of all substance). Alchemy appears to have addressed various aspects of QFT with terms such as: the philosopher's stone, *palingenesis, spiritus mundi,* and *aurum potabile*. Within this synthetic universe, the molecule is the seed of the future. This ancient "seed" (God) symbol is pervasive in mythology. Also, the seed principle plays a significant part in quantum research as it applies to genetics, physics, artificial intelligence (AI), robotics, holonomics, cognition, and mass media technology.

Quantum fields are dynamically interactive—not merely a medium for observing particles—and this interactivity constitutes intelligence that corresponds to animal and human choice-making. "Sentient" interactivity is a fundamental aspect of both cell research and dream analysis upon which the entire Jungian theory depends; i.e. dream symbols that constitute the "environment" change kaleidoscopically as a dreamer's understanding changes. Interactivity in cell research equates to "Intelligent behavior" that can be attributed to "a functional membrane with both receptor (awareness) and effector (action) proteins. "These protein complexes are the fundamental units of cellular intelligence," (Lipton, 2005, p. 87). Moreover, cell research has established the interactivity between cell membranes and silicon chips, "The chip and cell membrane shared the same technical definition," and "The cell membrane was indeed a structural and functional equivalent (homologue) of a silicon chip," (Lipton, 2005, p. 91).

Interactivity may be extended to social dynamics. QFs have the paradoxical characteristics of distinctiveness and pervasiveness that may be applicable to exploration of New Order morality in social contexts that are intrinsically both individual and collective—and in a universal field of sentience. Essentially, QF is the incipient, "theory of everything" beginning with propagation: "If a field quantum transfers its energy to an atom it must do so as a unit and disappear from all space, no matter how spread out it is. Even if it transfers only part of its energy, it must collapse every time energy is transferred," (Brooks, 2016, p. 93). Among other things, this suggests that individuality (and Jungian Individuation) are comprehensively transitory (Heraclitus)[1] and evolutionary events that are unique and unrepeatable.

This reality of physics (as we presently understand it) must be associated with the idea of "metaphorical resonance" and the structure of language (rhetoric). The Quantum Field Theory introduces a scientifically proven concept that goes beyond even quantum mechanics and sets the stage for metaphorical exploration and re-evaluation of universal sentience—a concept that would completely alter the trajectory of research in physics by incorporating a practical cognitive-moral dimension provided by rhetoric.

Renormalization, a mathematical term in physics, is based on the contention that electrons act upon themselves and that a vacuum does not exist in space—only fields. (If all charges contribute to making a single common field, and if that common field acts back on all the charges, then each charge must act back upon itself. (Brooks, 2016, p. 93) If this is not sentient dream-likeness, what is? Valuable as it is, the paradoxes of QM obscure its relevance to real life. On the other-hand QFT, based on the well-founded Schwinger theory, provides a consistent, paradox-free picture of nature that is contrary to Feynman's view of particle reality and resolves QM paradoxes such as the uncertainty principle. QFT explains "spooky action at a distance," (Brooks, 2016, pp. 98, 124) and allows humans to explore the invisible ranges of sentience in morally meaningful ways.

Super-Human Sentience of the Soul Projected in the Dreamscape

The current imperative is to go beyond prejudice and to research the existence of super-human fields of sentience. It does not really matter what we call these super-human fields. What matters is that their coherent qualities can be accessed by humanitarians in order to amplify the clarity of direct communication. Given the scope of scientific research on sub-human dimensions of cosmic sentience (Norris, 2014Thincube, 2015; Hameroff & Penrose, 2014), discovery of super-human sentience is inevitable and past-due. So many philosopher-scientists have remarked upon or alluded to the correlations that exist between mystical ontology and modern science that an entire volume of references could be dedicated to their views. However, references are not the primary objective of Superconductive Sentience and The Mediated Dreamscape.

Rather, the objective is to invoke creative thinking and exploration in diverse fields that provide scientific evidence of this New Order mystic-scientific ontology. By "New Order" we mean a paradigm shift in research focus that unabashedly combines mystical and scientific realities. With-regard to the invisible force fields of thought and feeling, motivation and cognitive patterns, such a culture may be understood as superconductive. While this may sound like a prodigious undertaking, it is long overdue, and much of the new cultural organism is in the final stages of gestation before birth. One objective of this book is to correlate science and myth in- order to argue the pragmatism of applying Rhetorical ratio to technological data analysis. The maintenance of coherent states must characterize the cognitive structures upon which a global culture of conscience can be based.

In view of this QF complexity, how is one to correlate the field principles of physics in-order to develop a field theory of sentience? Three research fundamentals currently exist for this purpose:

1. Quantum level vibrations in microtubules inside brain neurons were discovered by Stuart Hameroff and Sir Roger Penrose (2014), and subsequently verified.[2] (Gutschner, 2014) It is postulated that these quantum vibrations are the source of consciousness.[3,4,5]
2. Coherent cognitive-functional human states as measured by heart-rate variability exist as measurable sine-wave frequencies (McCraty & Childre, 2010) that seem to represent far-ranging moral qualities that are consistent within a sentient cosmos. Quantum collapse (wave function reduction) may connect non-computational decision-making to experiences embedded in spacetime's fundamental geometry.
3. Jungian Psyche is based on the "energetic totality" (Jacoby, 1973, p. 2) cognitive-functional states which, today, may be translated into the "framing" language of the cognitive unconscious (Lakoff, George, 2008 & 2002)[6,7] and the mathematics of QFT (Brooks, 2016) thereby providing a framework for research on superconductive sentience.

Long ago, in *The Tempest*, Shakespeare observed that, "We are such stuff as dreams are made on." More recently, Sir Arthur Eddington said, "The stuff of the world is mind-stuff," (Foster, 1985, pp. 7-12) Deepak Chopra has argued that, "We are experiencing the final stages of the overthrow of the superstition of materialism," (Chopra, 1995) and Lewis Carroll asked, "Is all our life, then, but a dream? Taken together, these statements sound a keynote not only for study, but for survival in the emergent age of virtual illusions. Survival has to do with the development of a New Order human responsibility that is suggested by the emergent humanitarian organism.

Because appreciation for the cognitive value of dreams—as understood by the Priest-crafts and adepts—was universal in the great cultures of the past and because modern media technology provides a perfect metaphor relative to study of a sentient cultural Media-Dream, dream analysis according to principles of Carl Jung may be a common denominator for better understanding of the practical morality of such a Dreamscape which appears to have the characteristics of a quantum field (QF). The QF is based fundamentally on Fourier (recursive) mathematics as is the media-sphere, so—if this recursive mathematics is applied to the Jungian principle of archetypal representations (projections of subconscious quantum patterns), the link between physical fields, dreams, and fields of the media-sphere—a measurable upgrade of ancient subjective science becomes manifest as humanitarian agency.

The hypothetical approach taken here is that the salient features of QF are sentience and interactivity (superconductivity); i.e. superconductivity may arise from the existence of integrated elemental-sentient fields such as mineral, vegetable, animal, and spiritual fields that have the potential for coherent entrainment.

BACKGROUND

The theoretical mystic-scientific integrity of PSYCHecology video games has been detailed in the books, *Gaming Cultural Atonement* (Schafer, in press) and *Generating a Superconductive Culture of Conscience* (Schafer, 2017). For the purposes of the present publication, PSYCHecology is defined as an ecology of Psyche that is consistent with principles of dream analysis developed by Carl Jung. The successful dynamics of Jungian therapy have now been verified not only with decades of clinical success, but (perhaps unknowingly or without attribution) by modern cognitive researchers including George Lakoff, Bernard Baars, Rollin McCraty, Ken Wilbur, Deepak Chopra, Maharishi Mahesh Yogi, Stuart Hameroff and Sir Roger Penrose, Luca Maria Aiello, David Bohm, Karl Pribram, Sir Arthur Eddington, Alfred North Whitehead, Sir James Jeans, and many others in fields ranging through cognition, consciousness studies, optics, quantum computing, video game design, and information and communication technologies (ICTs).

The term PSYCHecology can also be closely associated with vast environmental research and extended to cognitive-sentient dimensions in plants, animals, minerals, and artificial intelligence. PSYCHecology refers to the cosmic system of "living psyche" to which environmental science has contributed extensively in terms of information and research methodology. Perhaps the most accurate way to understand the term PSYCHecology is with the term soul, but this will not be immediately sensible to many moderns who think of themselves as atheists. Mystics and theologians have debated the ontology of the Soul from time immemorial, but modern scientists have avoided the term based on the false assumption that such ontology is not scientifically valid. This posture has produced very unfortunate consequences that have supported immorality in all human choices and have brought the planet to a suicidal crisis. As Carl Jung observed about the modern shift to materialism in *Modern Man is Search of a Soul* (1933):

Let no one suppose that so radical a change in man's outlook could be brought about by reasoning and reflection for no chain of reasoning can prove or disprove the existence of either mind or matter. Both these concepts, as every intelligent man today may ascertain for himself, are mere symbols that stand for something unknown and unexplored, and this something is postulated or denied according to man's mood and disposition or as the spirit of the age dictates. There is nothing to prevent the speculative intellect from treating the psyche, on the one hand, as a complicated biochemical phenomenon, and at bottom a mere play of electrons, or, on the other, from regarding the unpredictable behavior of electrons as the sign of mental life even in them. The fact that a metaphysics of the mind was supplanted in the nineteenth century by a metaphysics of matter, is a mere trick if we consider it as a question for the intellect; yet regarded from the standpoint of psychology, it is an unexampled revolution in man's outlook upon the world. Other-worldliness is converted into matter-of-factness; empirical boundaries are set to man's discussion of every problem, to his choice of purposes, and even to what he calls "meaning." Intangible inner happenings seem to have to yield place to things in the external, tangible world, and no value exists if it is not founded on a so-called fact. Just as formerly the assumption was unquestionable that everything that exists takes its rise from the creative will of a God who is spirit, so the nineteenth century discovered the equally unquestionable truth that everything arises from material causes. Today the psyche does not build itself a body, but on the contrary, matter, by chemical action, produces the psyche (Jung, Carl, 1933, p. 174-175).

Without a doubt, Carl Jung's scientific understanding of this quintessential global human psychology and its applicability to a mediated age positions him as the greatest scientist-philosopher of our time. Relative to the soul, another philosopher-scientist, Ken Wilbur, is concise and consistent with both the Ancient Mysteries and modern mystical-science.

When it comes to understanding the multidimensional nature of our body, it's far more rich and profound than meets the eye. We have a waking (gross) body, a dream (subtle) body, and a dreamless (causal) body. And each of these bodies is intrinsically woven into corresponding states of consciousness (Wilbur, 2018).

In Wilbur's quote, the Soul equates to the Dream Body. For many, this statement may seem strange, but it should not be so for esotericists, quantum field theorists, or cognitive researchers. This body-image in which both the physical body and its quantum context are like a river of changing quantum field forces constitutes the basic theoretical framework for creating PEGs. According to game dynamics, PEGs can foster cultural transcendence beyond the state of personality selfishness, fearfulness, and malignance to a state of the collectively coherent and inclusive Soul. In the language of physics, the quantum fields of the Soul, its energy, spin, and forces, provides the best mystic-scientific significator for universal interactive sentience, but in mythic terms, perhaps the Hindu God of Fire, Agni is the best expression. Use of the fire metaphor would emphasize fluid dynamics in the algorithmic dynamics of PEG.

There is no doubt that many scientific "philosophers" will disagree about the details of the book's theme or about the validity of the entire concept of universal sentience and the correlation of media dreams with Jungian dreams. The editor welcomes informed disagreement but asks the question: "Upon what is the disagreement based?" It is fairly certain that the answer to this question will be "closed mindedness' that forecloses all progress or outright "pre-judgment" that has no place in authentic science. Nevertheless, fomenting disagreement is good if it contributes to adjustment of the current amoral and atheistic scientific research trajectory.

Assumptions Intrinsic to Research Methodology

- The dynamics of quantum field theory (QFT) may be applicable to a unified field of super-human sentience (Psyche or Divinity—the Absolute, the Ineffable) and both realities may exist as synchronized individualized beings.
- The ancient axiom, "As it is above, so it is below." Applies to the recursive correlation between the macrocosm and the microcosm. One example of this is the relationship between God and Humanity, and this correspondence may be researched in terms of the human functions (mind, emotion, physical senses, & intuition) known symbolically as the elements air, water, earth, and fire. So, sentience of the Absolute may be researched in terms of the sentience of the human microcosm and its agency including biological and virtual realms of metaphorical and computational Arts and Sciences.
- Unity (non-duality)—God Transcendent—finds expression as the virtual illusion of reality that appears to be a duality—God Immanent (correlated with principles of field distinctiveness and pervasiveness?)
- In current cognitive terms, God Transcendent and God Imminent may be understood as the collective unconscious (Transcendence) and the collective conscious (Immanence).
- The cognitive unconscious (God Transcendent) known to psychiatrists and cognitive researchers not only exists but is a powerful generative dynamic in human perceptions of so-called reality.
- The Jungian process of analytical psychology is uniquely authoritative for psychoanalysis with the use of dream symbolism that exists as archetypal patterns in the personal and collective unconscious and is projected as archetypal representations that are perceived as mediated images or virtual realities.
- The Jungian Amplification Method for psychological healing of individuals and collectives using analysis of dreams has been authenticated with decades of research and clinical experience. (At sub-atomic levels is amplification correlated with the mathematics of field quanta as defined by Brooks? (p. 47))

- Research on Jungian principles of psychological healing may be directly correlated with current cognitive theory and research and used to study ethical behavior in virtual realities.
- Whether viewed from the perspective of sub-atomic particles, Einstein's relativity, or Hermetic alchemy, the integrity of the universe may be stipulated. Modern science verifies universal integrity with robust research in every field ranging from genetics to extra-planetary research with optics. A comprehensive illustration of this manifest universal integrity is the Media Dream which would not be possible without the integrity of Fourier transforms and the complex algorithms of the modern media. The Holographic universe (David Bohm), the Holographic brain (Karl Pribram), the Gaia principle, and theological monotheism[8] are dependent on the assumption of universal integrity.
- The Media-sphere exists as a holographic Jungian Psyche in which archetypal representations exist as projected content; i.e., algorithmic content is unconscious until projected into conscious mediated programs.
- The Media-sphere can be interrogated as a collective dreamscape according to Jung's principles of dream analysis and healing dynamics.
- Language grounds thought, so the nature of symbolic language, language in general, and how it influences human reality perceptions can be understood as the seed ground for ethics.
- As researched by the Institute of HeartMath, the quality of coherence can be correlated with both Jungian Compensation (leading to psychological harmony) and with measurable ethical (coherent) behavior.
- The recent evolution in computer science and technologies, artificial intelligence, video game technology (Togelius, 2016), and robotics allows for use of PSYCHecology video games (PEG) to do robust research on the cognitive unconscious, for verification of Jungian dynamics, for understanding ethics in virtual realities, and for healing of contextual Collective Personae.
- Algorithms employing the dynamics of Fourier transforms for psychological (functional) analysis of the media-sphere can be used to analyze the media dreams of contextual collectives according to authentic Jungian principles for analysis of dreams. "A people [collective], in the doctor's eyes, presents only a somewhat more complex picture of psychic life than the individual (Jung, 1933, pp. 209-210).
- Current and historical study of ethics strongly suggests that ethics are based on coherent thoughts, feelings, and behavior that can be fostered with PSYCHecology video games (PEGs) that are programmed according to Jungian principles of personality types. The principles of coherence, Jungian Compensation, and mystical Transcendence—as well as the Flow experience—are directly correlated.
- PEG may be employed as a research instrument for both psychological healing and ongoing research relative to the nature of cultural ethics in virtual realities.
- The paradigm shift into a media age could herald the projection of the World-Soul (Archetypal Representation) into a manifest culture of conscience.

Research citations supporting various elements of the hypotheses exist in the tens of thousands—many of which are cited in my own works which are included in the reference list for this chapter.

Outdated scientific perspectives on scientific materialism have long since given way to models based on relativity, Quantum Field Theory (QFT), and communication and information technologies (ICTs). References to Jungian dynamics of dream analysis certify the fact that the QFT is tantamount to Jungian

Psyche. The QFT perspective is literally "dreamlike." What is not commonly credited is that Jungian research is entirely consistent with fundamental findings in the fields of Artificial Intelligence (AI), Robotics, Genetics, and advanced imaging technologies such as fMRI and MEG (Schafer, 2016, 2016a, 2016b, 2016c, 2016d; Schafer, 2012a; Schafer, 2012b; Schafer & Palamides, 2011; Schafer, 2006). These technologies have given rise to robust research on the neurobiological cognitive unconscious, and these advances have thrown new light on the dream research of Carl G. Jung. For example, decades of research on coherence-based heart-rate-variability (HRV) can be directly correlated with the Jungian principle of Compensation, and the media-sphere can be correlated with dynamics of Jungian Psyche. Diverse research already provides the keys for synchronizing with the World-Soul, and the keys have already verified fundamental dynamics of cosmic sentience upon which the thesis of this paper depend.

However, though modern research has made profound discoveries relative to what may be defined as a "sentient" cosmos, a re-evaluation of metaphysical principles would speed the process of healing. Though research on the phenomenon of "consciousness" is longstanding, modern research on "sentience" is only beginning to be taken seriously by academic researchers. Incipient research includes study of sentience in water, plants, and animals, and recent research on human cognition substantiates the overwhelming influence of unconscious sentience upon human choices and behaviors. For one example, it is now known that gut bacteria influence human dietary tastes (Norris, 2014). Also, the frequencies of intentions, thoughts, and words create human reality (Direct Action 2017, 2013; Thincube, 2015). Consciousness is estimated to produce only about 5% of human function. (Lakoff, George, 2002), but most important is the fact that—most likely—consciousness has its foundation in quantum neurobiological fields of sentience (Hameroff & Penrose, 2014; Chopra, 1995).

Meanwhile, in advanced research in the field of signal transduction within biological proteins that activate cell behavior, we find such statements as the following which begin the process of correlating human sentience with sub-human molecular sentience:

Because [protein] receptors can read energy fields, the notion that only physical molecules can impact cell physiology is outmoded. Biological behavior can be controlled by invisible forces, including thought.

While the receptor provides an awareness of environmental signals, the cell still has to engage in an appropriate, life-sustaining response, that is the venue of the effector proteins.

The membrane's receptors are the equivalent of sensory nerves, and the effector proteins are the equivalent of action-generating motor nerves. Together, the receptor-effector complex acts as a switch, translating environmental signals into cellular behavior.

Signal transduction scientists are busily classifying hundreds of complex information pathways that lie between the membrane's reception of environmental signals and the activation of the cell's behavior proteins.

To exhibit "intelligent" [sentient] cells need a functioning membrane with both receptor (awareness) and effector (action) behavior proteins (Lipton, 2005, pp. 84-87).

SUGGESTIONS FOR FURTHER RESEARCH

All the above hypotheses require ongoing research because the evolution of human consciousness has a long way yet to go. People do not yet realize what precious vessels of energetic sentience they are. They will not admit that they represent condensers and transformers of energy. When such apparatuses are damaged or misdirected, the cosmic distribution of energy is disturbed, and this perilous manifestation undoubtedly creates disruption not only in human affairs but in climatic conditions and beyond. Global warming constitutes implicit proof of this fact of a sentient cosmic interaction, but such ontological issues are not presently being addressed by science. Manifesting as a sentient creation of humans, the Media-sphere with its powerful impacts on planetary life is a good place to begin research, and Robots as sentient human creations offer a rich field for an extension of research. For example, the following hypotheses would advance research on sentience:

1. Archetypes & sub-atomic fields are directly related.
2. A definitive relationship exists between archetypes & archetypal representations.
3. A definitive relationship exists between Jungian Contextuality and the distinctiveness of fields.
4. Quantum fields have rhetorical structure.
5. Quantum collapse is related to coherent Individuation and to the concept of a superconductive culture.

GAMING CULTURAL ATONEMENT

Hypothetically, propagation of fields according to quantum field theory (QFT) can be correlated with cognitive neurobiological dynamics in-order to better understand and establish the scientific basis for universal "sentience." In the incipient age of mediated illusion characterized by the internet, artificial intelligence, robotics, political hacking, the neuro-marketing psychology of the unconscious, and the proven integrity of species, it would be foolhardy to ignore the dynamics of "sentience" which lead to an exploration of universal mystic-scientific morality.

Since the seventeenth century, science has become increasingly subjective and susceptible to an ontological paradigm shift that closely approximates mystical worldviews based on Divinity, Logos, or God. Richard Feynman's title, *The God Particle* (referring to the Higgs boson), was certainly not intended as a treatise on theology, but the title is more prophetic than he suspected. The title (actually chosen by the publisher) is suggestive of the emergent integrity of science and myth rather than an atheistic reality.

Masterfully enumerated by Rodney A. Brooks in *Fields of Color: The Theory that escaped Einstein* (2016), the process of discovery for QFT has been long, very complex, and incomplete; it has included a brilliant sequence of scientific discoveries ranging from Newton's gravity to Einstein's gravity through electromagnetic fields, force and matter fields, self-fields and renormalization, J. Schwinger's final solution, and Special Relativity.

The difference between quantum field (QF) theory and quantum mechanics (QM) is that—though fields often behave like particles—they are "fields" that are paradoxically discreet and pervasive whereas QM is "a theory in which reality is made of particles, described by a wave function that gives the probability of finding the particle at a given point," (p. 93). Quantum fields are dynamic—not merely a medium for observing particles. QFs have the paradoxical characteristics of distinctiveness and pervasiveness—that may be applicable to exploration of New Order morality in a universal field of sentience. "If a field quantum transfers its energy to an atom it must do so as a unit and disappear from all space, no matter how spread out it is. Even if it transfers only part of its energy, it must collapse every time energy is transferred," (p. 93).

Quantum Field Theory introduces a mathematically proven concept that goes beyond even quantum mechanics and sets the stage for exploration of an ontology of universal sentience—a concept that would completely alter the trajectory of research in physics to incorporate a practical moral-cognitive dimension. This statement is based on the pragmatism that—at least—QFT is not full of paradoxes as is QM; i.e., in QFT there is no particle-wave, no uncertainty principle (Schrödinger's cat), and no measurement problem (Brooks, p. 128). Such a "practical" mathematical approach supports exploration of "sentience" because sentience appears to be both discreet and pervasive. In the human dimension, this discrete pervasiveness can be understood in terms of individual and collective contextuality.

In the Jungian sense contextuality may be defined in terms of, QF "discretization" which is measurable in terms of quantum collapse (entanglement) (p. 127). In other words, "spooky-action-at-a-distance" becomes less spooky. "Quantum fields are not described by simple numbers. They are described by vectors in what mathematicians call Hilbert space and their dynamics are described by operators that obey partial differential equations," (p. 132). Also, QF resolution can solve the uncertainty principle with the use of "*Fourier's theorem* that relates the spatial spread of a field to the spread of its wavelengths," (p. 125). Beyond the intuitive stipulation that moral action is practical, it is also measurable in terms of "coherence" or harmonious entrainment of human functions—physical, emotional, mental, and spiritual. Coherence is measurable using heart-rate-variability (HRV). Considered as quantum fields, discreet degrees of sentience should be measurable and their discreet functions within the Field defined.

The first phenomenon which attracted our attention as a manifestation of field reactions was the electromagnetic mass of the electron. The electron, having a charge, produces an electromagnetic field around itself. In turn, this field, the so-called self-field of the electron, interacts with the electron. We call this interaction the field reaction. Because of the field reaction, the apparent mass of the electron differs from the original mass. The excess mass due to this field reaction is called the electromagnetic mass of the electron and the experimentally observed mass is the sum of the original mass and this electromagnetic mass." Mathematical calculations were—at first—discouraging because the influence of the field reaction becomes infinite. However, renormalization in calculations provides the solution. Renormalization takes the influence of the field reaction into account; i.e., As observed by Sin-Itiro Tomonaga, instead of calculating from the original mass and charge, calculating the experimental mass and charge as modified by field reactions becomes finite (p. 94).

Renormalization is based on the contention that electrons act upon themselves and that a vacuum does not exist in space—only fields. (If all charges contribute to making a single common field, and if that common field acts back on all the charges, then each charge must act back upon itself (p. 93).

Field theory, based on the well-founded Schwinger theory, provides a consistent, paradox-free (particle vs. field) picture of nature that is contrary to Feynman's view of particle reality (p. 98).

In QF, the QM dynamic of spin becomes "helicity" because no axis exists upon which to spin "Helicity gives rise to angular momentum" (p. 134) that may prove commensurate with the geometrical vectors in astrology and, therefore, to cognitive measurement. However far-fetched this may sound, the QF exclusion principle (if it is applied to human scale and experience) might provide insight as to why humans appear as separate entities. Two matter quanta cannot be in the same quantum state and are seen only as separate entities (p. 134).

The new imperative is to go beyond prejudice and to research the existence of super-human fields of sentience. Given the scope of scientific research on sub-human dimensions of cosmic sentience (Norris, 2014Thincube, 2015; Hameroff, & Penrose, 2014), discovery of super-human sentience is inevitable. So many philosopher-scientists have remarked upon or alluded to the correlations that exist between mystical ontology and modern science that an entire volume could be dedicated only to references. However, references are not the primary objective of *Superconductive Sentience & The Mediated Dreamscape.* Rather, the objective is to invoke creative thinking and exploration in diverse fields that provide scientific evidence of his New Order mystic-scientific ontology. By "New Order" we mean a paradigm shift on research focus that unabashedly combines mystical and scientific realities.

Because appreciation for the cognitive value of dreams—as understood by the Priest-crafts and adepts—was universal in the great cultures of the past and because modern media technology provides a perfect metaphor relative to study of a sentient cultural Media-Dream, dream analysis according to principles of Carl Jung may be a common denominator for better understanding the practical morality of such a dreamscape which appears to have the characteristics of a quantum field (QF).

The hypothetical approach taken in the book is that the salient features of QF are sentience and superconductivity; i.e. superconductivity may arise from the existence of integrated elemental-sentient fields such as mineral, vegetable, animal, and spiritual fields that have the potential for coherent entrainment.[9]

The Pythagorean focus on mathematics and geometry creates a bridge between alchemy, astrology, and quantum field theory (QFT). The vectors of astrological geometry are meant to measure the sentient relationship between heavenly bodies and the personal-social realities of an individual's astrological chart. Astrological vectors could be compared with the vectors of quantum fields[10] (Brooks, 2016, p. 132). In fact, they already have been compared. A fractal evolutionary process has already been defined according to the "intelligence" of cell membranes (Lipton, 2005, pp. 195-198). For all we know, the quantum collapse of QF may be correlated with contextual (mass, charge, and spin?) psychological evolution ensconced within the astrological chart. In such a case, "transcendent" states of being on the human scale would be correlated with superconductive states of being at the atomic scale.

Within a vast array of mystical theory, Pythagorean thought provides a bridge between Western and Eastern schools of mysticism. Pythagoras spent fifteen years in Egypt being initiated into the secret mysteries, and most of what we know about him comes from the Pythagorean school, some writings by Plato and Aristotle, and the subsequent Neoplatonism (Hall, 2003, pp. 191-221). The Pythagorean aphorisms correlate geometrical shape with cognitive meaning in a way very similar to Jungian dream analysis with its resulting "healing" compensation. The aphorisms, like Jungian symbols, must be read intuitively and are never susceptible to rational understanding. We must remember that much of the emergent new science is not susceptible to rational understanding. The reason for this is that all rational thought depends upon a valid hypothesis, and given the emergent state of science, valid hypotheses are difficult to frame. Perhaps this is why we must refer to mysticism for valid hypotheses such as the transcendence/immanence of God in which the part (potentially) is greater than the whole. For example, Pythagoras taught that the dot (".") symbolized the power of the number one:

The wind blowing, adore the sound. "Pythagoras here reminds his disciples that the fiat of God is heard in the voice of the elements, and that all things in nature manifest through harmony, rhythm, order, or procedure—the attributes of the Deity," (Hall, 2003, p. 201).

So, Pythagorean reality is based on geometry, numbers, words, and their metaphorical resonances. It may be mystically stipulated that cosmic reality is "sentient"—a concept ignored by modern science until the advent of algorithms used in modern research on artificial intelligence (AI). Algorithmic entrainment has become so sophisticated that AI is expected to surpass human intelligence (Eagleman, 2009; Lewis, 2014; McDonald, 2015) and robots are being viewed as a new species (Saffell, 2015).

Pythagorean genius is a primary example of the modern relevance of "mystic" science which pervades mythology, theology, and their esoteric schools throughout the world. What today would be researched in terms of quantum electrodynamic, optics, and neuro-cognitive dynamics has been the subject of millennial research on astrology, the science of the chakras, and meditation. So, a new ontology is needed—an ontology of superconductive sentience. Yet, ontology is like Gregory Bateson's "map," that constitutes our best guess about the nature of the territory, but it is not the territory which always evades complete understanding. Ontology ranges in validity from the fairly accurate to the ridiculous—like a space-probe or moon landing that generates claims that humans are the "masters of the universe" and know all the mysteries of cosmos (Eisley, 1970). Even highly successful quantum physics only describes how fields behave, not what a field is or like a personality profile that is useful but limited. If science can get beyond prejudice, so-called mysticism tends to extend human understanding of the territory and of human responsibility therein. This is an important contribution because such responsibility is blatantly disregarded by most modern science.

In-order to grasp the critical scale of the present planetary complex of crises, science must review mystical precedents without prejudice and incorporate them among the fields of science. Based on vast mystical-ontological precedent as well as robust modern research in many scientific fields, the existence of super-human dimensions of sentience is more than probable. In-order to deal with human problems of the present and the future, planetary issues must be based on advanced understanding of mystic-scientific-ontology. At a time when some politicians are advocating a ludicrous repudiation of all science, science must address the potential for practical wisdom proffered by super-human dimensions of Soul sentience—both personal and collective. Proven ways to address a sentient reality are Carl Jung's principles of dream analysis and psychiatric healing as-well-as current research on coherence by the Institute of Heartmath. Jungian principles and coherence are directly correlated with current research on superconductive sentience in a multiverse. Just as the term "coherence" is defined differently in different scientific fields, the terms used to describe cognitive "superconductivity" may include transcendence, flow, coherence, compensation, harmony, the absolute, "0", "1", Logos, quantum collapse, "at-one-ment," PSYCHecology, God, and others. Pre-judgments asserted about the rightness of any one of these terms are neither open-minded nor scientific.

The concept of a scientific cosmic sentience must go beyond theory and must be applied, and—as a rule—application often proves the rule. Perhaps the best illustration of practical application is the environmental movement that understands reality as a sentient ecology—a living matrix of sentient—mostly biological--organisms. Another name for a sentient ecology is PSYCHecology, and this book recommends a practical instrument for researching the sentient PSYCHecology. It has been taught that, "Even each passing shadow of a man leaves an ineradicable trace," (Heart, 1932, p. 160). "What cannot burn away?" The answer to this old riddle is "thought." What, then, shall one say about thoughts and words used in the sentient media-sphere? The answer is daunting and requires a perspective on human responsibility inspired by the Gaia Principle of cosmic reality that has universal precedent in the myths of all ages and is supported by robust science in multiple fields.

Media Models to Foster Human Transcendence Into a Coherent PSYCHecology

This book recommends a New-Order approach to scientific research using PSYCHecology games (PEGs) for research on more efficient access to super-human bandwidths of Cosmic Sentience. The concept behind the development of a PEG prototype is that authentic game play—based on the laws of physics, the most current cognitive research, and the principles of dream analysis identified by Carl Jung—can promote individual and collective psychological atonement, and this "at-one-ment" constitutes a coherent state of being that equates to the functionality of the Soul. In other words, PEGs can promote entrainment in all the dimensions of human functional choices (McCraty & Childre, 2010) and that entrainment can be measured in terms of coherent sentience. The dynamic for achieving "at-one-ment" is grounded in the scientific hypothesis that universal reality is a coherent quantum electrodynamic (QED)[11] field. Jung—with his research on "Psyche"—was ahead of the learning curve. According to Jung, "unified energy" QED patterns of the archetypal unconscious are projected as images that have intrinsic recursive dramatic structure. Images tell stories, and stories contain imbedded psychological meaning. Therefore, contextual meaning imbedded in story images can be discovered with the application of narrative architecture. Images abound in human experience: Dreams, memories, fantasies, the mind's eye, television, technological monitors (Information & Communication Technologies—ICTs), and real-life (waking dreams). Therefore, images can be understood and researched as a common denominator in a unified QED field of sentience or a Jungian dreamscape. (Schafer, 2017; Schafer, 2017; Schafer, 2012).[12]

As we experience a paradigm shift into a media age, ICTs are altering the psychological parameters of human reality. The premise of this book is that the psychological dynamics of interactive images projected as Media Dreams correspond with the psychological dynamics of dream images as defined by Carl G. Jung (Jacobi, 1973). Encryption and steganography (Cheddad, Condell, Curran, & McKevitt, 2009) illustrate technological dynamics comparable to the Jungian concept of meaningful symbolism in dream images.[13] In other words, meaningful information can be imbedded in images, so media dream images can mirror patterns of energy and information in what Jung called the collective unconscious. Jung calls images archetypal representations or projections of archetypal energy patterns that are structured as metaphorical narrative. The most recent cognitive research (Lakoff, 2002, 2008) verifies that—indeed—the cognitive unconscious has the framework of metaphorical narrative and that these story patterns correlate with energy patterns in the nervous system and the cognitive unconscious.

Jungian compensation is essentially the same thing as coherence, and recent research on coherence confirms that coherent states can be evoked with specific biofeedback technologies (Schafer, 2018). So, the Soul may be collectively evoked. Jung also knew that dreams "have a purpose," and that discovery of purpose may compensate for psychological incoherence—better known as psychosis, neurosis, and a plethora of lately defined syndromes such as bipolar disorder and PTSD. Discovery of meaning has the power to harmonize conscious and unconscious psyche. Harmony—discussed as measurable coherence—must necessarily be the fundamental quality of super-human soul sentience. Moreover, coherent psychological states increase emotional and perceptual stability as well as alignment (entrainment) among human functional systems: the physical, cognitive, and emotional systems. The hypothesis treated in the book is that the images projected by Information and Communication Technologies (ICTs)—the media dreams of a population—are subject to psychological analysis in-order to disclose unconscious sources of psychological stress in contextual collectives and to provide healing insight to cultures with coherent mediated biofeedback.

Sub-Human and Super-Human Sentience

Emergent science has already extended the concept of "sentience". Though the idea that cosmos is "conscious" has many precedents—both philosophical and scientific—Dean Radin (2006, 1997) has made a comprehensive argument for the scientific proof of psychic phenomena, and his latest book, *Real Magic* (2018) has correlated Ancient wisdom with modern science in-order to discuss "the secret power of the universe from the perspective of physics." Francisco DiBiasi (2013) has developed a cognitive model "based in the holonomic neural networks of Karl Pribram (Pribram, 1991, 1971), in the holographic quantum theory developed by David Bohm, and in the non-locality property of the quantum field described by Hiroomi Umezawa." This model—an extension of the interactive dualism of Sir John Eccles--propounds an interconnection between brain and spirit by means of quantum microsites named dendrons and psychons. This is a dynamic concept of "sentience" that is perceived as a holoinformational flux interconnecting the holonomic informational quantum brain dynamics with the quantum informational holographic nature of the universe.

Meanwhile, Stuart Hameroff and Sir Roger Penrose (2014) have proven that the origins of "sentience" in neural synapses, and robust research on coherent "sentience" is being done by the *Institute of Heartmath* and its global network of colleagues (McCraty et al., 2017; McCraty & Childre, 2010; McCraty & Tomasino, 2006; McCraty, 2005; McCraty, 2002). The research of this network is directly correlated with DiBiasi's self-organizing flux of neuronal information that can be optimized through practices of deep meditation, prayer, and others states of higher sentience that underlie the coherence of cerebral waves. "In brain mapping studies performed during the occurrence of these harmonic states we can see the spectral array of brain waves highly synchronized and perfectly ordered like a unique harmonic wave, as if all frequencies of all neurons from all cerebral centers played the same symphony. DiBiasi observes that this highly coherent brain state generates the non-local holographic informational cortical field of sentience that interconnects the human brain and holographic cosmos.

Similar cognitive models extend from Bateson (1972) to Baars (1997) and Lakoff (2008, 2002), and this is only the tip of the iceberg. Many philosopher-researchers have entertained essentially the same idea from diverse perspective on quantum-based human-cosmic sentience (Schafer, 2017; Schafer, 2017; Gackenbach & Hakopdjanian, 2015; Aiello, 2014; Khodarahimi, 2009; Institute of Heartmath, 2007; Pobachenko, Kolesnik, Borodin, & Kalyuzhin, 2006; Ho, 2005; Pert & Lipton, 2005, 2003; Jenkins, 2005; Chopra, 1995; Zohar & Marshall, 1993; Talbot, 1991; Zohar, 1990; Hinton, McClelland, Rumelhart et al., 1986; Maharishi's Programmes, n.d.). It is high time for science to divest itself of old habits and to adopt the counter-intuitive holonomic quantum informational nature of brain-consciousness-universe interconnectedness that, "allows us to solve the old mind-matter cartesian hard problem, unifying science, philosophy, and spiritual traditions in a more transdisciplinary, holistic, integrated paradigm" (DiBiasi, 2013).

Research on cognitive models is vast, and it includes hard academic and metaphysical research long thought to be anomalous. Synthesizing the updates on hard-scientific research on scalar sentience—due to its scope—is nearly impossible. Yet, the scope, dynamics and language of these numerous cognitive models can be suggested (Direct Action 2017, 2013; von Foerster, Heinz, 2013; Luhmann, 2012; Sheldon, 2011; Pruchnic, 2008; Papadopolous & Byng-Hall, 2007; Luhmann, 2002; Pruchnic, 2008; Peters, 1999; Nikolaev, Rudakov, Mansurov, and Mansurova, 1976; von Foerster, 1973, 1962; Weaver, 1949; Shiga, 2013). Diverse and robust research has enhanced our knowledge of:

- Sentience in plants (Streedhar, 2017; Documentary TV, 2014), animals (Clark, 2014; Norris, 2014; Burnett, 2012)
- Sentience in the elements (Thincube, 2014; Eagleman, 2009)
- Sentience induced with hallucinogens (Strassman, 2001; Langlitz, 1972; Lilly, 1985; Lilly, 1967; Lilly, 1974)
- Sentience in robots, and artificial intelligence (Rogers, 2018; Galeon, 2017, 2016; Ohmori, 2016; McDonald, 2015; Saffell, 2015; Simonite, 2015; Knight, 2015; Jeffrey, 2015; Lewis, 2014; Taylor, 2014; Trafton, 2014; Kirk, 2014; Schafer, 2012a, 2012b; Schafer, & Yu, 2011; Twist, 2004; McCarthy, Minsky & Rochester, 1955; Emerging Technology from the arXiv, 2016)

Amplification

The fact that a metaphysics of the mind was supplanted in the nineteenth century by a metaphysics of matter, is a mere trick if we consider it as a question for the intellect; yet regarded from the standpoint of psychology, it is an unexampled revolution in man's outlook upon the world. Other-worldliness is converted into matter-of-factness; empirical boundaries are set to man's discussion of every problem, to his choice of purposes, and even to what he calls 'meaning.' Intangible inner happenings seem to have to yield place to things in the external, tangible world, and no value exists if it is not founded on a so-called fact. Just as formerly the assumption was unquestionable that everything that exists takes its rise from the creative will of a God who is spirit, so the nineteenth century discovered the equally unquestionable truth that everything arises from material causes. Today the psyche does not build itself a body, but on the contrary, matter, by chemical action, produces the psyche (Jung, 1933, p. 174-175).

Has not a poet spoken of the 'nations' of his soul? And quite correctly, as it seems to me, for in one of its aspects the psyche is not individual, but is derived from the nation, from collectivity, or from humanity even. In some way or other we are part of an all-embracing psychic life, of a single 'greatest' man, to quote Swedenborg. And so we can draw a parallel: just as in me, a single human being, the darkness calls forth the helpful light, so does it also in the psychic life of a people (Jung, 1933, p. 209-210).

Carl Jung discovered that assisting patients to understand the symbols in their dream images led to insight as to the causes of subconscious problems. Because images are projections of subconscious energy and information patterns, such insight has the capacity to harmonize cognitive energy patterns, to compensate for dissonance, and to heal patients by helping them achieve a degree of psychic coherence. More recently, cognitive research has verified Jung's theories and empirical observations from the perspective of neurobiological patterns and cognitive framing—Jung's archetypes of the unconscious are frames of the cognitive unconscious. Jung based his method of dream analysis on his observation that dreams have dramatic structure and—like drama—have a purpose or exposition. George Lakoff, researching cognitive frames, has demonstrated that the narrative-metaphorical architecture of the cognitive unconscious frames human values. Lakoff defines the two fundamental family metaphors that determine conservative values (The Strict Father Metaphor) and progressive values (The Nurturant Parent Metaphor). Such narratives and metaphors with which we identify at subconscious levels determine our value systems, our worldview, and our actions. Lakoff observes that without reference to cognitive narratives, humans cannot know others. Without these unconscious narrative-metaphorical reference points, people cannot even know themselves. Essentially, we refer our identity and our very existence to our dreamlike cognitive narratives.

Most interesting is that cognitive dream dynamics of the microcosm correspond to the dynamics of the macrocosm. Jung said that, "Disruption in the spiritual life of an age shows the same pattern as radical change in an individual," (Jung, 1933, p. 202). Also, "A people presents only a somewhat more complex picture of psychic life than the individual," (Jung, 1933, p. 210). That-is to say, archetypal patterns are shared at individual and collective scales. Archetypes are projected as images embedded with the potentials for meaningful insight. Their recursive symbolic content spans the "gap" between unconscious and conscious (sentient) energy patterns of both personal and collective psyche. This suggests that, in-order to achieve curative insight, the key to navigation of the unconscious psyche resides in the correspondences (based on algorithmic recursions of isometric form) that may be mediated between digital software and the archetypal images that the software projects. Jung also observed that because the psychic process has a quantitative aspect, the conservation of energy in the physical world is analogous to "compensation" by the psyche. Accordingly, "No psychic value can disappear without being replaced by another of equivalent intensity," (Jung, 1933, p. 209).

These ideas raise questions as to the existence of structural parallels between Jungian psyche and the emergent media sphere. The media sphere has been defined as the collective ecology of the world's media including newspapers, journals, television, radio, books, novels, advertising, press releases, publicity and the blogosphere—any and-all media both broadcast and published. If such parallels exist, it is possible to analyze the images of the media sphere according to Jungian methods in-order to diagnose and heal "contextual" collectives (Schafer & Yu, 2011). This chapter addresses this possibility and argues that the possibility of fostering coherent psychological states with ICTs should provide the foundation upon which all serious research on information and communication technologies is based (Schafer, 2016, 2012).

CONCLUSION AND DISCUSSION

In a sentient universe characterized by rules of QFT, the dream dynamics defined by Carl Jung, dynamics of the cognitive unconscious averred by George Lakoff, video game design, and the metaphor of the holographic media-sphere, hypothetical PEGs designed as dream analogs could serve as research instruments to explore the cognitive unconscious and the cosmic-sentient reality. Research hypotheses include the following:

1. The unified field of physics is essentially the same as the unified field of Psyche or Divinity—the Absolute, the Ineffable—and both realities exist as vital sentient beings. Though this assertion has not yet been "scientifically proven", quantum field theory (QFT) has advanced to a stage where it becomes probable.
2. The ancient axiom, "As it is above, so it is below." Applies to the recursive correlation between the macrocosm and the microcosm. One example of this is the relationship between God and Humanity, and this correspondence may be researched in terms of the human functions (mind, emotion, physical senses, intuition) known symbolically as the elements air, water, earth, and fire. So, the Absolute

may be researched from the perspective of diverse scientific fields—especially everything known about the virtual realms of metaphor from the perspective of the human arts and sciences.

3. Unity (non-duality)—God Transcendent—finds expression as the virtual illusion of reality that appears to be a duality—God Immanent.
4. In cognitive terms, God Transcendent and God Imminent may be understood as the collective unconscious (transcendence) and the collective conscious (Immanence).
5. The cognitive unconscious known to psychiatrists and cognitive researchers not only exists but is a powerful generative dynamic in human perceptions of so-called reality.
6. The Jungian process of Analytical Psychology is uniquely authoritative for psychoanalysis with the use of dream symbolism that exists as archetypal patterns in the personal and collective unconscious and is projected as archetypal representations that are perceived as mediated images or virtual realities.
7. The Jungian Amplification Method for psychological healing of individuals and collectives using analysis of dreams has been authenticated with decades of research and clinical experience.
8. Research on Jungian principles of psychological healing may be directly correlated with current cognitive theory and research and used to study ethical behavior in virtual realities.
9. The universe of physics is holographic (David Bohm) and the brain is holographic (Karl Pribram).
10. The Media-sphere exists as a holographic Jungian archetypal representation.
11. The Media-sphere can be interrogated as a collective dreamscape according to Jung's principles of dream analysis and healing dynamics.
12. Language grounds thought, so the nature of symbolic language, language in general, and how it influences human reality perceptions can be understood as the seed ground for ethics.
13. As researched by the Institute of HeartMath, the quality of coherence can be correlated with both Jungian Compensation (leading to psychological harmony) and with measurable ethical behavior.
14. The recent evolution in computer science and technologies, artificial intelligence, video game technology (Togelius, 2016), and robotics allows for use of PSYCHecology video games (PEG) to do robust research on the cognitive unconscious, for verification of Jungian dynamics, for understanding ethics in virtual realities, and for healing of contextual collective personae.
15. Algorithms employing the dynamics of Fourier transforms for psychological analysis of the media-sphere can be used to analyze the media dreams of contextual collectives using authentic Jungian principles for analysis of dreams.
16. Current and historical study of ethics strongly suggests that ethics are based on coherent thoughts, feelings, and behavior that can be fostered with PSYCHecology video games (PEG) that are programmed according to Jungian principles of personality types. The principles of coherence, Jungian Compensation, and mystical transcendence—as well as the Flow experience—are directly correlated.
17. PEG may be employed as a research instrument for both psychological healing and ongoing research relative to the nature of ethics in virtual realities.
18. The paradigm shift into a media age could herald the projection of the World-Soul into a manifest superconductive culture of conscience.

PEG Video Games: Resolving the Global Problematique

This proposal for developing a genre of PSYCHecology video games (PEG) defines an unprecedented model for addressing human transition into the alternative reality of holographic mediation—what we call the media-sphere. The PEG model is a practical approach to resolving the global problematique and advancing robust research on the virtual reality that we call the PSYCHecology—the Ecology of Psyche (ΨΘ)—that is having critical impact on the earth's Geomagnetic fields and beyond. These fields are quantum electrodynamic (QED) in nature, and they are carriers of biologically relevant information that connects all living systems and extends to other cosmic dimensions of sentience. The author uses the term "bandwidth" to describe the infinite dynamic for defining this quantum interactivity—not because it adheres to every definition of reality, but because it is understandable in terms of much current scientific research and technological mediation. Understanding how "electrodynamic bandwidths" function in the Media-sphere should cause a radical alteration in human perceptions of sentience and of their human function and responsibility not only to life on earth but to cosmic harmony. While there is much yet to learn about "QED bandwidths", the conceptual map is not the territory, but it serves our present purposes (Korzybski's Dictum).

The gigantic video game industry has only recently seen the value of psychological narrative relative to the success of games. Though the concepts addressed in the book has been a work in progress for the last thirty years, video game technology, imaging technology, and computational advances have only recently achieved the requisite technological capabilities to realize the PEG concept. When this proposal was first expressed, there was little interest in quality game stories on the part of an incipient game industry. The industry cliché was that writers were not highly valued and only needed after the game was ready to be produced. Now, finally, there is a realization that the game-narrative is the essence and focus around which popular games are designed. Though it was also obvious that emotion constitutes the raison d'être of video games, game popularity, and the industry in general—people play games because they are fun, and fun is an emotional response—the industry was not much interested in the psychological sources of emotion. Only within the last few years has game-writing and research on emotion become paramount in game design.

During the last several years, research and the dynamics of perceived reality have changed radically. Radical change has made the subject of this chapter and its proposals a primary concern not only of the game industry, but of education, media law, and the ethics of such issues as global political hacking and social media. The scientific research from which chapters are drawn includes QED physics & chemistry, Computer Science, Mathematics, Government Policy, and many more. It provides an unprecedented perspective on diverse fields. Though the chapter targets an academic audience, its content is accessible to educated people in many professions.

REFERENCES

Academy Science USSR. (1976). *Reprint N 17a*. Moscow: IZMIRAN.

Aiello, L. M. (2014, August 1). How Yahoo Research Labs studies culture as a formal computational concept. *MIT Technology Review*. Retrieved from http://www.technologyreview.com/view/529521/how-yahoo-research-labs-studies-culture-as-a-formal-computational-concept

Baars, B. J. (1997). In the theatre of consciousness: Global workspace theory, a rigorous scientific theory of consciousness. *Journal of Consciousness Studies*, *4*(4), 292–309.

Bateson, G. (1972). *Steps to an ecology of mind: collected essays in anthropology, psychiatry, evolution, and epistemology*. Chicago: University Of Chicago Press.

Brooks, R. A. (2016). *Fields of Color: The theory that escaped Einstein*. Rodney A. Brooks.

Burnett, D. G. (2012). *The Sounding of the Whale: Science and Cetaceans in the Twentieth Century*. Chicago: University of Chicago Press.

Cheddad, A., Condell, J., Curran, K., & McKevitt, P. (2010). Digital image steganography: Survey and analysis of current methods. *Signal Processing*, *90*(3), 727–752.

Chopra, D. (1995). *Body, mind, and soul. Turner Home Entertainment*. Atlanta, GA: PBS Home Video.

Clark, B. (2014). *John Lilly, the mind of the dolphin, and communication out of bounds*. USA: Texas Tech University.

DiBiasi, F. (2013). *A holoinformational model of the physical observer*. Universidade Geraldo Di Biase.

Direct Action 2017. (2013). Dr. Emoto Thoughts/Intentions Shape Physical World - Amazing... SHARE! [YouTube video]. Retrieved from https://www.youtube.com/watch?v=Is8FE0RQo8A

Documentary TV. (2014, September 1). How Plants Communicate & Think – Amazing Nature Documentary [YouTube video].

Eagleman, D. M. (2009). Silicon immortality: Downloading consciousness into computers. What Will Change Everything. Retrieved from https://www.eagleman.com/blog/silicon-immortality-downloading-consciousness-into-computers

Eisley, L. (1970). *The cosmic prison*. (S. Buscemi & S. Charlotte, Eds.). Boston, MA: McGraw Hill.

Emerging Technology from the arXiv. (2015). The emerging science of human-data interaction. *MIT Technology Review*. Retrieved from https://www.linkedin.com/grp/post/1039687-5900533809526378496

Emerging Technology from the arXiv. (2016, November 7). Neural networks can read book covers. *Kyushu University, Japan* MIT Technology Review. Retrieved from https://www.technologyreview.com/s/602807/deep-neural-network-learns-to-judge-books-by-their-covers/?utm_campaign=newsletters&utm_source=newsletter-weekly-computing&utm_medium=email&utm_content=201611

Gackenbach, J., & Hakopdjanian, S. (2015). Breaking the frame of digital, dream, and waking realities. In S. Schafer (Ed.), *Exploring the Collective Unconscious in a Digital* Age (pp. 79-106). Hershey, PA: IGI Global. Retrieved from http://www.igi-global.com/book/exploring-collective-unconscious-age-digital/140975

Galeon, D. (2016). For the first time in history, we are seeing alien worlds (with spectroscopy). *Futurism*. Retrieved from https://futurism.com/for-the-first-time-in-history-we-are-seeing-alien-worlds/

Galeon, D. (2017). New research shows that the universe might once have been a hologram. *Futurism*. Retrieved from https://futurism.com/new-research-shows-the-universe-may-have-once-been-a-hologram/

Gutschner, M. (2014). Discovery of Quantum Vibrations in "Microtubules" Inside Brain Neurons Corroborates Controversial 20-Year-Old Theory of Consciousness [Press release]. *Elsevier*. Retrieved from https://www.elsevier.com/about/press-releases/research-and-journals/discovery-of-quantum-vibrations-in-microtubules-inside-brain-neurons-corroborates-controversial-20-year-old-theory-of-consciousness

Hameroff, S., & Penrose, S. R. (2014). Discovery of quantum vibrations in 'mocrotubules' corroborates theory of consciousness. *PhysOrg*. Retrieved from https://phys.org/news/2014-01-discovery-quantum-vibrations-microtubules-corroborates.html

Hinton, G. E., McClelland, J. L., Rumelhart, D. E., & Rnmelhart, D. E. (1986). *Parallel distributed processing: Explorations in the microstructure of cognition*. Cambridge, MA: MIT Press.

Ho, M. W. (2005). *The rainbow and the worm: the physics of organisms*. Singapore: World Scientific Publishing Co.

Institute of Heartmath. (2007). New research project: global coherence monitoring system. IHM Summer 2007 Newsletter. Retrieved from http://www.heartmath.org/templates/ihm/e-newsletteer/2007/Summer-2007/summer_2007_newsletter.htm

Jeffrey, C. (2015). First-ever QED device that detects and corrects its own errors. *New Atlas*. Retrieved from https://newatlas.com/quantum-device-detects-corrects-errors/36404/

Jenkins, H. (2005). *Game design as narrative architecture*. Henry Jenkins. Retrieved from http://homes.lmc.gatech.edu/~bogost/courses/spring07/lcc3710/readings/jenkins_game-design.pdf

Joye, S. R. (2016). *The holoflux theory of consciousness: An integral interpretation of the theories of Karl Pribram, David Bohm, and Pierre Teilhard de Chardin. California*. San Francisco, CA: Institute of Integral Studies.

Jung, C. G. (1933). *Modern Man in Search of a Soul*. New York: Harcourt Brace Jovanovich.

Khodarahimi, S. (2009, October-December). Dreams in Jungian psychology: The use of dreams as an instrument for research diagnosis and treatment of social phobia. *The Malaysian Journal of Medical Sciences, 16*(4), 42–49. PMID:22135511

Kirk, T. (2014). Superconducting spintronics pave way for next-generation computing. *PhysOrg*. Retrieved from http://phys.org/news/2014-01-superconducting-spintronics-pave-next-generation.html

Knight, W. (2015, November 2). A new tool for analyzing academic papers uses cutting-edge AI to find meaning in billions of words. *MIT Technology Review*. Retrieved from http://www.technologyreview.com/news/542981/academic-search-engine-grasps-for-meaning/

Lakoff, G. (2002). *Moral politics*. Chicago, IL: The University of Chicago Press. doi:10.7208/chicago/9780226471006.001.0001

Lakoff, G. (2008). *The political mind*. USA: Viking Penguin.

Langlitz, N. (1972). Neuropsychedelia: The Revival of Hallucinogen Research since the Decade of the Brain. Berkeley, CA: University of California Press.

Langlitz, N. (2006). Tripping in Solitude: Introducing Honza Samotar by Way of John Lilly. In K. Solhdju (Ed.), *Introspective Self-Rapports: Shaping Ethical and Aesthetic Concepts 1850–2006* (pp. 81–92). Berlin: Max Planck Institute for the History of Science.

Lewis, T. (2014). This computer chip can think like a human brain. *Live Science*. Retrieved from https://www.livescience.com/47240-ibm-computer-chip-simulates-brain.html

Lilly, J. C. (1967). *The Mind of the Dolphin: A Nonhuman Intelligence*. Garden City, NY: Doubleday.

Lilly, J. C. (1974). *Programming and Metaprogramming in the Human Biocomputer: Theory and Experiments*. New York: Bantam.

Lilly, J. C. (1985). *The center of the cyclone: An autobiography of inner space*. Three Rivers Pr.

Lipton, B. (2005). The biology of belief. Santa Rosa, CA: Mountain of Love/Elite Books.

Luhmann, N. (2002). The Paradox of Observing Systems. In W. Rasch (Ed.), *Theories of Distinction: Redescribing the Descriptions of Modernity* (pp. 73–93). Stanford: Stanford University Press.

Luhmann, N. (2012). *Theory of Society* (Vol. 1). (R. Barrett, Trans.). Stanford: Stanford University Press.

Maharishi Effect. (n.d.). Maharishi's Programmes. Retrieved from maharishi-programmes.globalgoodnews.com/maharishi-effect

McCarthy, J., Minsky, M. L., & Rochester, N. (1955). Dartmouth summer research project on artificial intelligence. *BBC*. Retrieved from http://www.bbc.com/future/story/20161114-how-we-built-machines-that-can-think-for-themselves?ocid=ww.social.link.linkedin

McCraty, R. (2002). Heart rhythm coherence: And emerging area of biofeedback. *Biofeedback*, *30*(1), 17–19.

McCraty, R. (2005). *Enhancing emotional, social, and academic learning with heart rhythm coherence feedback*. Heartmath Research Center, Institute of Heartmath.

McCraty, R., Atkinson, M., Stolc, V., Alabdulgader, A. A., Vainoras, A., & Ragulskis, M. (2017). Synchronization of Human Autonomic Nervous System Rhythms with Geomagnetic Activity in Human Subjects. *International Journal of Environmental Research and Public Health*, *14*(7), 770. doi:10.3390/ijerph14070770

McCraty, R., & Childre, D. (2010). Coherence: bridging personal, social, and global health. *Alternative Therapies*, *16*(4). Retrieved from https://www.heartmath.org/assets/uploads/2015/01/coherence-bridging-personal-social-global-health.pdf

McCraty, R., & Tomasino, D. (2006). Coherence-building techniques and heart rhythm coherence feedback: New tools for stress reduction, disease prevention, and rehabilitation. In E. Molinari, A. Compare, & G. Parati (Eds.), *Clinical Psychology and Heart Disease*. Milan, Italy: Springer-Verlag. doi:10.1007/978-88-470-0378-1_26

McDonald, C. (2015). The good, the bad, and the robot: Experts are trying to make robots moral. *Cal Alumni Association*. Retrieved from https://alumni.berkeley.edu/california-magazine/just-in/2015-06-08/good-bad-and-robot-experts-are-trying-make-machines-be-moral

Mindvalley Talks. (2011). How to change your frequency to change your reality [YouTube video].

Nikolaev, Y. S., Rudakov, Y. Y., Mansurov, S. M., & Mansurova, L. G. (1976). Interplanetary magnetic field sector structure and disturbances of the central nervous system activity.

Norris, J. (2014). Do gut bacteria rule our minds. *UCSF*. Retrieved from https://www.ucsf.edu/news/2014/08/116526/do-gut-bacteria-rule-our-minds

Ohmori, K. (2016). World's fastest quantum simulator. *Institute for Molecular Science, National Institutes of Natural Sciences, Japan, Nature Communications*. Retrieved from http://phys.org/news/2016-11-world-fastest-quantum-simulator-atomic.html

Papadopolous, R., & Byng-Hall, J. (Eds.). (2007). *Narrative Based Medicine: A narrative approach to medicine*.

Pert, C. B. (2003). *Molecules of emotion*. New York: Scribner.

Peters, J. D. (1999). *Speaking into the Air: A History of the Idea of Communication*. Chicago: University of Chicago Press. doi:10.7208/chicago/9780226922638.001.0001

Pobachenko, S. V., Kolesnik, A. G., Borodin, A. S., & Kalyuzhin, V. V. (2006). The contingency of parameters of human encephalograms and Schumann Resonance electromagnetic fields revealed in monitoring studies. *Complex Systems Biophysics.*, *51*(3), 480–483.

Pribram, K. (1969). The neurophysiology of remembering. *Scientific American*, *220*(1), 73–87.

Pribram, K. (1971). *Languages of the brain; experimental paradoxes and principles in neuropsychology*. Englewood Cliffs, NJ: Prentice-Hall.

Pribram, K. (1977). *Languages of the Brain. Monterey*, CA: Wadsworth Publishing.

Pribram, K. (1991). *Brain and perception: holonomy and structure in figural processing*. Hillsdale, NJ: Lawrence Erlbaum Associates.

Pruchnic, J. (2008). Neurorhetorics: Cybernetics, Psychotropics, and the Materiality of Persuasion. *Configurations*, *16*(2), 2. doi:10.1353/con.0.0057

Radin, D. (1997). *The conscious universe: Scientific proof of psychic phenomena*. USA: Harper One.

Radin, D. (2006). *Entangled Minds: Extrasensory Experiences in a Quantum Reality; · Supernormal: Science, Yoga, and the Evidence for Extraordinary Psychic Abilities*. USA: Paraview Pocket Books.

Radin, D. (2018). *Real magic: Ancient wisdom, modern science, and a guide to the secret power of the universe*. USA: Penguin/Random House.

Rogers, K. (2018). *MIT Device Uses Wi-Fi to 'See' Through Walls and Track Your Movements*. Vice.

Saffell, N. (2015). On the origin of robot species: robots building robots by "natural selection." University of Cambridge. Retrieved from http://www.cam.ac.uk/research/news/on-the-origin-of-robot-species?sthash.zMSyp5OS.mjjo

Schafer, S. (2012a). Optimizing Cognitive Coherence, Learning, and Psychological Healing with Drama-based Games. *Video Game Play and Consciousness*. Nova Science Publishers. Retrieved from http://www.facebook.com/insights/?sk=po_261980910540561#!/pages/Video-Game-Play-and-Consciousness/261980910540561?sk=wall

Schafer, S. (Ed.). (2017b). *Exploring the collective unconscious in a digital age*. Hershey, PA: IGI Global.

Schafer, S., & Yu, G. (2011). *Meaningful video games: drama-based video games as transformational experience*. Hershey, PA: IGI Global. doi:10.4018/978-1-60960-567-4.ch019

Schafer, S. B. (2012b). Articulating the paradigm shift: Serious games for psychological healing of the collective persona. In *Handbook of Research on Serious Games as Educational, Business and Research Tools* (pp. 352–377). Hershey, PA: IGI Global.

Schafer, S. B. (2016e). Employing dynamics of Jungian compensation to foster psychological coherence with ICTs. *International Journal of Privacy and Health Information Management*, *4*(1), 23–49.

Schafer, S. B. (2017a). *Generating a superconductive culture of conscience*. Mauritius: Lambert Academic Publishing.

Schafer, S. B. (2018). Synergy with the World Soul: Establishing the humanitarian organism. International Journal of Civic Engagement and Social Change, 4(3). doi:10.4018/IJCESC.20170701

Shiga, J. (2013). Of Other Networks. *Amodern 2: Network Archaeologies*. Retrieved from http://amodern.net/article/of-other-networks/

Simonite, T. (2015). A better way to design brain inspired chips. *MIT Technology Review*. retrieved from http://www.technologyreview.com/news/537211/a-better-way-to-build-brain-inspired-chips/

Strassman, R. (2001). DMT, the spirit molecule. Rochester, VT: Park Street Press.

Streedhar, N. (2017). Micro-organisms and their influence in the human body. *Financial Express*. Retrieved from https://www.financialexpress.com/lifestyle/health/micro-organisms-and-their-influence-in-the-human-body-find-out-how/547433/

Talbot, M. (1991). *The holographic universe*. New York: Harper.

Taylor, T. (2014). Outline of the history of artificial life. *LinkedIn*. Retrieved from https://www.linkedin.com/grp/post/1039687-5900533809526378496

Thincube. (2015). Dr. Masaru Emoto's Water Experiment - Words are Alive! [YouTube video]. Retrieved from https://www.youtube.com/watch?v=au4qx_l8KEU

Trafton, A. (2014). New brain-scanning technique allows scientists to see when and where the brain processes visual information. *Medical Press*. Retrieved from http://medicalxpress.com/news/2014-01-brain-scanning-technique-scientists-brain-visual.html

Twist, J. (2004). Brainwaves control video game. *BBC*. Retrieved from http://news.bbc.co.uk/2/hi/technology/3485918.stm

Von Foerster, H. (1962). Bio-Logic. In E. E. Bernard & M. R. Kare (Eds.), *Biological Prototypes and Synthetic Systems* (Vol. 1, pp. 1-12). New York: Plenum Press. doi:10.1007/978-1-4684-1716-6_1

von Foerster, H. (1973). On Constructing a Reality. In *Understanding Understanding*. New York: Springer.

Von Foerster, H., Müller, A., Müller, K. H., Rooks, E., & Kasenbacher, M. (2014). *The Beginning of heaven and earth has no name: Seven days with second-order cybernetics*. Fordham University; doi:10.2307/j.ctt13wzw8d

Weaver, W. (1949). Recent Contributions to the Mathematical Theory of Communication. In *C. Shannon (Ed.), The Mathematical Theory of Communication* (pp. 3–28). Urbana, IL: University of Illinois Press.

WilburK. (2018). Cultivating your superhuman body.

WilburK. (2018). Cultivating your superhuman body.

Zohar, D. (1990). *The quantum self: Human nature and consciousness defined by the new physics*. New York: William Morrow and Company, Inc.

Zohar, D., & Marshall, I. (1993). The quantum society: Mind, physics, and a new social vision. New York: Bloomsbury Publishing.

ENDNOTES

1 Heraclitus: "No man ever steps in the same river twice, for it's not the same river and he's not the same man"

2 The recent discovery of warm temperature quantum vibrations in microtubules inside brain neurons by the research group led by Anirban Bandyopadhyay, PhD, at the National Institute of Material Sciences in Tsukuba, Japan (and now at MIT), corroborates the pair's theory and suggests that EEG rhythms also derive from deeper level microtubule vibrations. In addition, work from the laboratory of Roderick G. Eckenhoff, MD, at the University of Pennsylvania, suggests that anesthesia, which selectively erases consciousness while sparing non-conscious brain activities, acts via microtubules in brain neurons.

3 Orchestrated objective reduction is a hypothesis that consciousness in the brain originates from processes inside neurons, rather than from connections between neurons. The mechanism is held to be a quantum physics process called objective reduction that is orchestrated by molecular structures called microtubules. It is proposed that objective reduction is influenced by non-computable factors embedded in spacetime geometry which thus may account for the Hard Problem of Consciousness.

4 The hard problem of consciousness is the problem of explaining how and why we have qualia or phenomenal experiences—how sensations acquire characteristics, such as colors and tastes.

5 The recent discovery of warm temperature quantum vibrations in microtubules inside brain neurons by the research group led by Anirban Bandyopadhyay, PhD, at the National Institute of Material Sciences in Tsukuba, Japan (and now at MIT), corroborates the pair's theory and suggests that EEG rhythms also derive from deeper level microtubule vibrations.

6 Reminiscent of Jungian amplification, Lakoff describes radial conceptual categories as the most common form of metaphorical structure. "The radial categories show how the coherent ideologies

in each category fit together and what the relationships among them are." (Lakoff, 2002, p. 14) Radial categories are structured according to variations on a central model. As an example, he uses the category mother and explains that the term is characterized by four metaphorical sub-models: The birth model, the genetic model, the nurturance model, and the marriage model. (Lakoff, 2002, p. 8) All of the frames in the radial categories are couched in little metaphorical stories within stories. Presumably, if "mother" is the prototype and sub-model variants of the "mother metaphor" constitute radial categories or variations on the central model, discovering how the categories fit together and how they are related in a contextual field would result in discovery of meaning. (Schafer, 2012)

7 "Neural binding circuitry consists just of neurons and connections but has a special effect: it creates new experiences." (Lakoff, 2008, p. 26) We understand events in the world in terms of what our bodies can do. As a result, every simple narrative has such an 'event structure.' In the field of computational neural modeling, the technical term is an 'executing schema,' or 'X-schema' for short. (p. 27)

8 Monotheism is a universal concept espoused by all the Great Religions and is implied in so-called primitive religions. The existence of numerological hierarchies of sentience is fundamentally mystical. The One-God principle of "Logos" is both mystical and scientific (ranging from Gnostic, Greek, Islamic, Hindu, and Buddhist interpretations to Carl Jung and Gottfried Wilhelm Leibniz. Monotheism is not the province of Judeo-Christian philosophy only.

9 Entrainment, within the study of chronobiology, occurs when rhythmic physiological or behavioral events match their period and phase to that of an environmental oscillation. A common example is the entrainment of circadian rhythms to the daily light–dark cycle, which ultimately is determined by the Earth's rotation.

10 "Instead of the ordinary numbers used in conventional mathematics, the quantum nature of the fields requires the use of Hilbert algebra, in which the physical properties are described by "vectors" in an abstract Hilbert space and by "operators" that act on those vectors."

11 Abundant and robust research attests to this fact—far too much to cite, but the following references provide good examples: (Alabdulgader, McCraty, Atkinson, Dobyns, Vainoras, Ragulskis, and Stolc, 2018; Garling, & Cully, 2015; Mortier, Haddadi, Henderson, McAuley, & Crowcroft, 2015; McCraty, 2004)

12 From the unpublished document, *Gaming Cultural Atonement,* Stephen Brock Schafer.

13 Steganography is the science that involves communicating secret data in an appropriate multimedia carrier, e.g., image, audio, and video files.

Compilation of References

(1959). C. Psychophysiological approaches to the evaluation of psychotherapeutic process and outcome. InRubinstein, E., & Parloff, M. (Eds.), *Research in Psychotherapy* (pp. 160–208). Washington, DC: American Psychological Association. doi:10.1037/10036-010

Abbott, L. F., & Nelson, S. B. (2000). Synaptic plasticity: Taming the beast. *Nature Neuroscience*, *3*(S11), 1178–1183. doi:10.1038/81453 PMID:11127835

Abdi, A. A., & Shultz, L. (2008). *Educating for human rights and global citizenship*. Albany, NY: State University of New York Press.

Abram, D. (1997). *The spell of sensuous perception and language in a more than human world.* New York: Vintage Books, A Division of Random House Book, Inc.

Academy Science USSR. (1976). *Reprint N 17a*. Moscow: IZMIRAN.

Adams, M. V. (2002). African American dreaming and the beast of racism. The cultural unconscious in Jungian analysis. *Psychoanalytic Psychology*, *19*(1), 182–198. doi:10.1037/0736-9735.19.1.182

Aggleton, J. P. (Ed.). (1992). *The Amygdala: Neurobiological Aspects of Emotion, Memory and Mental Dysfunction.* New York: Wiley-Liss.

Aharonov, Y., Popescu, S., Rohrlich, D., & Skrzypczyk, P. (2015). Quantum Cheshire cat. *New Journal of Volume Physics, 15*, 1-5. Retrieved from http_iopscience.iop.org_1367_2630_15_11_113015_art.pdf

Ahn, D., Jeon, S., & Yoo, B. (2014). *What shows your age? An empirical analysis of identity fraud in online game classification systems.* GCCR Working Paper Series 14-8. Retrieved from https://ssrn.com/abstract=2442683

Aiello, L. M. (2014). How Yahoo Research Labs studies culture as a formal computational concept. *MIT Technology Review*. Retrieved from http://www.technologyreview.com/view/529521/how-yahoo-research-labs-studies-culture-as-a-formal-computational-concept

Aiello, L. M. (2014, August 1). How Yahoo Research Labs studies culture as a formal computational concept. *MIT Technology Review*. Retrieved from http://www.technologyreview.com/view/529521/how-yahoo-research-labs-studies-culture-as-a-formal-computational-concept

Akhter, R. (2014). *Existentialism and its relevance to the contemporary system of education in India: Existentialism and present educational scenario*. Hamburg, Germany: Anchor Academic Publishing.

Alabdulgader, A., McCraty, R., Atkinson, A., Dobyns, Y., Vainoras, A., Ragulskis, M., & Stolc, V. (2018). Long-Term study of heart rate variability responses to changes in the solar and geomagnetic environment. *Scientific Reports.* Retrieved from https://www.heartmath.org/assets/uploads/2018/03/long-term-study-of-hrv-responses-to-changes-in-solar-geomagnetic-enviroment.pdf

Alabdulgader, A., McCraty, R., Atkinson, M., Dobyns, Y., Vainoras, A., Ragulskis, M., & Stolc, V. (2018). Long-Term study of heart rate variability responses to changes in the solar and geomagnetic environment. *Scientific Reports.* Retrieved from https://www.heartmath.org/assets/uploads/2018/03/long-term-study-of-hrv-responses-to-changes-in-solar-geomagnetic-enviroment.pdf

Albala-Bertrand, J. M. (2000). Complex emergencies versus natural disasters: An analytical comparison of causes and effects. *Oxford Development Studies, 28*(2), 187–204. doi:10.1080/713688308

Aloni, N. (2007). *Enhancing humanity: The philosophical foundations of humanistic education.* Dordrecht, The Netherlands: Springer.

Anaxagoras. (2015). Anaxagoras of Clazomenae. In *Stanford Encyclopedia of Philosophy.* Retrieved from http://plato.stanford.eddu/entries/anaxagoras.

Andrews, P. L. R. (1986). Vagal afferent innervation of the gastrointestinal tract. *Progress in Brain Research, 67*, 65–86. doi:10.1016/S0079-6123(08)62757-0 PMID:3823483

Annette, B. S., Oernbo, E. K., Stoica, A., Gerkau, N. J., Barbuskaite, D., Tritsaris, K., ... MacAulay, N. (2018). Cotransporter-mediated water transport underlying cerebrospinal fluid formation. *Nature Communications, 9*(1). PMID:29867199

Anonymous. (2009). *What Einstein Thought About Thermodynamics, The Second Law of Life.* Retrieved from https://secondlawoflife.wordpress.com/2009/07/12/what-einstein-thought-about-thermodynamics/

Anonymous. (2014). The curious evolution of artificial life. *MIT Technology Review.* Retrieved 7/23/2017 from https://www.linkedin.com/grp/post/1039687-5900533809526378496

Anonymous. (2016a). Google Unveils Neural Network with "Superhuman" Ability to Determine the Location of Almost Any Image. *MIT Technology Research.* Retrieved 7/23/2017 from https://www.technologyreview.com/s/600889/google-unveils-neural-network-with-superhuman-ability-to-determine-the-location-of-almost/

Anonymous. (2016b). Neural networks learn to identify criminals by their faces. *MIT Technology Review.* Retrieved 7/23/2017 from https://www.technologyreview.com/s/602955/neural-network-learns-to-identify-criminals-by-their-faces/?utm_campaign=newsletters&utm_source=newsletter-weekly-computing&utm_medium=email&utm_content=20161

Anonymous. (2017). How we build machines that can think for themselves. *BBC.* Retrieved from http://www.bbc.com/future/story/20161114-how-we-built-machines-that-can-think-for-themselves

Anonymous. (2018). The machine vision challenge to better analyze satellite images of Earth. *MIT Technology Review.* Retrieved from https://www.technologyreview.com/s/611198/the-machine-vision-challenge-to-better-analyze-satellite-images-of-earth/

Anonymous. (2018a). The energetic heart. CA: *Institute of Heartmath.* Retrieved from https://www.heartmath.org/articles-of-the-heart/science-of-the-heart/the-energetic-heart-is-unfolding/

Anonymous. (2018b). Study shows that geomagnetic fields and solar activity affects human autonomic nervous system functions. *Heartmath Institute.* Retrieved from https://www.heartmath.org/articles-of-the-heart/study-shows-geomagnetic-fields-solar-activity-affect-human-autonomic-nervous-system-functions/

Anonymous. (2018c). Global coherence initiative. *Heartmath Institute.* Retrieved from https://www.heartmath.org/gci/

Aquinas, T. (1992). Summa theologica, question 94, on the natural law. In M. L. Morgan (Ed.), Classics of moral and political theory (pp. 482–483). Indianapolis, IN: Hackett.

Aquinas, T. (1986). The order of learning the sciences (A. Maurer, Trans.). In *The division and methods of the sciences* (pp. 99–102). Toronto: Pontifical Institute of Mediaeval Studies.

Aristotle. (1947). Analytica Posteriora (G. R. G. Mure, Trans.). In McKeonR. (Ed.), *Introduction to Aristotle* (pp. 9–109). New York: Modern Library.

Aristotle. (2004). *Nicomachean Ethics (C. Rowe, Trans.).* Oxford, UK: Oxford University Press.

Aristotle. (2010). *The Politics: translated and with the introduction, notes and glossary by Carnes Lord.* The University of Chicago Press.

Armour, J.A. (2003). *Neurocardiology—Anatomical and functional principles.* Boulder Creek, CA: HeartMath Research Center, Institute of HeartMath, Publication No. 03-011.

Armour, J. A. (1994). Peripheral autonomic neuronal interactions in cardiac regulation. In J. A. Armour & J. L. Ardell (Eds.), *Neurocardiology* (pp. 219–244). New York: Oxford University Press.

Ascham, R. (1970). Report of the affair and state of Germany. In W. A. Wright (Ed.), *English Works* (pp. 123–169). Cambridge, UK: Cambridge University Press.

Atkinson, R. L., Atkinson, R. C., & Smith, E. E. (1990). *Introduction to Psychology.* Orlando, FL: Harcourt Brace Jovanovich.

Attanasio, A. A. (2016). *Science Fiction.* Retrieved from http://dreadfuljoy.blogspot.com/2016/12/welive-in-world-not-of-science-but-of.html

Augustine. (1998). *The city of God against the pagans* (R. W. Dyson, Trans. & Ed.). Cambridge, UK: Cambridge University Press.

Ax, A. F. (1953). The physiological differentiation between fear and anger in humans. *Psychosomatic Medicine, 15*(5), 433–442. doi:10.1097/00006842-195309000-00007 PMID:13100546

Baars, B. J. (1997). In the theatre of consciousness: Global workspace theory, a rigorous scientific theory of consciousness. *Journal of Consciousness Studies, 4*(4), 292–309.

Bae, B.-I., Jayaraman, D., & Walsh, C. A. (2015). Genetic Changes shaping the Human Brain. *Developmental Cell, 32*(4), 423–434. doi:10.1016/j.devcel.2015.01.035 PMID:25710529

Ball, P. (2001). *Life's matrix: A biography of water.* Berkeley, CA: University of California Press. Retrieved from https://www.nasa.gov/vision/universe/solarsystem/Water:_Molecule_of_Life.html

Baloyannis, S. J. (2018). Anaxagoras on mind. *J Neural Stroke, 8*(1), 00269. doi:10.15406/jnsk.2017.08.00269

Banafa, A. (2016). The Last Mile of IoT. *Open Mind.* Retrieved 11/15/2016 from: https://www.bbvaopenmind.com/en/the-last-mile-of-iot-artificial-intelligence-ai/#.WCMHOfLODIA.linkedin

Bandura, A. (1986). *Social foundations of thought and action.* Englewood Cliffs, NJ: Prentice-Hall.

Barfield, O. (1957). Saving the Appearances. London: Faber and Faber.

Baron-Cohen, S. (1995). *Mindblindness: An essay on autism and theory of mind.* Boston, MA: MIT Press/Bradford Book. doi:10.7551/mitpress/4635.001.0001

Baron-Cohen, S., Leslie, A. M., & Frith, U. (1985). Does the autistic child have a 'theory of mind'? *Cognition, 21*(1), 37–46. doi:10.1016/0010-0277(85)90022-8 PMID:2934210

Bartholomew, Y. B. (2018). The evolution of consciousness. *Environment and Ecology Research, 6*(4), 259-269, http://www.hrpub.org

Bateson, G. (1972). *Steps to an ecology of mind: collected essays in anthropology, psychiatry, evolution, and epistemology.* University Of Chicago Press.

Baturin, D. A. (2016). Fentezijnyj neomif v kontekste teorii arhetipov K.G. Junga [Fantasy & neomyth in the context of C.G. Jung's theory of archetypes]. Novaja nauka: opyt, tradicii, innovacii, 2016(2), 177-180.

Baudrillard, J. (1998). When Bataille attacked the metaphysical principles of economy. In F. Bottig & S. Wilson (Eds.), *Bataille: A critical reader.* Oxford, UK: Blackwell.

Bauman, Z. (2000). *Liquid modernity.* Cambridge, UK: Polity Press.

Bergethon, K. (n.d.). *Cyber currency investing for beginners.* Loveland, CO: Legendary Ventures. Retrieved from https://thefranklinsociety.com/cyber-currency-investing-for-beginners/

Bernstein, D. A. (2010). *Essentials of Psychology.* Boston: Cengage Learning.

Bisht, D. B. (1985). *The spiritual dimension of health.* New Delhi: Directorate General of Health Services, Government of India.

Black, R. (2001). *Humanism and education in Medieval and Renaissance Italy: Tradition and innovation in Latin schools from the twelfth to the fifteenth century.* New York: Cambridge University Press. doi:10.1017/CBO9780511496684

Bohm, D. (1987). Hidden variables and the implicate order. Quantum Implications.

Bohr, N. (1976). Niels Bohr, Collected Works: Vol. 3. *The Correspondence Principle (1918-1923).* Amsterdam: North-Holland.

Bonvallet, M., & Allen, M. B. Jr. (1963). Prolonged spontaneous and evoked reticular activation following discrete bulbar lesions. *Electroencephalography and Clinical Neurophysiology, 15*(6), 969–988. doi:10.1016/0013-4694(63)90141-X PMID:14132294

Bourdieu, P. (1977). *Outline of a theory of practice.* Cambridge, UK: Cambridge University Press. doi:10.1017/CBO9780511812507

Bourdieu, P. (1990). *The logic of practice.* Cambridge, UK: Polity Press.

Bourdieu, P. (2004). *The science of science and reflexivity.* Chicago, IL: The University Of Chicago Press.

Bower, G. H. (1990). Mood-congruity of social judgements. In J. Forgas (Ed.), *Emotions and Social Judgements.* Oxford, UK: Pergamon.

Bower, G. H. (1992). How might emotions affect learning? In S. A. Christianson (Ed.), *The Handbook of Emotion and Memory: Research and Theory* (pp. 3–31). Hillsdale, NJ: Lawrence Erlbaum.

Bradley, R. T. (2006). The psychophysiology of entrepreneurial intuition: A quantum-holographic theory. *Institute for Whole Sciences; Institute of Heartmath; e-motion Institute.* Retrieved from https://www.heartmath.org/research/research-library/intuition/psychophysiology-of-entrepreneurial-intuition-a-quantum-holographic-theory/

Bradley, R. T. (2006). The psychophysiology of entrepreneurial intuition: A quantum-holographic theory. *Proceedings of the third AGSE International Entrepreneurship Research Exchange.*

Bradley, R. T. (2006). *The psychophysiology of entrepreneurial intuition.* Auckland, New Zealand: Institute for Whole Social Science.

Brennan, L., Binney, W., Parker, L., Aleti, T., & Nguyen, D. (2014). *Social marketing and behavior change. Models, theory and applications.* Cheltenham, UK: Edward Elgar Publishing.

Brock, S. S. (2018). Gaming Cultural Atonement. Hershey, PA: IGI Global.

Brooks, R. A. (2016). Fields of color: The theory that escaped Einstein. Silver Springs, MD: Universal Printing, LLC.

Brooks, R. A. (2010). *Fields of color.* Silver Springs, MD: Universal Printing, LLC.

Brooks, R. A. (2016). *Fields of Color* (3rd ed.). Silver Springs, MD: Universal Printing, LLC.

Brooks, R. A. (2016). *Fields of Color: The theory that escaped Einstein.* Rodney A. Brooks.

Brown, J. K., & Zinnert, J. C. (2018). Mechanisms of surviving burial: Dune grass interspecific differences drive resource allocation after sand deposition. *Ecosphere, 9*(3), e02162. doi:10.1002/ecs2.2162

Brugnoli, M. P., Pesce, G., Pasin, E., Basile, M. F., Tamburin, S., & Polati, E. (2015). The role of clinical hypnosis and self-hypnosis to relief pain and anxiety in severe chronic diseases in palliative care: a 2-year long-term follow-up of treatment in a nonrandomized clinical trial. *Ann. Palliat. Med.* doi:10.21037/apm.2017.10.03

Burnett, D. G. (2012). *The sounding of the whale: Science and cetaceans in the twentieth century.* Chicago: University of Chicago Press.

Burnett, D. G. (2012). *The Sounding of the Whale: Science and Cetaceans in the Twentieth Century.* Chicago: University of Chicago Press.

Burton, W. K., Cabrera, N., & Frank, F. C. (1951). The growth of crystals and the equilibrium structure of their surfaces. In Philosophical Transactions of the Royal Society A, 243(866), 299. doi:10.1098/rsta.1951.0006

Butler, A. B., & Hodos, W. (1996). *Comparative Vertebrate Neuroanatomy: Evolution and Adaptation.* New York: Wiley-Liss.

Byrne, P. (2010). *The Many Worlds of Hugh Everett III: Multiple Universes, Mutual Assured Destruction, and the Meltdown of a Nuclear Family.* Oxford University Press.

Bzdok, D., Schilbach, L., Vogeley, K., Schneider, K., Laird, A. R., Langner, R., & Eickhoff, S. B. (2012). Parsing the neural correlates of moral cognition: ALE meta-analysis on morality, theory of mind, and empathy. *Brain Structure & Function, 217*(4), 783–796. doi:10.100700429-012-0380-y PMID:22270812

Campbell, J. (1968). The hero with a thousand faces (2nd ed.). Princeton University Press.

Cannon, W. B. (1927). The James-Lange theory of emotion: A critical examination and an alternative theory. *The American Journal of Psychology, 39*(1/4), 106–124. doi:10.2307/1415404

Cannon, W. B. (1929). *Bodily Changes in Pain, Hunger, Fear and Rage: An Account of Recent Researches into the Function of Emotional Excitement* (2nd ed.). New York: D. Appleton & Company. doi:10.1097/00007611-192909000-00037

Cantin, M., & Genest, J. (1986). The heart as an endocrine gland. *Scientific American, 254*(2), 76–81. doi:10.1038cien tificamerican0286-76 PMID:2935933

Cara, C. (2009). Creating a coherent culture: Building a culture that enables innovation and change. *International Journal of Knowledge, Culture and Change Management, 9*(4), 15-34.

Carroll, S. (2016). *The big picture: On the origins of life, meaning, and the universe itself.* New York: Dutton.

Carvalho, R. (2014). Synchronicity, the infinite unrepressed, dissociation and the interpersonal. *The Journal of Analytical Psychology*, *59*(3), 366–384. doi:10.1111/1468-5922.12085 PMID:24919629

Cashford, J. (2018). 'Who is my Jung?' The progressive, though sometimes ambivalent, expansion of Jung's idea of the collective unconscious: From an 'unconscious humanity' to - in all but name - the soul of the world. *The Journal of Analytical Psychology*, *63*(3), 322–335. doi:10.1111/1468-5922.12413 PMID:29750347

Caughill, P. (2017). Boeing Just Took a Major Step Toward Autonomous Electric Flights. *Futurism.* Retrieved from https://futurism.com/boeing-just-took-a-major-step-toward-autonomous-electric-flights/

Certeau, M. de. (1984). Practices of everyday life. Berkeley, CA: University of California Press.

Chalidze, V. (1991). *Ierarkhicheskii chelovek [The hierarchical man].* Moscow: Terra.

Chalmers, D. (1995). Facing up to the problem of consciousness. *Journal of Consciousness Studies*, *2*(3), 200–219.

Cheddad, A., Condell, J., Curran, K., & McKevitt, P. (2010). Digital image steganography: Survey and analysis of current methods. *Signal Processing*, *90*(3), 727–752.

Childre, D. (2006). *De-stress kit for the changing times*. HeartMath LLC.

Childre, D., & Martin, H. (1999). *The HeartMath Solution.* San Francisco: HarperSanFrancisco.

Childre, D., & Rozman, D. (2002). *Overcoming Emotional Chaos: Eliminate Anxiety, Lift Depression and Create Security in Your Life*. San Diego: Jodere Group.

Chopra, D. (1995). *Body, mind, and soul. Turner Home Entertainment.* Atlanta, GA: PBS Home Video.

Chzhan, Ts., & Tolkachev, S. P. (2016). Issledovanie arkhetipov v kitayskoy telereklame [The study of the archetypes in the Chinese TV commercials]. Obrazovanie i nauka v sovremennykh usloviyakh, 2(2), 29-31.

Clark, A. (2000). *The theory of sentience.* Oxford Press. doi:10.1093/acprof:oso/9780198238515.001.0001

Clark, B. (2014). *John Lilly, the mind of the dolphin, and communication out of bounds*. USA: Texas Tech University.

Clore, G. C. (1994). Why emotions are felt. In P. Ekman & R. J. Davidson (Eds.), *The Nature of Emotion* (pp. 103–111). New York: Oxford University Press.

Collet, C., Vernet-Maury, E., Delhomme, G., & Dittmar, A. (1997). Autonomic nervous system response patterns specificity to basic emotions. *Journal of the Autonomic Nervous System*, *62*(1-2), 45–57. doi:10.1016/S0165-1838(96)00108-7 PMID:9021649

Collins, S. (2014). What Is Your Gut Telling You? *WEBMED.* Retrieved from https://www.webmd.com/digestive-disorders/news/20140820/your-gut-bacteria#1

Connoly, T. (2019). *Plato: Phaedo. The International Encyclopedia of Philosophy: Socrates' intellectual history* (96a-102a). Retrieved from https.//www.iep.upm.edu//Phaedo/

Connoly, T. (2019). *Plato: Phaedo. The International Encyclopedia of Philosophy: Socrates' intellectual history* (96a-102a). Retrieved from https://www.iep.upm.edu//Phaedo/

Coon, D., & Mitterer, J. O. (2008). *Introduction to Psychology: Gateways to Mind and Behavior*. Boston, MA: Cengage Learning.

Corsini, R. J. (2002). *The Dictionary of Psychology*. Florence, KY: Psychology Press.

Crompton, P. (2017). *Mary, Jesus' Mother' is new name for UAE mosque: Mohammad Bin Zayed orders name change to 'consolidate bonds of humanity between followers of different religions'*. Retrieved from https://gulfnews.com/news/uae/culture/mary-jesus-mother-is-new-name-for-uae-mosque-1.2043544

Cummins, D. D. (2014). Neural correlates of causal power judgments. *Frontiers in Human Neuroscience*, *8*(1014). PMID:25566033

Cuny, F. (1983). *Disasters and Development*. New York: Oxford University Press.

Curtis, A. (2009). *Century of the self: Happiness machines for free*. British Broadcasting Corporation. Retrieved from http://freedocumentaries.org/film.php?id=140

Dainton, B. (2003). *Stream of Consciousness: Unity and continuity in Conscious Experience*. London: Routledge, International Library of Philosophy. Retrieved from http://psyche.cs.monash.edu.au/symposia/dainton/prec.html

Damasio, A. (2003). *Looking for Spinoza, sorrow, and the feeling brain*. Orlando, FL: Harcourt.

Damasio, A. R. (1994). *Descartes' Error: Emotion, Reason and the Human Brain*. New York: G.P. Putnam's Sons.

Damasio, A. R. (1999). *The Feeling of What Happens*. Orlando, FL: Harcourt.

Darwin, C. (1871) The Descent of Man, and Selection in Relation to Sex. New York: Appleton.

Davis, P. (1983). *God and the New Physics*. Simon and Schuster.

Davydova, T.R., Kasimov, R.Kh. (2015). Kategoriya vseedinstva kak arkhetip russkoy filosofii i sotsiologii [Category of unity as the archetype of Russian philosophy and sociology]. *Elektronniy nauchno-prakticheskiy zhurnal kul'tura i obrazovanie, 1*.

De Wied, M., Goudena, P. P., & Matthys, W. (2005). Empathy in boys with disruptive behavior disorders. *Journal of Child Psychology and Psychiatry, and Allied Disciplines*, *46*(8), 867–880. doi:10.1111/j.1469-7610.2004.00389.x PMID:16033635

Deep Neural Networks Learns to Judge Books by their Covers. (2016). *MIT Review*. Retrieved 11/15/2016 from https://www.technologyreview.com/s/602807/deep-neural-network-learns-to-judge-books-by-their-covers/?utm_campaign=newsletters&utm_source=newsletter-weekly-computing&utm_medium=email&utm_content=201611

Denkmayr, T., Geppert, H., Sponar, S., Lemmel, H., Matzkin, A., Tollaksen, J., & Hasegawa, Y. (2013). *Observation of a quantum Cheshire Cat in a matter wave interferometer experiment*. Retrieved from http://www.arxiv.org/pdf/1312.3775

Descartes, R. (1649). Treatise on the Passions of the Soul (E. Haldane & G. Ross, Trans.). In *The Philosophical Works of Descartes*. Cambridge, UK: Cambridge University Press.

DiBiasi, F. (2013). *A holoinformational model of the physical observer*. Universidade Geraldo Di Biase.

Dinan, T. G., Stilling, R. M., Stanton, C., & Cryan, J. F. (2015). Collective unconscious: How gut microbes shape human behavior. *Journal of Psychiatric Research*, *63*, 1–9. doi:10.1016/j.jpsychires.2015.02.021 PMID:25772005

Direct Action 2017. (2013). Dr. Emoto Thoughts/Intentions Shape Physical World - Amazing... SHARE! [YouTube video]. Retrieved from https://www.youtube.com/watch?v=Is8FE0RQo8A

Disalle, R. (2009). Space and Time: Inertial Frames. In E. N. Zalta (Ed.), *The Stanford Encyclopedia of Philosophy*. Retrieved from http://plato.stanford.edu/archives/sum2002/entries/spacetime-iframes/#Oth

Dobzhansky, T. (1973). Nothing in biology makes sense except in the light of evolution. *The American Biology Teacher*, *35*(3), 125–129. doi:10.2307/4444260

Documentary TV. (2014, September 1). How Plants Communicate & Think – Amazing Nature Documentary [YouTube video].

Documentary, W. how your consciousness affects the universe, Dr. Emoto. (n.d.). In *YouTube*. Retrieved from https://www.bing.com/videos/search?q=consciousness+of+water&view=detail&mid=431C0918B170F7B18E81431C0918B170F7B18E81&FORM=VIRE

Doeller, C. F., Barry, C., & Burgess, N. (2010). Evidence for grid cells in a human memory network. *Nature*, *463*(7281), 657–661. doi:10.1038/nature08704 PMID:20090680

Dr Masaru Emoto's Water Experiment - Words are Alive! (2014). In *YouTube*. Retrieved from https://www.bing.com/videos/search?q=consciousness+of+water&&view=detail&mid=3C404585CAAEB8A0536F3C404585CAAEB8A0536F&&FORM=VDRVRV

Dr Masaru Emoto's Water Experiment - Words are Alive! (2014). *YouTube*. Retrieved from https://www.bing.com/videos/search?q=consciousness+of+water&&view=detail&mid=3C404585CAAEB8A0536F3C404585CAAEB8A0536F&&FORM=VDRVRV

Dr. Emoto Thoughts/Intentions Shape Physical World - Amazing... SHARE! (2013). In *YouTube*. Retrieved 6/01/2016 from https://www.bing.com/videos/search?q=consciousness+of+water&&view=detail&mid=EECD4295AEE23D260AC4EECD4295AEE23D260AC4&&FORM=VDRVRV

Dretzin, R., Goodman, B., & Rushkoff, D. (Producers). (2004). *The Persuaders* [Video file]. Retrieved from https://www.pbs.org/wgbh/frontline/film/showspersuaders/

Dreyfus, H. L. (2005). Heidegger's Ontology of Art. In H. L. Dreyfus & M. A. Wrathall (Eds.), *A Companion to Heidegger*. Malden, MA: Blackwell Publishing. doi:10.1002/9780470996492.ch25

Drinkhill, M. J., & Mary, D. A. (1989). The effect of stimulation of the atrial receptors on plasma cortisol level in the dog. *The Journal of Physiology*, *413*(1), 299–313. doi:10.1113/jphysiol.1989.sp017655 PMID:2600852

Drozdovski, A. K. (2015). The connection between typological complexes of properties of the nervous system, temperaments, and personality types in the professions and sports. *Open Access Journal of Sports Medicine*, *6*, 161–172. doi:10.2147/OAJSM.S75612 PMID:26056499

Duffield, M., & Prendergast, J. (1994). *Without troops and tanks: The emergency humanitarian desk and the cross border operation into Eritrea and Tigray*. Lawrenceville, NJ: The Red Sea Press.

Dulcan, C. D. (2009). *Inteligenta materiei [Intelligence of matter]* (3rd ed.). Cluj, Romania: EIKON.

Dunbar, R. I. M. (2003). The social brain: Mind, language, and society in evolutionary perspective. *Annual Review of Anthropology*, *32*(1), 163–181. doi:10.1146/annurev.anthro.32.061002.093158

Durham, W. H. (1991). *Coevolution: Genes, culture, and human diversity*. Stanford, CA: Stanford University Press.

Eagleman, D. (n.d.). Silicon Immortality: Downloading Consciousness into Computers. *David Eagleman*. Retrieved from https://www.eagleman.com/blog/silicon-immortality-downloading-consciousness-into-computers

Eagleman, D. M. (2009). Silicon immortality: Downloading consciousness into computers. What Will Change Everything. Retrieved from https://www.eagleman.com/blog/silicon-immortality-downloading-consciousness-into-computers

Eichenbaum, H. (2014). Time cells in the hippocampus: A new dimension for mapping memories. *Nature Reviews. Neuroscience*, *15*(11), 732–744. doi:10.1038/nrn3827 PMID:25269553

Einstein, A. (2001). Relativity: The Special and the General Theory (R. W. Lawson, Trans.). Florence, KY: Routledge. (Originally published 1920)

Einstein, A. (1950). *Out of My Later Years*. New York: Philosophical Library.

Eisley, L. (1970). *The cosmic prison.* (S. Buscemi & S. Charlotte, Eds.). Boston, MA: McGraw Hill.

Ekman, P., Levenson, R. W., & Friesen, W. V. (1983). Autonomic nervous system activity distinguishes among emotions. *Science*, *221*(4616), 1208–1210. doi:10.1126cience.6612338 PMID:6612338

Ellsworth, P. C. (1994). William James and emotion: Is a century of fame worth a century of misunderstanding? *Psychological Review*, *101*(2), 222–229. doi:10.1037/0033-295X.101.2.222 PMID:8022957

Emel'kina, I.V. (2016). Spetsifika zakona i prava v rossiyskom (russkom) arkhetipe [The specifics of the law and the rights in the Russian archetype]. *Mir nauki i obrazovaniya, 1.*

Emerging Technology from the arXiv. (2015). The emerging science of human-data interaction. *MIT Technology Review.* Retrieved from https://www.linkedin.com/grp/post/1039687-5900533809526378496

Emerging Technology from the arXiv. (2016, November 7). Neural networks can read book covers. *Kyushu University, Japan* MIT Technology Review. Retrieved from https://www.technologyreview.com/s/602807/deep-neural-network-learns-to-judge-books-by-their-covers/?utm_campaign=newsletters&utm_source=newsletter-weekly-computing&utm_medium=email&utm_content=201611

Enriquez-Geppert, S., Huster, R. J., Figge, C., & Herrmann, C. S. (2014). Self-regulation of frontal-midline theta facilitates memory updating and mental set shifting. *Frontiers in Behavioral Neuroscience*, *8*(420). PMID:25538585

Equilibrium. (n.d.). Burbank, CA: Buena Vista Home Entertainment, Inc.

Erasmus, D. (1965b). Christian Humanism and the Reformation (J. C. Olin, Ed. & Trans.). New York: Harper and Row.

Erasmus, D. (1965a). *The colloquies of Erasmus (C. R. Thompson, Trans.).* Chicago: University of Chicago Press.

Erasmus, D. (1985a). De pueris instituendis / De recta pronuntiatione. In *Collected Works of Erasmus* (Vol. 4). Toronto: University of Toronto Press.

Erasmus, D. (1985b). A Declamation on the Subject of Early Liberal Education for Children. Beert C. Verstraete (Trans.). In J. K. Sowards (Ed.), *Collected Works of Erasmus* (Vol. 26, pp. 291–346). Toronto: University of Toronto Press.

Erasmus, D. (1985c). On Good Manners for Boys. Brian McGregor (Trans.). In J. K. Sowards (Ed.), *Collected Works of Erasmus* (Vol. 25, pp. 269–289). Toronto: University of Toronto Press.

Erokhina, T. I. (2016). Gendernye arkhetipy «serebryanogo veka» v massovoy kul'ture [Gender archetypes of "Silver Age" in mass culture]. Yaroslavskiy pedagogicheskiy vestnik, 5, 329-334.

Everett, H. (1957). 'Relative state' formulation of quantum mechanics. *Reviews of Modern Physics*, *29*(3), 454–462. doi:10.1103/RevModPhys.29.454

F World. (1933). New York: Agni Yoga Society.

Fernandez, Y. M., & Marshall, W. L. (2003). Victim empathy, social self-esteem, and psychopathy in rapists. *Sexual Abuse, 15*(1), 11–26. doi:10.1177/107906320301500102 PMID:12616926

Fet, A. I. (2005). *Instinkt i sotsial'noe povedenie [Instinct and social behavior]*. Novosibirsk: Sova.

Feynman, R. P. (1998). *Six Not-So-Easy Pieces: Einstein's Relativity, Symmetry, and Space-Time*. New York: Basic Books.

Field, J. O. (Ed.). (1993). *The challenge of famine: Recent experiences, lessons learned*. West Hartford: Kumarian Press.

Fillmore, C. (1975). An alternative to checklist theories of meaning. In *Proceedings of the First Annual Meeting of the Berkeley Linguistic Society*. Berkeley, CA: Berkeley Linguistics Society.

Fishbein, M. (2009). An integrative model for behavioral prediction and its application to health promotion. In R. J. Diclemente, R. A. Crosby, & M. C. Kegler (Eds.), *Emerging theories in health promotion practice and research* (2nd ed.; pp. 215–234). San Francisco: Jossey-Bass.

Flanigan, S., Gabrieli, E. R., & MacLean, P. D. (1957). Cerebral changes revealed by radioautography with S35-labeled l-methionine. *Archives of Neurology and Psychiatry, 77*(6), 588–594. doi:10.1001/archneurpsyc.1957.02330360046003 PMID:13423931

Flay, B. R., & Schure, M. (2012). The theory of triadic influence. In A. C. Wagenaar & S. C. Burris (Eds.), *Public health law research: Theory and methods* (pp. 169–192). San Francisco: Jossey-Bass.

Floyd, K. (2011). *Interpersonal communication*. McGraw-Hill Education.

Foerster, V. (2003). On constructing a reality. In *Understanding Understanding: Essays on Cybernetics and Cognition* (pp. 211–227). New York: Springer. doi:10.1007/0-387-21722-3_8

Fourier, J. B. J. (1878). *The Analytical Theory of Heat* (A. Freeman, Trans.). The University Press. (Original publication 1822)

Fracasso, C., & Friedman, H. (2011). Near-death experiences and the possibility of disembodied consciousness: Challenges to prevailing neurobiological and psychosocial theories. *NeuroQuantology: An Interdisciplinary Journal of Neuroscience and Quantum Physics, 9*(1), 41–53. doi:10.14704/nq.2011.9.1.389

Freeman, D. (2004). *Creating emotion in games*. Boston: New Riders Publishing. doi:10.1145/1027154.1027179

Freeman, M. (2016). From the collective unconscious to the narrative unconscious: Re-imagining the sources of selfhood. *Europe's Journal of Psychology, 12*(4), 513–522. doi:10.5964/ejop.v12i4.1325 PMID:27872663

Freud, S. (1894). The neuro-psychoses of defence. SE, 3, 41-61.

Freud, S. (1923). The ego and the id. SE, 19, 1-66.

Fromm, E. (2007). *Begstvo ot svobody [Escape from freedom]*. Moscow: Academic Project.

Frye, N. (1971). *The critical path: An essay on the social context of literary criticism*. Bloomington: Indiana University Press.

Frysinger, R. C., & Harper, R. M. (1990). Cardiac and respiratory correlations with unit discharge in epileptic human temporal lobe. *Epilepsia, 31*(2), 162–171. doi:10.1111/j.1528-1167.1990.tb06301.x PMID:2318169

Fuchs, C. (2016). *Critical Theory of Communication*. London: University of Westminster Press.

Gabbriellini, S. (2014). The Evolution of Online Forums as Communication Networks: An Agent-Based Model. *Revue Francaise de Sociologie, 55*(4), 805–825. doi:10.3917/rfs.554.0805

Gabriele, J. D., & Desmond, J. E. (1998). The role of left prefrontal cortex in learning and memory. *Proceedings of the National Academy of Sciences of the United States of America*, *95*(3), 906–913. doi:10.1073/pnas.95.3.906 PMID:9448258

Gabriele, J. D., & Kao, Y. (2007). Development of the declarative memory system in the human brain. *Nature Neuroscience*, *10*(9), 1198–1205. doi:10.1038/nn1950 PMID:17676059

Gabrieljan, T. O. (2016). Rol' arhetipa v vizual'no-graficheskoj tipologii brenda [The role of the archetype in the visual-graphic brand typology]. Vestnik kemerovskogo gosudarstvennogo universiteta kul'tury i iskusstv, 1, 88-94.

Gackenbach, J., & Hakopdjanian, S. (2015). Breaking the frame of digital, dream, and waking realities. In *Exploring the Collective Unconscious in a Digital Age.* IGI Global. Retrieved from http://www.igi-global.com/book/exploring-collective-unconscious-age-digital/140975

Gackenbach, J., & Hakopdjanian, S. (2015). Breaking the frame of digital, dream, and waking realities. In S. Schafer (Ed.), *Exploring the Collective Unconscious in a Digital* Age (pp. 79-106). Hershey, PA: IGI Global. Retrieved from http://www.igi-global.com/book/exploring-collective-unconscious-age-digital/140975

Gahery, Y., & Vigier, D. (1974). Inhibitory effects in the cuneate nucleus produced by vago-aortic afferent fibers. *Brain Research*, *75*(2), 241–246. doi:10.1016/0006-8993(74)90744-6 PMID:4366859

Gaiseanu, F. (2016b). Informational subsystems of the consciousness. *The Science of Consciousness, Abstract Book, TSC 2016 TUCSON*, *9*, 29. Retrieved from http://www.consciousness.arizona.edu/documents/TSC2016_BOOK_of_Abstracts_for_emailWeb.pdf

Gaiseanu, F. (2017b). Quantum-assisted process of disembody under near-death conditions: an informational-field support model. *NeuroQuantology*, *15*(1), 4-9. Retrieved fromhttp://www.neuroquantology.com/index.php/journal/article/view/971

Gaiseanu, F. (2017c). *An information based model of consciousness fully explaining the mind normal/paranormal properties*. NeuroQuantology 15(2),132-140. Retrieved from https://www.neuroquantology.com/index.php/journal/article/view/1040

Gaiseanu, F. (2018a). Near-Death experiences and immortality from the Perspective of an informational modeling of consciousness. Gerontology & Geriatrics Studies, 2(3). 1-4. Retrieved from https://crimsonpublishers.com/ggs/pdf/GGS.000538.pdf

Gaiseanu, F. (2018b). An informational modeling of consciousness and cognitive centers. *Proc. of the Human Project Brain (HPB) International Conference: Understanding Consciousness, A scientific quest for the 21-st century*. Barcelona, June 21-22. Retrieved from http://hbp-ic.com/scientific-information/programme/

Gaiseanu, F., & Graur, A. (2018). Cognitive centers related attitude: Application for an iterative evaluation method in music-based therapy process. *Abstract Book of the Conference on Science of Consciousness*. Tucson (Arizona) April 2-7. http://www.consciousness.arizona.edu/documents/TSC2018AbstractBookfinal3.pdf

Gaiseanu, F. (2013). Contributions to the modelling and simulation of the atomic transport processes in silicon and polysilicon and applications. *Proceedings of the Romanian Academy, Series A*, *4*(4), 376–384. Retrieved from http://www.acad.ro/sectii2002/proceedings/doc2013-4/15-Gaiseanu.pdf

Gaiseanu, F. (2016a). Consciousness as informational system of the human body: Consciousness and life physics. *Cosmology and Astrophysics Journal*, *16*(1), 14–25. Retrieved from http://physics.socionic.info/index.php/physics/article/view/227/182

Gaiseanu, F. (2017a). Modelling and simulation of the impurity diffusion and related phenomena in silicon and polysilicon systems in microfabrication and micromachining technologies. *Annals of the Academy of Romanian Scientists, 10*(1), 41–78. Retrieved from http://aos.ro/wp-content/anale/IFVol10Nr1Art.4.pdf

Gaiseanu, F. (2018c). Destiny or free will decision? A life overview from the perspective of an informational modeling of consciousness Part II: Attitude and decision criteria, free will and destiny. *Gerontology & Geriatric Studies, 4*(1), 1–7. doi:10.31031/GGS.2018.04.000576

Gaiseanu, F. (2018d). Information: From philosophic to physics concepts for an informational modeling of consciousness. *Philosophy Study J., 8*(8), 68–82. doi: 10.17265/2159-5313/2018.08.004.Retrieved from http://www.davidpublisher.org/Public/uploads/Contribute/5c06323653cd2.pdf

Gaiseanu, F. (2019a). Destiny or free will decision? A life overview from the perspective of an informational modeling of consciousness Part I: Information, consciousness and life cycle. *Gerontol & Geriatric Stud., 4*(3), 1–6. Retrieved from https://crimsonpublishers.com/ggs/pdf/GGS.000586.pdf

Gaiseanu, F. (2019b). The informational model of consciousness: Mechanisms of embodiment/disembodiment of information. *NeuroQuantology: An Interdisciplinary Journal of Neuroscience and Quantum Physics.* (in press) 17(4)

Galarneau, L. L. (2005). *Spontaneous Communities of Learning: Learning Ecosystems in Massively Multiplayer Online Gaming Environments.* Retrieved from https://ssrn.com/abstract=810064

Galeon, D. (2016). For the first time in history, we are seeing alien worlds (with spectroscopy). *Futurism.* Retrieved 7/11/2017 from https://futurism.com/for-the-first-time-in-history-we-are-seeing-alien-worlds/

Galeon, D. (2016). For the first time in history, we are seeing alien worlds (with spectroscopy). *Futurism.* Retrieved from https://futurism.com/for-the-first-time-in-history-we-are-seeing-alien-worlds/

Galeon, D. (2017). New research shows that the universe might once have been a hologram. *Futurism.* Retrieved 7/11/2017 from https://futurism.com/new-research-shows-the-universe-may-have-once-been-a-hologram/

Galeon, D. (2017). New research shows that the universe might once have been a hologram. *Futurism.* Retrieved from https://futurism.com/new-research-shows-the-universe-may-have-once-been-a-hologram/

Garin, E. (Ed.). (1957). L'educazione in Europa (1400–1600). Rome: Bari.

Garin, E. (1965). *Italian humanism (P. Munz, Trans.).* Oxford, UK: Blackwell.

Garling, C., & Cully, A. (2015). Robots walk off Injuries: Robots can now keep moving after an injury without human intervention. *MIT Technology Review.* Retrieved from http://www.evolvingai.org/files/mit_tech_review_robot_adapts.pdf

Gaudry, J. D. (2018). Liking what's good for you: Evolution, Subjectivity and purpose. *Cosmos and History: The Journal of Natural and Social Philosophy, 14*(3), 359–375.

Gay, P. (1995). *The Enlightenment: The rise of modern paganism.* New York: W. W. Norton & Company. (Original work published 1966)

Gellhorn, E., & Loofbourrow, G. N. (1963). *Emotions and Emotional Disorders: A Neurophysiogical Study.* New York: Harper & Row.

Geril, P. (n.d.). *Smart dust is getting smarter.* EUROSIS—The European Multidisciplinary Society for Modeling and Simulation Theory. Retrieved 7/11/2017 from https://www.linkedin.com/groups/1039687/1039687-5922629195124473858

Gillin, L.M., Lapira, F., McCraty, R., Bradlley, R., Atkinson, M., Simpson, D., & Scicluna, P. (n.d.). *Before cognition: The active contribution of the heart/ANS to intuitive decision making as measured on repeat entrepreneurs in the Cambridge Technopol.* Australian Graduate School of Entrepreneurship.

Gillin, M., La Pira, F., McCraty, R., Bradley, R., Atkinson, M., Simpson, D., & Scicluna, P. (2007). Before cognition: The active contribution of the heart/ANS to intuitive decision making as measured on repeat entrepreneurs in the Cambridge Technopol. *Australian Graduate School of Entrepreneurship & Heartmath Institute.* Retrieved from https://www.heartmath.org/research/research-library/intuition/before-cognition-the-active-contribution-of-the-heart/

Ginsberg, J. P., Berry, M. E., & Powell, D. A. (2010). Cardiac coherence and PTSD in combat veterans. *Alternative Therapies in Health and Medicine*, *16*(4), 52–60. PMID:20653296

Giorbran, G. (2006). Everything forever: Learning to see timelessness. *Gevin.*, *ISBN10*, 0979186102.

Glassner, A. (2004). *Interactive storytelling*. Natick, MA: A.K. Peters.

Glassner, A. (2004). *Interractive storytelling*. Natick, MA: A.K. Peters.

Glazunova, V. V. (2016a). Primenenie arkhetipov sistem v formirovanii innovatsionnogo potentsiala organizatsii [The use of systems archetypes in shaping the innovative potential of organization]. Naukovedenie, 8(2).

Gleick, J. (1987). *Chaos*. Viking Penguin Inc.

Goffman, E. (1974). *Frame analysis; an essay on the organization of experience*. New York: Harper & Row.

Goffman, I. (1974). *Frame analysis: An essay on the organization of experience*. New York: Harper and Row.

Goldstein, E. R. (2009). *Sensation and Perception*. Boston, MA: Cengage Learning.

Goleman, D. (1998). *Working with Emotional Intelligence*. New York: Bantam Books.

Grafton, A. (2004). *Bring Out Your Dead: The Past as Revelation*. Cambridge, MA: Harvard University Press.

Graham, D. T., Stern, J. A., & Winokur, G. (1958). Experimental investigation of the specificity of attitude hypothesis in psychosomatic disease. *Psychosomatic Medicine*, *20*(6), 446–457. doi:10.1097/00006842-195811000-00002 PMID:13614604

Grasse, R. (1996). *The waking dream*. Wheaton, IL: The Theosophical Publishing House.

Gray, P. O. (2006). *Psychology* (5th ed.). New York: Worth.

Grayson, P. (n.d.). *The enunciation of the virgin dead.* Retrieved from: https://www.google.com/search?q=Grayson+Perry,+The+Annunciation+of+the+Virgin+Deal&rlz=1C1KMZB_enUS577US671&tbm=isch&source=iu&ictx=1&fir=mWcwHLNjqNGzyM%253A%252COceTlgB_IOeLyM%252C_&usg=AI4_-kSNO--aTpOlSYUyOF6jHLxVtP0V9w&sa=X&ved=2ahUKEwi396iN1uHfAhUfwMQHHVuZD6cQ9QEwAnoECAUQBA#imgdii=Pvug-5M-3bzRgM:&imgrc=mWcwHLNjqNGzyM

Grazer, B. (Producer). (2005). *Apollo 13 anniversary edition.* Universal Studios.

Green, M., Schwarz, J. H., & Witten, E. (1988). *Superstring Theory* (Vol. 1). Cambridge University Press.

Gregary, R. (1987). Oxford Companion to the Mind. New York: Oxford.

Gribunin, V. V. (2016b). Arkhetip stradaniya v russkoy kul'ture: religioznyy aspekt [The archetype of suffering in Russian culture: religious aspect]. Osnovnye tendentsii gosudarstvennogo i obshchestvennogo razvitiya rossii: istoriya i sovremennost', 1, 31-38.

Gribunin, V. V., & Kanarskiy, D. I. (2016). Transformatsiya russkogo pravoslavnogo arkhetipa v sovetskiy period [Transformation of Russian Orthodox archetype in the Soviet period]. Osnovnye tendentsii gosudarstvennogo i obshchestvennogo razvitiya rossii: istoriya i sovremennost', 1, 91-95.

Gritskov, Yu. V. (2006). *Fenomen stradaniia i sposoby ego osvoeniia v kul'ture* [The phenomenon of suffering and ways of its learning in culture] (Unpublished doctoral dissertation). Omsk State Pedagogical University, Omsk, Russia.

Gritskov, Yu. V. (2004). *Obrazy stradaniia v stradaiushchem soznanii [The images of suffering in suffering consciousness].* Krasnoyarsk State University.

Gritskov, Yu. V., & Lvov, D. V. (2015). *Interaction of In-Group Archetypes and Hierarchy while Forming in-Group Community. Journal of Siberian Federal University. Humanities & Social Sciences, 11,* 2325–2330.

Grujic, P. V. (2001). *The Concept of Fractal Universe Anaxagoras.* Belgrade Institute of Physics.

Gullatz, S. (2010). Constructing the collective unconscious. *The Journal of Analytical Psychology, 55*(5), 691–714. doi:10.1111/j.1468-5922.2010.01878.x PMID:20961321

Gutkowska, J., Jankowski, M., Mukaddam-Daher, S., & McCann, S. M. (2000). Oxytocin is a cardiovascular hormone. *Brazilian Journal of Medical and Biological Research, 33*(6), 625–633. doi:10.1590/S0100-879X2000000600003 PMID:10829090

Gutov, E. V. (2016). Transformatsiya arkhetipov sotsial'noy telesnosti [Transformation of social archetypes of corporeality]. Kul'tura, nauka, obrazovanie: problemy i perspektivy, 329-332.

Gutschner, M. (2014). Discovery of Quantum Vibrations in "Microtubules" Inside Brain Neurons Corroborates Controversial 20-Year-Old Theory of Consciousness [Press release]. *Elsevier.* Retrieved from https://www.elsevier.com/about/press-releases/research-and-journals/discovery-of-quantum-vibrations-in-microtubules-inside-brain-neurons-corroborates-controversial-20-year-old-theory-of-consciousness

Hafting, T., Fyhn, M., Molden, S., Moser, M.-B., & Moser, E. I. (2005). Microstructure of a spatial map in the entorhinal cortex. *Nature, 436*(7052), 801–806. doi:10.1038/nature03721 PMID:15965463

Hajdukovic, D. (2013). Can observations inside the solar system reveal the gravitational properties of the quantum vacuum? *Astrophysics and Space Science, 343*(2), 505–509. doi:10.100710509-012-1303-3

Hall, M. (2001). *The secret teaching of all ages: An encyclopedia outline of Masonic, Hermitic, Qabalistic and Rosicrucian symbolical philosophy.* Retrieved from http://www.sacred-texts.com/eso/sta/index.htm (Original publication 1928)

Hall, M. P. (2003). *The secret teachings of all ages.* New York: Tarcher/Penguin. Retrieved from http://www.sacred-texts.com/eso/sta/sta36.htm

Hall, M. P. (2003). *The Secret Teachings of All Ages.* New York: Penguin.

Halpern, E. H., & Wolf, A. A. (1972). Speculations of Superconductivity in Biological and Organic Systems. In K. D. Timmerhaus (Ed.), *Advances in Cryogenic Engineering. Advances in Cryogenic Engineering* (Vol. 17). Boston, MA: Springer. doi:10.1007/978-1-4684-7826-6_11

Hameroff, S., & Penrose, R. (2014). *Discovery of quantum vibrations in 'mocrotubules' corroborates theory of consciousness.* Retrieved 1/16/2014 http://phys.org/news/2014-01-discovery-quantum-vibrations-microtubules-corroborates.html

Hameroff, S., & Penrose, S. R. (2014). Discovery of quantum vibrations in 'mocrotubules' corroborates theory of consciousness. *PhysOrg.* Retrieved from https://phys.org/news/2014-01-discovery-quantum-vibrations-microtubules-corroborates.html

Hameroff, S., & Penrose, S. R. (2014). *Discovery of quantum vibrations in "microtubules" corroborates theory of consciousness.* Retrieved 1/16/2014 http://phys.org/news/2014-01-discovery-quantum-vibrations-microtubules-corroborates.html)

Hameroff, S. (1998). Quantum computation in brain microtubules? The Pennrose - Hameroff". Orch OR" model of consciousness. *Philos. Trans. Royal Society London, Ser.A. Math. Phys. Sci., 356*, 1869–1898. doi:10.1098/rsta.1998.0254

Hameroff, S., & Penrose, R. (1996). Orchestred reduction of quantum coherence in brain microtubules: A model for consciousness. *Mathematics and Computers in Simulation, 40*(3-4), 453–480. doi:10.1016/0378-4754(96)80476-9

Hameroff, S., & Penrose, R. (2014). Consciousness in the universe: A review of the 'Orch. OR' Theory. *Physics of Life Reviews, 1*(1), 39–79. doi:10.1016/j.plrev.2013.08.002 PMID:24070914

Harenski, C. L., Antonenko, O., Shane, M. S., & Kiehl, K. A. (2010). A functional imaging investigation of moral deliberation and moral intention. *NeuroImage, 49*(3), 2707–2716. doi:10.1016/j.neuroimage.2009.10.062 PMID:19878727

Harriss, J. (Ed.). (1995). *The politics of humanitarian intervention.* London: Pinter Publishers.

Hawking, S. W., & Penrose, R. (1970). The Singularities of Gravitational Colllapse and Cosmology. *Proceedings of the Royal Society of London. Series A, Mathematical and Physical Sciences, 314*(1519), 529–548. doi:10.1098/rspa.1970.0021

Health for All. (2001). *Kundalini energy yoga: Asanas, pranayama, relaxation, and meditation* [DVD]. Omaha, NE: Health for all, Inc. Available from https://www.youtube.com/watch?v=VH1UFVzQwkg https://www.youtube.com/watch?v=XdvwVz-qI1w&t=9s https://www.youtube.com/watch?v=Y3qnDDlSydE&t=16s

Heidegger, M. (1978). Basic Writings (D. F. Krell, Ed.). London: Routledge & Kegan Paul.

Heidegger, M. (1993). "Only a God can save us": Der Spiegel's Interview with Martin Heidegger (1966). In R. Wolin (Ed.), *The Heidegger controversy: A critical reader* (pp. 105–106). Cambridge, MA: The MIT Press.

Heidegger, M. (2002). The Origin of the Work of Art. In *Off the Beaten Track.* Cambridge University Press.

Helfer, L. R., & Austin, G. W. (2011). *Human rights and intellectual property mapping the global interface.* New York: Cambridge University Press.

Henderson, G. E., & Brown, C. (1997). *Glossary of Literary Theory.* Toronto: University of Toronto Press.

Henrich, J., & McElreath, R. (2003). The evolution of cultural evolution. *Evolutionary Anthropology: Issues. News Review (Melbourne), 12*(3), 123–135.

Hess, W. R. (1954). *Diencephalon: Autonomic and Extrapyramidal Functions.* New York: Grune & Stratton.

Heymans, C., & Neil, E. (1958). *Reflexogenic Areas of the Cardiovascular System.* Boston: Little, Brown.

Heyse, L. (2006). *Choosing the lesser evil: Understanding decision making in humanitarian aid NGOs.* Hampshire, UK: Ashgate Publishing Limited.

Hill, K. (2004). *War, humanitarian crises, population, displacement, and fertility: A review of evidence.* Washington, DC: The National Academies Press.

Hinton, G. E., McClelland, J. L., Rumelhart, D. E., & Rnmelhart, D. E. (1986). *Parallel distributed processing: Explorations in the microstructure of cognition.* Cambridge, MA: MIT Press.

Hodges, D. B., Aylet, K., & Reeves, S. (2008). Analytic research - Discourse Analysis. *BMJ (Clinical Research Ed.), 337*, a879. doi:10.1136/bmj.a879 PMID:18687729

Holmgren, J. (1981). Hypothesis on consciousness. Academic Press. doi:10.13140/RG.2.1.1611.5049

Holmgren, J. (2017). Human consciousness - a parsimonious approach. Principles for concept formation, for abstraction, and for the foundations of logic, mathematics, physics, and biology. Academic Press. doi:10.13140/RG.2.2.35901.67041

Holmgren, J. (2014). Natural Evolution and Human Consciousness. *Mens Sana Monographs, 12*(1), 127–138. doi:10.4103/0973-1229.130322 PMID:24891802

Ho, M. W. (2005). *The rainbow and the worm: the physics of organisms*. Singapore: World Scientific Publishing Co.

Hossenfelder, S. (2015). Lost in thought. How important to physics was Einstein's imaginings? *Scientific American, 313*(3). PMID:26455101

Houran, J., Lange, R., & Hooper, D. (2018). Cross-examining the case for precognition: Comment on Mossbridge and Radin (2018). *Psychology of Consciousness, 5*(1), 98109. doi:10.1037/cns0000126

How Plants Communicate & Think. (n.d.). *BBC & YouTube*. Retrieved from https://www.bing.com/videos/search?q=national+geographic+plants+video&&view=detail&mid=ABD11A5AADE89DE64EC6ABD11A5AADE89DE64EC6&&FORM=VDRVRV

How we build machines that can think for themselves. (2016). *BBC*. Retrieved 11/15/2016 from http://www.bbc.com/future/story/20161114-how-we-built-machines-that-can-think-for-themselves?ocid=ww.social.link.linkedin

How your consciousness affects the universe, Dr. Emoto. (n.d.). In *YouTube*. Retrieved from https://www.bing.com/videos/search?q=consciousness+of+water&view=detail&mid=431C0918B170F7B18E81431C0918B170F7B18E81&FORM=VIRE

Huang, M.-H., Friend, D. S., Sunday, M. E., Singh, K., Haley, K., Austen, K. F., ... Smith, T. W. (1996). An intrinsic adrenergic system in mammalian heart. *The Journal of Clinical Investigation, 98*(6), 1298–1303. doi:10.1172/JCI118916 PMID:8823294

Hubert, W., & de Jong-Meyer, R. D. (1991). Psychophysiological response patterns to positive and negative film stimuli. *Biological Psychology, 31*(1), 73–93. doi:10.1016/0301-0511(90)90079-C PMID:2021681

Huizinga, J. (1984). *Erasmus and the Age of Reformation*. Princeton, NJ: Princeton University Press. (Original work published 1924) doi:10.1515/9781400858071

Hunt, H. T. (2012). A collective unconscious reconsidered: Jung's archetypal imagination in the light of contemporary psychology and social science. *The Journal of Analytical Psychology, 57*(1), 76–98. doi:10.1111/j.1468-5922.2011.01952.x PMID:22288542

Hurford, R. J. (2008). The evolution of human communication and language. *In Cap 14*. In P. d'Ettorre & D. P. Hughes (Eds.), *Sociobiology of Communication, An Interdisciplinary Perspective* (pp. 249–264). Oxford University Press. doi:10.1093/acprof:oso/9780199216840.003.0014

Huxley, A. (2016). The perennial philosophy quotes. *Goodreads*. Retrieved 10/30/2016 from https://www.goodreads.com/work/quotes/889435-the-perennial-philosophy

Ignatenko, A. (1994). Cum sa devii un fenomen [How to become a phenomenon]. Bucharest: Ed. Biodova.

Institute of Heartmath. (2007). New research project: global coherence monitoring system. IHM Summer 2007 Newsletter. Retrieved from http://www.heartmath.org/templates/ihm/e-newsletteer/2007/Summer-2007/summer_2007_newsletter.htm

Introduction and Research Review. (n.d.). *Maharishi International University Fairfield, Iowa, U.S.A*. Retrieved from https://www.mum.edu/pdf_msvs/v01/alexander.pdf

Isbister, K. (2006). *Better game characters by design, a psychological approach.* San Francisco: Morgan Kaufmann Publishers—Elsevier.

Israel, N. R. (1968). *Cognitive control and pattern of autonomic response*. Paper presented at meeting of the Eastern Psychological Association, Washington, DC.

Israel, N. R. (1969). Leveling-sharpening and anticipatory cardiac response. *Psychosomatic Medicine*, *31*(6), 499–509. doi:10.1097/00006842-196911000-00003 PMID:5361745

Jacobi, J. (1943). *The psychology of Jung*. New Haven, CT: Yale University Press.

Jacobi, J. (1973). *The psychology of CG Jung*. London: Yale University Press.

Jacoby, J. (1959). *Complex, archetype, symbol in the psychology of C. G. Jung*. New York, NY: Princeton University Press.

Jacoby, J. (1973). *C. G. Jung*. Yale University Press.

Jaeger, G. (2014). What in the (quantum) world is macroscopic? *American Journal of Physics*, *82*(9), 896–905. doi:10.1119/1.4878358

James, W. (1884). What is an emotion? *Mind*, *9*(34), 188–205. doi:10.1093/mind/os-IX.34.188

James, W. (1890). *Principles of Psychology*. New York: Holt.

Janowski, J. S., Shimamura, A. P., & Squire, L. R. (1989). Source memory impairment in patients with frontal lobe lesions. *Neuropsychologia*, *27*(8), 1043–1056. doi:10.1016/0028-3932(89)90184-X PMID:2797412

Jedlicka, P. (2017). Revisiting the Quantum Brain Hypothesis: Toward Quantum (Neuro)biology? *Frontiers in Molecular Neuroscience*, *10*, 366. doi:10.3389/fnmol.2017.00366 PMID:29163041

Jeffery, K. (2007). Integration of sensory inputs to place cells: What, where, why, and how? *Hippocampus*, *17*(9), 775–785. doi:10.1002/hipo.20322 PMID:17615579

Jeffery, K., Anderson, M., Hayman, R., & Chakraborty, S. (2003). A proposed architecture for the neural representation of spatial context. *Neuroscience and Biobehavioral Reviews*, *28*(2), 201–218. doi:10.1016/j.neubiorev.2003.12.002 PMID:15172764

Jeffrey, C. (2015). First-ever QED device that detects and corrects its own errors. *New Atlas*. Retrieved from https://newatlas.com/quantum-device-detects-corrects-errors/36404/

Jeffrey, C. (2015). First-ever QED device that detects and corrects its own errors. *Physics*. Retrieved from https://newatlas.com/quantum-device-detects-corrects-errors/36404/

Jenkins, H. (2005). *Game design as narrative architecture*. Henry Jenkins. Retrieved from http://homes.lmc.gatech.edu/~bogost/courses/spring07/lcc3710/readings/jenkins_game-design.pdf

Jenkins, H. (2005). *Game design as narrative architecture*. Retrieved 11/7/2016 from http://www.electronicbookreview.com/thread/firstperson/lazzi-fair

John, P. J. (2009). Frontal-temporal dysfunction in schizophrenia: A selective review. *Indian Journal of Psychiatry*, *51*(3), 180–190. doi:10.4103/0019-5545.55084 PMID:19881045

Johnson, J. D., Suzuki, M., & Rugg, M. D. (2013). Recollection, familiarity and content sensitivity in lateral parietal cortex: A high-resolution fMRI study. *Frontiers in Human Neuroscience*, *7*(209). PMID:23734122

Jordan, B. (2009). Blurring Boundaries: The “Real and the “Virtual” in Hybrid Spaces. Introduction to Special Section on Knowledge Flow in Online and Offline Spaces. *Human Organization, 68*(2), 181–193. doi:10.17730/humo.68.2.7x4406g270801284

Joye, S. R. (2016). *The holoflux theory of consciousness: An integral interpretation of the theories of Karl Pribram, David Bohm, and Pierre Teilhard de Chardin. California.* San Francisco, CA: Institute of Integral Studies.

Jung C.G. (1991). *Arkhetip i simvol* [Archetype and symbol]. Moscow: Academic Press.

Jung C.G. (1998) *Psikhologicheskie aspekty arkhetipa materi* [Psychological aspects of the mother archetype]. Moscow: Academic Press.

Jung, C. (1933). *The collected works of C. G. Jung: The real and the surreal.* Retrieved from http://www.ihp.sinica.edu.tw/~medicine/book/Jung.html

Jung, C. G. (2009). The Red Book: Liber Novus (S. Shamdasani, M. Kyburz, J. Peck, & S. Shamdasani Trans. & Eds.). London: W. W. Norton.

Jung, C. G., von Franz, M.-L., Henderson, J., Jacobi, J., & Jaffe, A. (1964). Man and His Symbols, London: Aldus.

Jung, C. G. (1933). *Modern man in search of a soul.* Harcourt Brace & Company.

Jung, C. G. (1933). *Modern man in search of a Soul.* New York: Harcourt Brace & Co.

Jung, C. G. (1933). *Modern Man in Search of a Soul.* New York: Harcourt Brace Jovanovich.

Jung, C. G. (1968). *Part 1: The archetypes and the collective unconscious* (2nd ed.; Vol. 9; G. Adler & G. F. C. Hull, Trans. & Eds.). Princeton, NJ: Princeton University Press.

Jung, C. G. (1981). *The Archetypes and the Collective Unconscious* (2nd ed.). Princeton University Press.

Jung, C. G. (1989). *Memories, Dreams, Reflections* (A. Jaffé, Ed., WinstonR.WinstonC., Trans.). New York, NY: Vintage.

Kaada, B. R., Pribram, K. H., & Epstein, J. A. (1949). Respiratory and vascular responses in monkeys from temporal pole, insula, orbital surface and cingulate gyrus. *Journal of Neurophysiology, 12*(5), 347–356. doi:10.1152/jn.1949.12.5.347 PMID:18137711

Kahn, C. H. (1960). Anaximander and the Origins of Greek Cosmology. New York: Columbia University Press.

Katabi, D. (2018). RF-Pose, AI that can see through walls. *MIT Computer Science & Artificial Intelligence Lab.* Retrieved from MIT Technology Review, Futurism, & Motherboard (Kaleigh Rogers): https://motherboard.vice.com/en_us/article/a3aaqp/mit-device-uses-wifi-to-see-through-walls-and-track-your-movements;https://connect.xfinity.com/appsuite/#!!&app=io.ox/mail&folder=default0/INBOX

Kenji, I. B., & Uchida, S. (n.d.). *Neural networks can read book covers.* Kyushu University. Retrieved from MIT Technology Review: https://www.technologyreview.com/s/602807/deep-neural-network-learns-to-judge-books-by-their-covers/?utm_campaign=newsletters&utm_source=newsletter-weekly-computing&utm_medium=email&utm_content=201611

Khodarahimi, S. (2009, October-December). Dreams in Jungian psychology: The use of dreams as an instrument for research diagnosis and treatment of social phobia. *The Malaysian Journal of Medical Sciences, 16*(4), 42–49. PMID:22135511

Khoury, B., Sharma, M., Rush, S. E., & Fournier, C. (2015). Mindfulness-based stress reduction for healthy individuals: A meta-analysis. *Journal of Psychosomatic Research, 78*(6), 519–528. doi:10.1016/j.jpsychores.2015.03.009 PMID:25818837

Kim, M., & Lee, B. (2013). *Are There Too Many Superheroes? Analysis of the Social Distance in Massive Multiplayer Online Role Playing Game.* KAIST College of Business Working Paper Series No. 2013-019. Retrieved from https://ssrn.com/abstract=2330090

King, B. 2009. *Evolving God: A provocative View on the Origins of Religion.* Doubleday Publishing.

Kirk, T. (2014). Superconducting spintronics pave way for next-generation computing. *PhysOrg*. Retrieved from http://phys.org/news/2014-01-superconducting-spintronics-pave-next-generation.html

Kirk, T. (2014). Superconducting spintronics pave way for next-generation computing. *PhysOrg*. Retrieved January 15, 2014 from http://phys.org/news/2014-01-superconducting-spintronics-pave-next-generation.html

Kirmayer, L. J. (2018). Ethno and cultural psychiatry. *The International Encyclopedia of Anthropology.* Retrieved from https://connect.xfinity.com/appsuite/#!!&app=io.ox/mail&folder=default0/INBOX

Klose, J. (2009). Process Ontology from Whitehead to Quantum Physics. In H. Atmanspacher & H. Primas (Eds.), *Recasting Reality-Wolfgang Pauli's Philosophical Ideas and Contemporary Science* (pp. 151–170). Springer-Verlag. doi:10.1007/978-3-540-85198-1_8

Knight, W. (2015, November 2). A new tool for analyzing academic papers uses cutting-edge AI to find meaning in billions of words. *MIT Technology Review*. Retrieved from http://www.technologyreview.com/news/542981/academic-search-engine-grasps-for-meaning/

Koch, C. (2015). Two natural philosophers, centuries apart, converse about the mind. *Scientific American Mind, 26*(2), 28–31. doi:10.1038cientificamericanmind0315-28

Koch, C. (2016). Constructing the modern mind. *Scientific American Mind, 27*(3), 22–27. doi:10.1038cientificamericanmind0516-22

Kochetkov, V.V. (2015). Konstitutsionalizm i arkhetipy russkoy vlasti (casus 1906 goda) [Constitutionalism and the archetypes of Russian power (casus 1906)]. *Rossiyskiy zhurnal pravovykh issledovaniy, 1,* 155-163.

Koriat, A., & Goldsmith, M. (1996). Memory metaphors and the real-life/ laboratory controversy: Correspondence versus storehouse conceptions of memory. USA. *Behavioral and Brain Sciences, 19*(02), 167–228. doi:10.1017/S0140525X00042114

Koriath, J. J., & Lindholm, E. (1986). Cardiac-related cortical inhibition during a fixed foreperiod reaction time task. *International Journal of Psychophysiology, 4*(3), 183–195. doi:10.1016/0167-8760(86)90014-0 PMID:3793561

Kosov, A. A.,, Broderick, B. Y., Yorkin, C. S. (Producers) & Villeneuve, D. (Director). (2017). *Blade Runner 2049* [DVD]. USA: Alcon Media Group.

Kraye, J. (1996). *The Cambridge companion to Renaissance humanism.* Cambridge University Press. doi:10.1017/CCOL0521430380.016

Kreyszig, E. (2011). *Advanced Engineering Mathematics* (10th ed.). John Wiley and Sons Inc.

Krishnamurti, J. (n.d.). *Truth is a pathless land.* Retrieved from https://www.jkrishnamurti.org/about-dissolution-speech

Krishna, V., & John, M. (2004). The art of conversation: Eliciting information from experts through multi-stage communication. *Journal of Economic Theory, 117*(2), 147–179. doi:10.1016/j.jet.2003.09.008

Kristeller, P. O. (1979). Renaissance thought and its sources (M. Mooney, Ed.). New York: Columbia University Press.

Kushner, L. H. (2008). *Contrast in judgements of mental health.* Ann Arbor, MI: ProQuest.

Kuspa, A. (2018). *Lectins help social amoeba establish their own microbiome.* Baylor College of Medicine. Retrieved from https://phys.org/news/2018-07-lectins-social-amoeba-microbiome.html

Kwan, V. S. Y. (2008). Missing links in social cognition: The continuum from nonhuman agents to dehumanized humans. *Social Cognition, 26*(2), 125–128. doi:10.1521oco.2008.26.2.125

Lacey, B. C., & Lacey, J. I. (1974). Studies of heart rate and other bodily processes in sensorimotor behavior. In P. A. Obrist, A. H. Black, J. Brener, & L. V. DiCara (Eds.), *Cardiovascular Psychophysiology: Current Issues in Response Mechanisms, Biofeedback, and Methodology* (pp. 538–564). Chicago: Aldine.

Lacey, J. I. (1967). Somatic response patterning and stress: Some revisions of activation theory. In M. H. Appley & R. Trumbull (Eds.), *Psychological Stress: Issues in Research* (pp. 14–42). New York: Appleton-Century-Crofts.

Lacey, J. I., Kagan, J., Lacey, B. C., & Moss, H. A. (1963). The visceral level: Situational determinants and behavioral correlates of autonomic response patterns. In P. H. Knapp (Ed.), *Expression of the Emotions in Man* (pp. 161–196). New York: International Universities Press.

Lacey, J. I., & Lacey, B. C. (1970). Some autonomic-central nervous system interrelationships. In P. Black (Ed.), *Physiological Correlates of Emotion* (pp. 205–227). New York: Academic Press.

Lakoff, G. (2002). *Moral politics*. Chicago: The University of Chicago Press. doi:10.7208/chicago/9780226471006.001.0001

Lakoff, G. (2008). *The political mind.* Viking Penguin.

Lamborn, A. B. (2015). The Deus Absconditus and the post secular quest. In *Jung in the Academy and Beyond.* University of the South.

Langlitz, N. (1972). Neuropsychedelia: The Revival of Hallucinogen Research since the Decade of the Brain. Berkeley, CA: University of California Press.

Langlitz, N. (2006). Tripping in Solitude: Introducing Honza Samotar by Way of John Lilly. In K. Solhdju (Ed.), *Introspective Self-Rapports: Shaping Ethical and Aesthetic Concepts 1850–2006* (pp. 81–92). Berlin: Max Planck Institute for the History of Science.

Lara, A. H., & Wallis, J. D. (2015). The role of prefrontal cortex in working memory: A mini review. *Frontiers in Systems Neuroscience, 9*(173). PMID:26733825

Lashley, K. S. (1960). The thalamus and emotion. In F. A. Beach, D. O. Hebb, C. T. Morgan, & H. S. Nissen (Eds.), *The Neurospsychology of Lashley*. New York: McGraw-Hill.

Laurie, T. (2015). Becoming-animal is a trap for humans: Deleuze and Guattari in Madagascar. In H. Stark & J. Roffe (Eds.), *Deleuze and the Non-Human* (pp. 142–162). Hampshire, UK: Palgrave Macmillan. doi:10.1057/9781137453693_9

Lawrence, M., Davis, J., Dow, T., Godfry, W. (Producers) & Proyas, A. (Director). (2004). *I Robot* [DVD]. USA: Twentieth Century Fox.

Le Bon, G. (2002). The crowd: A study of the popular mind. Mineola, NY: Dover Publications (Originally published 1896)

LeDoux, J. (1996). *The Emotional Brain: The Mysterious Underpinnings of Emotional Life.* New York: Simon and Schuster.

LeDoux, J. E. (1994). Cognitive-emotional interactions in the brain. In P. Ekman & R. J. Davidson (Eds.), *The Nature of Emotion: Fundamental Questions* (pp. 216–223). New York: Oxford University Press.

Leek, B. F. (1972). Abdominal visceral receptors. In E. Neil (Ed.), Handbook of Sensory Physiology (Vol. 3, pp. 113–160). Heidelberg, Germany: Springer.

Lessmeier, T. J., Gamperling, D., Johnson-Liddon, V., Fromm, B. S., Steinman, R. T., Meissner, M. D., & Lehmann, M. H. (1997). Unrecognized paroxysmal supraventricular tachycardia: Potential for misdiagnosis as panic disorder. *Archives of Internal Medicine*, *157*(5), 537–543. doi:10.1001/archinte.1997.00440260085013 PMID:9066458

Levenson, R. W., Ekman, P., & Friesen, W. V. (1990). Voluntary facial action generates emotion-specific autonomic nervous system activity. *Psychophysiology*, *27*(4), 363–384. doi:10.1111/j.1469-8986.1990.tb02330.x PMID:2236440

Lewis, T. (2014). This computer chip can think like a human brain. *Live Science*. Retrieved from https://www.livescience.com/47240-ibm-computer-chip-simulates-brain.html

Lewis, S. L., & Maslin, M. A. (2015). Defining the Anthropocene. *Nature*, *519*(7542), 171–180. doi:10.1038/nature14258 PMID:25762280

Lilly, J. C. (1967). *The mind of the dolphin: A nonhuman intelligence*. Garden City, NY: Doubleday.

Lilly, J. C. (1967). *The Mind of the Dolphin: A Nonhuman Intelligence*. Garden City, NY: Doubleday.

Lilly, J. C. (1974). *Programming and metaprogramming in the human biocomputer: Theory and experiments*. New York: Bantam.

Lilly, J. C. (1974). *Programming and Metaprogramming in the Human Biocomputer: Theory and Experiments*. New York: Bantam.

Lilly, J. C. (1985). *The center of the cyclone: An autobiography of inner space*. Three Rivers Pr.

Lindsley, D. B. (1951). Emotion. In S. S. Stevens (Ed.), *Handbook of Experimental Psychology* (pp. 473–516). New York: Wiley.

Lipton, B. (2005). The biology of belief. Mountain of Love/Elite Books.

Lipton, B. (2005). The biology of belief. Santa Rosa, CA: Mountain of Love/Elite Books.

Lipton, B. (2016). Molecules of emotion. Santa Rosa, CA: Mountain of Love/Elite Books.

Lohrey, A. (2019). *The principles of nonlocal realism.* Retrieved from https://www.academia.edu/38239068/Nonlocal_Realism.docx

Long, A. A., & Sedley, D. N. (1987). *The Hellenistic Philosophers: Translations of the Principal Sources with Philosophical Commentary*. New York: Cambridge University Press.

Lucas, J. R., & Hodgson, P. E. (1985). *Space, Time and Causality: An Essay in Natural Philosophy*. New York: Oxford University Press.

Luhmann, N. (1995). *Social systems*. Stanford, CA: Stanford University Press.

Luhmann, N. (2002). The Paradox of Observing Systems. In W. Rasch (Ed.), *Theories of Distinction: Redescribing the Descriptions of Modernity* (pp. 73–93). Stanford: Stanford University Press.

Luhmann, N. (2012). *Theory of Society* (Vol. 1). (R. Barrett, Trans.). Stanford: Stanford University Press.

MacDonald, C. J., Lepage, K. Q., Eden, U. T., & Eichenbaum, H. (2011). Hippocampal "time cells" bridge the gap in memory for discontiguous events. *Neuron*, *71*(4), 737–749. doi:10.1016/j.neuron.2011.07.012 PMID:21867888

Machiavelli, N. (2008). *The Prince*. New York: Random House Modern Library.

Macintyre, A. (1981). *After virtue*. London: Duckworth.

MacLean, P. D. (1949). Psychosomatic disease and the "visceral brain": Recent developments bearing on the Papez theory of emotion. *Psychosomatic Medicine*, *11*(6), 338–353. doi:10.1097/00006842-194911000-00003 PMID:15410445

MacLean, P. D. (1952). Some psychiatric implications of physiological studies on frontotemporal portion of limbic system (visceral brain). *Electroencephalography and Clinical Neurophysiology*, *4*(4), 407–418. doi:10.1016/0013-4694(52)90073-4 PMID:12998590

MacLean, P. D. (1970). The triune brain, emotion and scientific bias. In F. O. Schmitt (Ed.), *The Neurosciences: Second Study Program* (pp. 336–349). New York: Rockefeller University Press.

MacLean, P. D. (1990). *The Triune Brain in Evolution: Role in Paleocerebral Functions*. New York: Plenum.

Mahan, G. P. (1961). *I-God (Secret Power of Kundalini)*. Madras, India: Universal Peace Sanctuary.

Mahan, G. P. (1981). *Kundalini yoga*. Tamil Nadu, India: Universal Peace Sanctuary.

Maharishi Effect. (n.d.). Maharishi's Programmes. Retrieved from maharishi-programmes.globalgoodnews.com/maharishi-effect

Maharishi, Y. V. (1986). *Karma yoga the holistic unity*. Madras, India: Veththiri Publications.

Maharishi, Y. V. (1989). *Yoga for modern age* (3rd ed.). Erode, India: Vethathiri Publications.

Maharishi, Y. V. (1992). *Journey of consciousness*. New Delhi: Macmillan India Limited.

Maharishi, Y. V. (1998). *Logical solutions for the problems of humanity*. Erode, India: Vethathiri Publications.

Malin, S. R. C., & Strivatsava, B. J. (1979). Correlation between heart attacks and magnetic activity. *Nature*, *277*(5698), 646–648. doi:10.1038/277646a0 PMID:423962

Malmo, R. B., Shagass, C., & Davis, F. H. (1950). Symptom specificity and bodily reactions during psychiatric interview. *Psychosomatic Medicine*, *12*(6), 362–376. doi:10.1097/00006842-195011000-00006 PMID:14797915

Mamychev, A. Yu. (2016). Arkhetipy gosudarstvennoy vlasti v pravosoznanii sovremennykh grazhdan RF [Archetypes of the government in the legal consciousness of modern citizens of the Russian Federation]. Aktual'nye problemy sovremennogo prava, 325-329.

Maniadakis, M., & Trahanias, P. (2011). Temporal cognition: A key ingredient of intelligent systems. *Frontiers in Neurorobotics*, *5*(2). PMID:21954384

Mann, M., & Brugge, P. J. (Producers) & Mann, M. (Director). (1999). *The Insider* [DVD]. USA: Touchstone Pictures.

Mansouri, F. A., Rosa, M. G. B., & Atapour, N. (2015). Working memory in the service of executive control functions. *Frontiers in Systems Neuroscience*, *9*(166). PMID:26696841

Marchetti, G., Giulio, M., & Alharbi, A. (Eds.). (2015). *Attention and meaning – The attentional basis of meaning*. New York: Nova Science Publishers, Inc.

Mariella, S., & Grech, V. (2016). Sentience in science fiction 101. *SFRA Review*, *315*, 14–18.

Marshall, L. H., & Magoun, H. W. (1998). *Discoveries in the Human Brain*. Totowa, NJ: Humana Press. doi:10.1007/978-1-4757-4997-7

Maslow, A. H. (1943). A theory of human motivation. *Psychological Review*, *50*(4), 370–396. doi:10.1037/h0054346

Mauco, O. (2008). Politicization of the On-Line Worlds as a Problematic of a Private Space Publicized: Comparative Study of MMORPG and Second Life. In *PCA ACA National conference*. Retrieved from https://ssrn.com/abstract=1329345

Maurer, A. P., Vanrhoads, S. R., Sutherland, G. R., Lipa, P., & McNaughton, B. I. (2005). Self-motion and the origin of differential spatial scaling along the septo-temporal axis of the hippocampus. *Hippocampus*, *15*(7), 841–852. doi:10.1002/hipo.20114 PMID:16145692

Maxwell, N. (2001). Evolution of sentience, consciousness and language viewed from a Darwinian and purposive perspective. In The Human World in the Physical Universe: Consciousness, Free Will and Evolution. Rowman and Littlefield.

McCarthy, J., Minsky, M. L., & Rochester, N. (1955). Dartmouth summer research project on artificial intelligence. *BBC*. Retrieved from http://www.bbc.com/future/story/20161114-how-we-built-machines-that-can-think-for-themselves?ocid=ww.social.link.linkedin

McCarthy, J., Minsky, M. L., & Rochester, N. (1955). *Dartmouth summer research project on artificial intelligence.* Retrieved from BBC Future: http://www.bbc.com/future/story/20161114-how-we-built-machines-that-can-think-for-themselves?ocid=ww.social.link.linkedin

McConkie, B. (1979). *Mormon doctrine* (2nd ed.). Amazon Prime.

McCraty, R. (2004). The Energetic Heart: Bioelectromagnetic Communication Within and Between People. *Heartmath Institute*. Retrieved from https://www.heartmath.org/research/research-library/energetics/energetic-heart-bioelectromagnetic-communication-within-and-between-people/

McCraty, R. (n.d.). *Coherence: Bridging personal, social, and global health.* Institute of Heartmath.

McCraty, R., & Atkinson, M. (2003). Psychophysiological coherence. Boulder Creek, CA: HeartMath Research Center, Institute of HeartMath, Publication 03-016.

McCraty, R., & Childre, D. (2002). The appreciative heart: The psychophysiology of positive emotions and optimal functioning. Boulder Creek, CA: HeartMath Research Center, Institute of HeartMath, Publication No. 02-026.

McCraty, R., & Childre, D. (2010). Coherence: bridging personal, social, and global health. Alternative Therapies, 16(4).

McCraty, R., & Childre, D. (2010). Coherence: Bridging personal, social, and global health. Alternative Therapies, 16(4).

McCraty, R., & Childre, D. (2010). Coherence: bridging personal, social, and global health. *Alternative Therapies*, *16*(4). Retrieved from https://www.heartmath.org/assets/uploads/2015/01/coherence-bridging-personal-social-global-health.pdf

McCraty, R., & Childre, D. (2010). Coherence: bridging personal, social, and global health. *Alternative Therapies*, *16*(4). Retrieved November 6, 2015 from https://www.heartmath.org/assets/uploads/2015/01/coherence-bridging-personal-social-global-health.pdf

McCraty, R., Atkinson, M., Stolc, V., Alabdulgader, A. A., Vainoras, A., & Ragulskis, M. (2017). Synchronization of Human Autonomic Nervous System Rhythms with Geomagnetic Activity in Human Subjects. *Int. J. Environ. Res. Public Health*, *14*(7), 770. Retrieved 10/08/2017 from http://www.mdpi.com/1660-4601/14/7/770

McCraty, R. (2002). Heart rhythm coherence: And emerging area of biofeedback. *Biofeedback*, *30*(1), 17–19.

McCraty, R. (2005). *Enhancing emotional, social, and academic learning with heart rhythm coherence feedback.* Boulder Creek, CA: Heartmath Research Center, Institute of Heartmath.

McCraty, R., Atkinson, M., Stolc, V., Alabdulgader, A. A., Vainoras, A., & Ragulskis, M. (2017). Synchronization of Human Autonomic Nervous System Rhythms with Geomagnetic Activity in Human Subjects. *International Journal of Environmental Research and Public Health*, *14*(7), 770. doi:10.3390/ijerph14070770

McCraty, R., Barrios-Choplin, B., Rozman, D., Atkinson, M., & Watkins, A. D. (1998). The impact of a new emotional self-management program on stress, emotions, heart rate variability, DHEA and cortisol. *Integrative Physiological and Behavioral Science*, *33*(2), 151–170. doi:10.1007/BF02688660 PMID:9737736

McCraty, R., & Schafer, S. (2016). *Exploring dimensions of the media dream: Functional context in collective personae.* Hershey, PA: IGI Global. doi:10.4018/978-1-4666-9891-8.ch001

McCraty, R., & Tomasino, D. (2006). Coherence-building techniques and heart rhythm coherence feedback: New tools for stress reduction, disease prevention, and rehabilitation. In E. Molinari, A. Compare, & G. Parati (Eds.), *Clinical Psychology and Heart Disease*. Milan, Italy: Springer-Verlag. doi:10.1007/978-88-470-0378-1_26

McDonald, C. (2015). The good, the bad, and the robot: Experts are trying to make robots moral. *Cal Alumni Association.* Retrieved from https://alumni.berkeley.edu/california-magazine/just-in/2015-06-08/good-bad-and-robot-experts-are-trying-make-machines-be-moral

McIntosh, K. (2008, Fall). The Social Construction of Virtual Space. *Michigan Sociological Review,* 196–214.

McNiell, Hardy, W. (1981). Discrepancies among the social sciences. *Conspectus of History*, *1*(7), 37–38.

McPheat, S. (2010). *Effective communication skills, training & venues*. Google/Scholar.

Meadows, D. H., Meadows, G., Randers, J., & Behrens, W. W. III. (1972). *The Limits to Growth.* New York: Universe Books.

Mehta, P., & Sharma, M. (2012-13). Evaluation of a social cognitive theory based yoga intervention to reduce anxiety. *International Quarterly of Community Health Education*, *32*(3), 205–217. doi:10.2190/IQ.32.3.d PMID:23353562

Meijer, D. (2013a). Information: What do you mean? On the formative element of our universe. *Syntropy*, *3*, 1–49.

Meijer, D. (2013b). Immortality: Myth or becoming reality? On the conservation of information. *Syntropy*, *3*, 168–207.

Melville, H., & Moncur, M. (2003-2012). Moby Dick chapter 132. *The Literature Page.* Retrieved from http://www.literaturepage.com/read/mobydick-565.html

Meredith-Owen, W. (2011). Winnicott on Jung: Destruction, creativity and the unrepressed unconscious. *The Journal of Analytical Psychology*, *56*(1), 56–75. doi:10.1111/j.1468-5922.2010.01890.x PMID:21241294

Micadei, K., Peterson, J. P., Souza, A. M., Sarthour, R. S., Oliveira, I. S., Landi, G. T., & Lutz, E. (2017). *Reversing the thermodynamic arrow of time using quantum correlations.* Ithaca, NY: Cornell University.

Miller, G. A., Galanter, E. H., & Pribram, K. H. (1960). *Plans and the Structure of Behavior*. New York: Henry Holt & Co. doi:10.1037/10039-000

Mindvalley Talks. (2011). How to change your frequency to change your reality [YouTube video].

Montoya P. (1994). *Herzwahrnehmung und hirnelektrische Aktivität. Eine Analyse der topographischen Verteilung von herzschlag-synchron evozierten Potentialen (HEP)*. Frankfurt/Main: Peter Lang.

Montoya, P., Schandry, R., & Muller, A. (1993). Heartbeat evoked potentials (HEP): Topography and influence of cardiac awareness and focus of attention. *Electroencephalography and Clinical Neurophysiology*, *88*(3), 163–172. doi:10.1016/0168-5597(93)90001-6 PMID:7684965

Morawski, S. (1994). The hopeless game of Flânerie. In K. Tester (Ed.), *The Flâneur* (p. 181). London: Routledge.

Morin, D. (2008). *Introduction to Classical Mechanics*. New York: Cambridge University Press. doi:10.1017/CBO9780511808951

Moritz, N. H., & Jaffe, T. (Producers) & Wiseman, L. (Director). (2012). *Total Recall* [DVD]. USA: Columbia Pictures Industries, Inc.

Morris, S. M. (2010). Achieving collective coherence: Group effects on heart rate variability coherence and heart rhythm synchronization. Alternative Therapies, 16(4).

Moser, M.-B., & Moser, E. I. (1998). Functional differentiation in the hippocampus. *Hippocampus*, *8*(6), 608–619. doi:10.1002/(SICI)1098-1063(1998)8:6<608::AID-HIPO3>3.0.CO;2-7 PMID:9882018

Mossbridge, J. A., & Radin, D. (2018). Plausibility, statistical interpretations, physical mechanisms and a new outlook: Response to commentaries on a precognition review. *American Psychological Association.* Retrieved from https://www.academia.edu/38151856/2018_Response_to_Commentaries_cns_5_1_110.pdf?email_work_card=view-paper

Moyers, B. (1985). *The Secret Government: Constitution in Crisis.* PBS.

Mukoyama, M., Nakao, K., Hosoda, K., Suga, S., Saito, Y., Ogawa, Y., ... Yasue, H. (1991). Brain natriuretic peptide as a novel cardiac hormone in humans. Evidence for an exquisite dual natriuretic peptide system, atrial natriuretic peptide and brain natriuretic peptide. *The Journal of Clinical Investigation*, *4*(4), 1402–1412. doi:10.1172/JCI115146 PMID:1849149

Müller, L. R. (1906). Ueber die Beziehung von seelischen Empfindungen zu Herzstörungen. *Munchener Medizinische Wochenschrift*, *53*, 14–16.

Munoz-Lopez, M. M., Mohedano-Moriano, A., & Insausti, R. (2010). Anatomical pathways for auditory memory in primates. *Frontiers in Neuroanatomy*, *4*(129). PMID:20976037

Myers, F. W. H. (1903). *Human personality and its survival of bodily death* (Vols. 1–2). London: Longmans.

Myss, C. (1996). Anatomy of the spirit: The seven stages of power and healing. Crown Publishers, Inc.

Näätänen, R. (1992). *Attention and Brain Function.* Hillsdale, NJ: Lawrence Erlbaum Associates.

National Institutes of Natural Sciences. (2016). World's fastest quantum simulator. *Institute for Molecular Science, National Institutes of Natural Sciences, Japan, Nature Communications.* Retrieved from PHYSORG: http://phys.org/news/2016-11-world-fastest-quantum-simulator-atomic.html

Needham, W. P., & Amado, C. A. (2007). Division of Natural Sciences. State University of New York at Purchase.

Neihardt, J. G. (1972). *Black Elk Speaks.* Pocket Books.

Nelson, S. M., McDermott, K. B., Wig, G. S., Schlagger, B. I., & Petersen, S. E. (2013). The critical roles of localization and physiology for understanding parietal contributions to memory retrieval. *The Neuroscientist*, *19*(6), 578–591. doi:10.1177/1073858413492389 PMID:23778789

Neundörfer, B., & Hilz, M. J. (1998). Ludwig Robert Müller (1870-1962) – A pioneer of autonomic nervous system research. *Clinical Autonomic Research*, *8*(1), 1–5. doi:10.1007/BF02267597 PMID:9532414

Nietzsche, F. (1909). Introduction. In *Thus Spake Zarathustra: A Book for All and None* (T. Common, Trans.). London: T.N. Foulis.

Nikolaev, Y. S., Rudakov, Y. Y., Mansurov, S. M., & Mansurova, L. G. (1976). Interplanetary magnetic field sector structure and disturbances of the central nervous system activity.

Nikolaev, Y. S., Rudakov, Y. Y., Mansurov, S. M., & Mansurova, L. G. (1976). Interplanetary magnetic field sector structure and disturbances of the central nervous system activity. *Acad Sci USSR. IZMIRAN, 1976*, 29.

Nikolenko, N.N. (2015). Traditsionnyy ornament nomadov kak vizual'nyy arkhetip [Traditional ornament of nomads as a visual archetype]. *Universum: filologiya i iskusstvovedenie, 9-10.*

Norris, J. (2014). Do gut bacteria rule our minds. *UCSF*. Retrieved from https://www.ucsf.edu/news/2014/08/116526/do-gut-bacteria-rule-our-minds

O'Keefe, J. (1976). Place units in the hippocampus of the freely moving rat. *Experimental Neurology*, *51*(1), 78–109. doi:10.1016/0014-4886(76)90055-8 PMID:1261644

O'Keefe, J., & Dostrovsky, J. (1971). The hippocampus as a spatial map. Preliminary evidence from unit activity in the freely-moving rat. *Brain Research*, *34*(1), 171–175. doi:10.1016/0006-8993(71)90358-1 PMID:5124915

O'Keefe, J., & Nadel, L. (1978). *The Hippocampus as a Cognitive Map*. London: Oxford University Press.

Odoul, M. (2014). *Ghidul complet al durerilor fizice - Terapii psihoenergetice, Prefață Dr. Medynski, Thierry* [The Complete Physical Pain Guide - Psycho-Therapeutic Therapies, Preface by Dr. Medynski, Thierry]. Bucuresti, Romania: Ed. Paralela 45. Retrieved from https://caleaechilibruluii.files.wordpress.com/2014/04/michel-odoul-ghidul-complet-al-durerilor-fizice.pdf

Ohmori, K. (2016). World's fastest quantum simulator. *Institute for Molecular Science, National Institutes of Natural Sciences, Japan, Nature Communications*. Retrieved from http://phys.org/news/2016-11-world-fastest-quantum-simulator-atomic.html

Orme-Johnson, D. W. (n.d.). Maharishi Effect. *Global Good News*. Retrieved November12, 2015 from maharishi-programmes.globalgoodnews.com/maharishi-effect

Oshchepkova, A. A. (2016). Arkhetip ryby v russkoy kul'ture [The archetype of the fish in Russian culture]. In VIII kirillo-mefodievskie chteniya "slavyanskoe slovo v kontekste vremeni", 101-106.

Panduro, A., Rivera-Iñigue, I., Sepulveda-Villegas, M., & Roman, S. (2017). Genes, emotions and gut microbiota: The next frontier for the gastroenterologist. *世界胃肠病学杂志:英文版*, *23*(17), 3030-3042. doi:10.3748/wjg.v23.i17.3030

Papadopolous, R., & Byng-Hall, J. (Eds.). (2007). Narrative Based Medicine: A narrative approach to medicine. Academic Press.

Papadopolous, R., & Byng-Hall, J. (Eds.). (2007). *Narrative Based Medicine: A narrative approach to medicine.*

Papez, J. W. (1937). A proposed mechanism of emotion. *Archives of Neurology and Psychiatry*, *38*(4), 725–743. doi:10.1001/archneurpsyc.1937.02260220069003

Parsons, T. (1951/1991). *The social system*. London: Routledge.

Penrose, R. (2005). *The Road to Reality*. London: Vintage Books.

Pereira, C., Reddy, J., & Shashi, K. (2018). The boy who grew a new brain: Understanding this miracle from a neuro-quantum perspective. *NeuroQuantology*, *16*(7), 39-48. Retrieved from https://www.researchgate.net/publication/326711377_The_Boy_Who_Grew_a_New_Brain_Understanding_this_Miracle_from_a_Neuro-Quantum_Perspective

Pereira, A. Jr, Nunn, C., Nixon, G. M., & Pregnolato, M. (2018). Consciousness and Cosmos: Building an Ontological Framework. *Journal of Consciousness Studies*, *25*(3-4), 151–205.

Pereira, C. (2016). Is it quantum sentience or quantum consciousness? A review of social behaviours observed in primitive and present-day microorganisms. *NeuroQuantology: An Interdisciplinary Journal of Neuroscience and Quantum Physics*, *1*, 16–27.

Permilovskaya, A. B. (2016). Derevyannyy pravoslavnyy khram kak voploshchenie etnokul'turnogo arkhetipa russkoy traditsionnoy kul'tury [Wooden orthodox temple as the embodiment of the ethno-cultural archetype of Russian traditional culture]. Yaroslavskiy pedagogicheskiy vestnik, 2, 245-250.

Personality Types Explained. (n.d.). Retrieved from http://www.humanmetrics.com/personality/type

Pert, C. B. (2003). *Molecules of emotion*. New York: Scribner.

Pert, C. B., Dreher, H. E., & Ruff, M. R. (1998). The psychosomatic network: Foundations of mind-body medicine. *Alternative Therapies in Health and Medicine*, *4*(4), 30–41. PMID:9656499

Petersen, E. (1988). The communication of the dead: notes on the *studia humanitatis* and the nature of humanist philology. In A. C. Dionisotti, A. Grafton, & J. Kraye (Eds.), *The Uses of Greek and Latin: historical essays* (pp. 57–69). London: Warburg Institute and University of London.

Peters, J. D. (1999). *Speaking into the Air: A History of the Idea of Communication*. Chicago: University of Chicago Press. doi:10.7208/chicago/9780226922638.001.0001

Petkov, V. (2010). *Minkowski Spacetime: A Hundred Years Later*. Berlin: Springer. doi:10.1007/978-90-481-3475-5

Piccolomini, A. S. (2002). The education of boys (C. W. Kallendorf, Trans. & Ed.). In *Humanist Educational Treatises* (p. 43). Cambridge, MA: Harvard University Press.

Pilant, W. (1979). *Elastic Waves in the Earth*. Elsevier Scientific Publishing Company.

Plato Political Philosophy. (n.d.). *Internet Encyclopedia of Philosophy IEP*. Retrieved from https://www.iep.utm.edu/platopol/

Pobachenko, S. V., Kolesnik, A. G., Borodin, A. S., & Kalyuzhin, V. V. (2006). The contingency of parameters of human encephalograms and Schumann Resonance electromagnetic fields revealed in monitoring studies. *Complex Systems Biophysics.*, *51*(3), 480–483.

Pokorny, M. (2012). Sentience, awareness consciousness. *Organon F*, *19*(Suppl.), 51–63. Retrieved from https://www.academia.edu/28728892/Sentience_Awareness_Consciousness

Polt, R. (2006). The Emergency of Being: On Heidegger's Contributions to Philosophy. Ithaca, NY: Cornell University Press.

Pope, R. (2019). Plato's map of the future omni artifact. *Science-Art Research Center of Austalia*. Retrieved from https://connect.xfinity.com/appsuite/#!!&app=io.ox/mail&folder=default0/INBOX

Pope, R. (2019). Plato's map of the future omni artifact. *Science-Art Research Center of Australia*. Retrieved from https://connect.xfinity.com/appsuite/#!!&app=io.ox/mail&folder=default0/INBOX

Popper, A. N. (2010). *Music Perception*. Berlin: Springer.

Postman, N. (1985). *Amusing ourselves to death*. New York: Penguin Books.

Pribram, K. (1969). The neurophysiology of remembering. *Scientific American, 220*, 75.

Pribram, K. (1963). Reinforcement revisited: A structural view. In *Nebraska Symposium on Motivation*. Lincoln, NE: University of Nebraska Press.

Pribram, K. (1969). The neurophysiology of remembering. *Scientific American*, *220*(1), 73–87.

Pribram, K. (1971). *Languages of the brain; experimental paradoxes and principles in neuropsychology*. Englewood Cliffs, NJ: Prentice-Hall.

Pribram, K. (1977). *Languages of the Brain. Monterey*, CA: Wadsworth Publishing.

Pribram, K. (1977). *Languages of the Brain.* Wadsworth Publishing.

Pribram, K. (1991). *Brain and perception: holonomy and structure in figural processing*. Hillsdale, NJ: Lawrence Erlbaum Associates.

Pribram, K. H. (1967). The new neurology and the biology of emotion: A structural approach. *The American Psychologist, 22*(10), 830–838. doi:10.1037/h0025031 PMID:4383680

Pribram, K. H. (1970). Feelings as monitors. In M. B. Arnold (Ed.), *Feelings and Emotions* (pp. 41–53). New York: Academic Press. doi:10.1016/B978-0-12-063550-4.50010-3

Pribram, K. H. (1971). *Languages of the Brain*. New York: Brandon House.

Pribram, K. H. (1984). Emotions: A neurobehavioral analysis. In K. R. Scherer & P. Ekman (Eds.), *Approaches to Emotion*. Hillsdale, NJ: Erlbaum.

Pribram, K. H., & Bagshaw, M. H. (1953). Further analysis of the temporal lobe syndrome utilizing fronto-temporal ablations. *The Journal of Comparative Neurology, 99*(2), 347–375. doi:10.1002/cne.900990208 PMID:13109029

Pribram, K. H., & McGuinness, D. (1975). Arousal, activation, and effort in the control of attention. *Psychological Review, 82*(2), 116–149. doi:10.1037/h0076780 PMID:1096213

Pribram, K. H., & Melges, F. T. (1969). Psychophysiological basis of emotion. In P. J. Vinken & G. W. Bruyn (Eds.), *Handbook of Clinical Neurology* (pp. 316–341). Amsterdam: North-Holland Publishing Company.

Privezentseva, V. A. (2016). Kul'turnye arkhetipy v angliyskom gumanizme vtoroy poloviny XVI - nachala XVII v [Cultural archetypes in English humanism in second half of XVI – beginning of XVII century]. Yaroslavskiy pedagogicheskiy vestnik, 2, 198-202.

Proctor, H., Carder, G., & Cornish, A. (2013). Searching for animal sentience: A systematic review of the scientific literature. *Animals, 3*, 882-906; doi:10.3390/ani3030882

Pruchnic, J. (2008). Neurorhetorics: Cybernetics, Psychotropics, and the Materiality of Persuasion. *Configurations, 16*(2), 2. doi:10.1353/con.0.0057

Pshenichnaya, V. V., & Gatiatullina, E. R. (2016). Problema proyavleniya arkhetipov v natsional'nom soznanii [The problem of the archetypes manifestation in the national consciousness]. Sovremennye problemy upravleniya prirodnymi resursami i razvitiem sotsial'no-ekonomicheskikh system, 307-312.

Radin, D. (2006). *Entangled minds, extrasensory experiences in a quantum theory*. Paraview Pocket Books, a division of Simon Shuster, Inc.

Radin, D. (1997). *The conscious universe: Scientific proof of psychic phenomena*. USA: Harper One.

Radin, D. (2006). *Entangled Minds: Extrasensory Experiences in a Quantum Reality; · Supernormal: Science, Yoga, and the Evidence for Extraordinary Psychic Abilities*. USA: Paraview Pocket Books.

Radin, D. (2018). *Real magic: Ancient wisdom, modern science, and a guide to the secret power of the universe*. New York: Ed. Harmony Books, Crown Publishing Group, Penguin Random House LLC.

Radin, D., Lund, N., Emoto, M., & Kizu, T. (2008). Effects of distant intention on water crystal formation: A triple-blind replication. *Journal of Scientific Exploration*, *22*(4), 481–493.

Randich, A., & Gebhart, G. F. (1992). Vagal afferent modulation of nociception. *Brain Research. Brain Research Reviews*, *17*(2), 77–99. doi:10.1016/0165-0173(92)90009-B PMID:1327371

Rau, H., Pauli, P., Brody, S., Elbert, T., & Birbaumer, N. (1993). Baroreceptor stimulation alters cortical activity. *Psychophysiology*, *30*(3), 322–325. doi:10.1111/j.1469-8986.1993.tb03359.x PMID:8497561

Reddy, J. (2018). Shashi Kiran (2018). The emergence of mind as a quantum field phenomenon. *NeuroQuantology: An Interdisciplinary Journal of Neuroscience and Quantum Physics*, *16*(11), 68–78. Retrieved from https://www.academia.edu/37922315/The_Emergence_of_Mind_as_a_Quantum_Field_Phenomenon?email_work_card=title

Reed, H. (2002). *Research ethics in complex humanitarian emergencies: Summary of a workshop*. Washington, DC: National Academy Press.

Reed, T. J. (2015). *Light in Germany: scenes from an unknown Enlightenment*. University of Chicago. doi:10.7208/chicago/9780226205243.001.0001

Reeves, B., & Nass, C. (1996). The media equation. Stanford, CA: Center For the study of Language and Information & Cambridge University Press.

Reeves, B., & Nass, C. (1996). *The media equation*. Stanford, CA: Center for the Study of Language and Information.

Rizzolatti, G., Fadiga, L., Gallese, V., & Fogassi, L. (1996). Premotor cortex and the recognition of motor actions. *Brain Research. Cognitive Brain Research*, *3*(2), 131–141. doi:10.1016/0926-6410(95)00038-0 PMID:8713554

Roberts, T., & Schleif, S. (2007). *What is the experimental basis of special relativity?* Usenet Physics FAQ.

Robinson, D. G. (2007). Plant neurology: No brain, no gain? *Elsevier: Science Direct*. Retrieved from http://www.linv.org/images/about_pdf/Trends%202007%20Alpi.pdf

Robinson, G. E., Fernald, R. D., & Clayton, D. F. (2008). Genes and social behavior. *Science*, *322*(5903), 896–899. doi:10.1126cience.1159277 PMID:18988841

Rogers, K. (2018). *MIT Device Uses Wi-Fi to 'See' Through Walls and Track Your Movements*. Vice.

Rolls, E. T. (2013). The mechanisms for pattern completion and pattern separation in the hippocampus. *Frontiers in Systems Neuroscience*, *7*(74). PMID:24198767

Romas, J. A., & Sharma, M. (2017). *Practical stress management. A comprehensive workbook*. London: Academic Press.

Rosenfeld, S. A. (1978). *Conversations between heart and brain*. Rockville, MD: National Institute of Mental Health.

Rossi, L. (2008). *MMORPG Guilds as Online Communities - Power, Space and Time: From Fun to Engagement in Virtual Worlds*. Retrieved from https://ssrn.com/abstract=2137594

Saffell, N. (2015). On the origin of robot species: robots building robots by "natural selection." University of Cambridge. Retrieved from http://www.cam.ac.uk/research/news/on-the-origin-of-robot-species?sthash.zMSyp5OS.mjjo

Saffell, N. (2015). *On the origin of robot species: robots building robots by "natural selection"*. University of Cambridge. Retrieved August 12, 2015 from http://www.cam.ac.uk/research/news/on-the-origin-of-robot-species?sthash.zMSyp5OS.mjjo

Sandman, C. A., Walker, B. B., & Berka, C. (1982). Influence of afferent cardiovascular feedback on behavior and the cortical evoked potential. In J. T. Cacioppo & R. E. Petty (Eds.), *Perspectives in Cardiovascular Psychophysiology* (pp. 189–222). New York: The Guilford Press.

Sartre, J.-P. (1946). *Existentialism is a humanism.* Yale University Press.

Sathya Sai International Organization-USA. (n.d.). Retrieved from https://sathyasai.us/devotion/discourse/purity-thought-word-deed)

Satsangi, A. K., & Brugnoli, M. P. (2018). Anxiety and psychosomatic symptoms in palliative care: From neuro-psychobiological response to stress, to symptoms' management with clinical hypnosis and meditative states. *Annals of Palliative Medicine*, *7*(1), 75–111. doi:10.21037/apm.2017.07.01 PMID:28866901

Schacter, D. L., Gilbert, D. T., & Wegner, D. M. (2011). *Semantic and episodic memory. In Psychology* (2nd ed.; pp. 240–241). New York: Worth Incorporated.

Schafer, S. (2006). Curriculum and the dream paradigm. *Gamasutra.* Retrieved from http://www.gmecareerguide.com/features/288/curriculum_and_the_dream_paradigm.php

Schafer, S. (2007). Premise: The key to interactive storytelling. *Gamasutra.* Retrieved from http://www.gamecareerguide.com/features/357/premise_the_key_to_interactive_.php?page=1

Schafer, S. (2007). Premise: The Key to Interactive Storytelling. *Game Career Guide.* Retrieved April 3, 2007 from http://www.gamecareerguide.com/features/357/premise_the_key_to_interactive_.php?page=1

Schafer, S. (2007). Premise: The Key to Interactive Storytelling. *Game Career Guide*: Retrieved April 3, 2007 from http://www.gamecareerguide.com/features/357/premise_the_key_to_interactive_.php?page=1

Schafer, S. (2008). *The Psychecology Game I-V; Using Story-based Video Games to Capture Veridical Data on Subliminal States of Being.* Consciousness Center & the University of Arizona. Retrieved from https://bandura.sbs.arizona.edu/login/consciousness/pubreport.aspx?aid=27

Schafer, S. (2008). The Psychecology Game I-V; Using Story-based Video Games to Capture Veridical Data on Subliminal States of Being. In *Toward a Science of Consciousness 2008.* Consciousness Center & the University of Arizona. Retrieved from https://bandura.sbs.arizona.edu/login/consciousness/pubreport.aspx?aid=2761

Schafer, S. (2008). *The Psychecology Game I-V; Using Story-based Video Games to Capture Veridical Data on Subliminal States of Being.* Toward a Science of Consciousness 2008, sponsored by the Consciousness Center & the University of Arizona. Retrieved from https://bandura.sbs.arizona.edu/login/consciousness/pubreport.aspx?aid=27

Schafer, S. (2011). Articulating the paradigm shift: Serious games for psychological healing of the collective person. In *The emerging science of human-data interaction—simulation of living systems: MIT Technology Review.* Retrieved July 31, 2014 from https://www.linkedin.com/grp/post/1039687-5900533809526378496

Schafer, S. (2011). Articulating the paradigm shift: Serious games for psychological healing of the collective persona. In *Business Social Networking: Organizational, Managerial, and Technological Dimensions.* IGI Global. Retrieved from http://www.igi-global.com/book/handbook-research-serious-games-educational/5

Schafer, S. (2011b). Articulating the paradigm shift: Serious games for psychological healing of the collective persona. In *Business Social Networking: Organizational, Managerial, and Technological Dimensions.* IGI Global. Retrieved from http://www.igi-global.com/book/handbook-research-serious-games-educational/5

Schafer, S. (2012). Optimizing Cognitive Coherence, Learning, and Psychological Healing with Drama-based Games. In *Video Game Play and Consciousness*. Nova Science Publishers. http://www.facebook.com/insights/?sk=po_261980910540561#!/pages/Video-Game-Play-and-Consciousness/261980910540561?sk=wall

Schafer, S. (2012). Optimizing Cognitive Coherence, Learning, and Psychological Healing with Drama-based Games. In *Video Game Play and Consciousness*. Nova Science Publishers. Retrieved from http://www.facebook.com/insights/?sk=po_261980910540561#!/pages/Video-Game-Play-and-Consciousness/261980910540561?sk=wall

Schafer, S. (2012a). Optimizing Cognitive Coherence, Learning, and Psychological Healing with Drama-based Games. *Video Game Play and Consciousness*. Nova Science Publishers. Retrieved from http://www.facebook.com/insights/?sk=po_261980910540561#!/pages/Video-Game-Play-and-Consciousness/261980910540561?sk=wall

Schafer, S. B. (2012e). Optimizing Cognitive Coherence, Learning, and Psychological Healing with Drama-based Games. In *Video Game Play and Consciousness*. Nova Science Publishers. Retrieved from http://www.facebook.com/insights/?sk=po_261980910540561#!/pages/Video-Game-Play-and-Consciousness/261980910540561?sk=wall

Schafer, S. B. (2016d). Fostering psychological coherence with ICTs. *International Journal of Reliable and Quality E-Healthcare*, *4*(4), 1-31. DOI doi:10.4018/IJRQEH.2015100101

Schafer, S. B. (2018). *Gaming cultural atonement*. Lambert Academic Publishers. Retrieved from https://www.morebooks.de/gb/search?utf8=%E2%9C%93&q=+978-613-9-89469-7+

Schafer, S. B. (2018a). *Generating a superconductive culture of conscience*. Dusseldorf, Germany: LAP Academic Publishers. Retrieved from: https://www.morebooks.de/store/gb/book/generating-a-superconductive-culture-of-conscience/isbn/978-620-2-07643-2

Schafer, S. B. (2018b). Synergy with the World Soul: Establishing the humanitarian organism. International Journal of Civic Engagement and Social Change, 4(3). Doi:10.4018/IJCESC.20170701

Schafer, S. B. (2018c). *Gaming cultural atonement*. Lambert Academic Publishers. Retrieved from https://www.morebooks.de/gb/search?utf8=%E2%9C%93&q=+978-613-9-89469-7+

Schafer, S., & Palamides, T. (2011). Designing an Ethical structure for Social Influence Marketing. In *Business Social Networking: Organizational, Managerial, and Technological Dimensions*. IGI Global. Retrieved from www.igi-global.com

Schafer, S., & Palamides, T. (2011c). Designing an Ethical structure for Social Influence Marketing. In *Business Social Networking: Organizational, Managerial, and Technological Dimensions*. IGI Global. www.igi-global.com

Schafer, B. S. (2016). *The Media-Sphere as Dream: Researching the Contextual Unconscious of Collectives. In Exploring the Collective Unconscious in the Age of Digital Media*. IGI Global.

Schafer, S. (Ed.). (2017b). *Exploring the collective unconscious in a digital age*. Hershey, PA: IGI Global.

Schafer, S. B. (2012b). Articulating the paradigm shift: Serious games for psychological healing of the collective persona. In *Handbook of Research on Serious Games as Educational, Business and Research Tools* (pp. 352–377). Hershey, PA: IGI Global.

Schafer, S. B. (2016a). *The dream analog: Psychecology video games. In Exploring the Collective Unconscious in a Digital Age*. Hershey, PA: IGI Global. doi:10.4018/978-1-4666-9891-8.ch002

Schafer, S. B. (2016d). *Employing dynamics of Jungian compensation to foster psychological coherence with ICTs* (Vol. 4). Hershey, PA: International Journal of Privacy and Health Information Management.

Schafer, S. B. (2016e). *Employing dynamics of Jungian compensation to foster psychocological coherence with ICTs. International Journal of Privacy and Health Information Management*, 4(1).

Schafer, S. B. (2016e). Employing dynamics of Jungian compensation to foster psychological coherence with ICTs. *International Journal of Privacy and Health Information Management*, *4*(1), 23–49.

Schafer, S. B. (2017a). *Generating a superconductive culture of conscience*. Mauritius: Lambert Academic Publishing.

Schafer, S. B. (2018b). *Generating the superconductive culture of conscience*. Dusseldorf, Germany: Lambert Academic Publishing. Retrieved from https://www.morebooks.de/store/gb/book/generating-a-superconductive-culture-of-conscience/isbn/978-620-2-07643-2

Schafer, S. B. (Ed.). (2016). *Exploring the collective unconscious in a digital age. In Exploring the Collective Unconscious in a Digital Age*. Hershey, PA: IGI Global. doi:10.4018/978-1-4666-9891-8

Schafer, S., & Yu, G. (2011). Meaningful video games: drama-based video games as transformational experience. In M. M. Cruz-Cunha, V. H. Valvalho, & P. Tavares (Eds.), *Business, Technological, and Social Dimensions of Computer Games: Multidisciplinary Developments* (pp. 312–329). Hershey, PA: IGI Global. doi:10.4018/978-1-60960-567-4.ch019

Schandry, R., & Leopold, C. (1997). The severity of diabetic autonomic neuropathy is reflected in the heartbeat evoked brain potential [abst.]. *Clinical Autonomic Research*, *7*(5), 249–250.

Schandry, R., & Montoya, P. (1996). Event-related brain potentials and the processing of cardiac activity. *Biological Psychology*, *42*(1-2), 75–85. doi:10.1016/0301-0511(95)05147-3 PMID:8770371

Schandry, R., Sparrer, B., & Weitkunat, R. (1986). From the heart to the brain: A study of heartbeat contingent scalp potentials. *The International Journal of Neuroscience*, *30*(4), 261–275. doi:10.3109/00207458608985677 PMID:3793380

Schultze, U. (2000). A confessional account of an ethnography about knowledge work. *Management Information Systems Quarterly*, *24*(1), 3–41. doi:10.2307/3250978

Sen, S., Mahendra, A., & Priyadarshi, P. (2015). *Communication and language skills*. Cambridge University Press.

Serino, S., & Riva, G. (2014). What is the role of spatial processing in the decline of episodic memory in Alzheimer's disease? The "mental frame syncing" hypothesis. *Frontiers in Aging Neuroscience*, *6*(33). PMID:24653696

Seung, H. S. (2000). Half a century of Hebb. *Nature Neuroscience*, *3*(S11), 1166. doi:10.1038/81430 PMID:11127829

Shannon, C. (1948). A mathematical heory of communication. *The Bell System Technical Journal*, *27*(3), 379–423. doi:10.1002/j.1538-7305.1948.tb01338.x

Sharma, M. (2001). Pilot test of a Kundalini yoga intervention developing mind-body connection. *International Journal of Yoga Therapy*, *11*, 85–91.

Sharma, M. (2005). Effects of a yoga intervention as a supportive therapy in arthritis. *Yoga Studies*, *4*, 12–16.

Sharma, M. (2014). Potential efficacy of yoga in various diseases. *Journal of Symptoms and Signs*, *3*(3), 194–199.

Sharma, M. (2015). Multi-theory model (MTM) for health behavior change. *WebmedCentral Behaviour*, *6*(9), WMC004982.

Sharma, M. (2017a). *Theoretical foundations of health education and health promotion* (3rd ed., pp. 250–262). Burlington, MA: Jones and Bartlett.

Sharma, M. (2017b). Trends and prospects in public health education: A commentary. [Editorial]. *Social Behavior Research & Health*, *1*(2), 67–72.

Sharma, M. (2018). *Introspective meditations for complete contentment (Santosha)*. Omaha, NE: Health for All.

Sharma, M., & Corbin, D. E. (2006). A preliminary study of the effects of an innovative social cognitive theory driven yoga-based behavioral intervention on smoking cessation. *Californian Journal of Health Promotion*, *4*(4), 63–76.

Sharpless, S., & Jasper, H. (1956). Habituation of the arousal reaction. *Brain*, *79*(4), 655–680. doi:10.1093/brain/79.4.655 PMID:13396069

Sheikh, K. (2017). How Gut Bacteria Tell Their Hosts What to Eat: By suppressing or increasing cravings, microbes help the brain decide what foods the body "needs". *Scientific American: Neuroscience*. Retrieved from https://www.scientificamerican.com/article/how-gut-bacteria-tell-their-hosts-what-to-eat/

Sheldon, C. M. (2011). How to change your frequency to change your reality. *YouTube*. Retrieved from https://www.bing.com/videos/search?q=consciousness+of+water&&view=detail&mid=E3E62E256E8BFBBA90C8E3E62E256E8BFBBA90C8&&FORM=VDRVRV

Shiga, J. (2013). Of Other Networks. *Amodern 2: Network Archaeologies*. Retrieved from http://amodern.net/article/of-other-networks/

Shiga, J. (n.d.). Of other networks: Closed-world and green-world networks in the work of John C. Lilly. *Amodern 2: Network Archaeologies*. Retrieved from http://amodern.net/article/of-other-networks/

Shin, C. (2016a). Introducing a function with plural derivatives. *Zeitschrift für Angewandte Mathematik und Physik*.

Shin, C. (2016b). *God before the Big Bang*. International Education & Research Journal.

Shin, C., & Cha, D. (2015). An Attempt towards Mathematical Description of the Human Mental State. *ADAM*, *5*(1), 1–8.

Shin, C., Ha, W., Chung, W., & Park, S. (2012). Unresolvable Human Mental States Based on a Parallel Universe Theory. *ADAM*, *1*(1), 18–29.

Shor, J. (1944). Review of the psychology of Jung. *Psychological Bulletin*, *41*(7), 496–497. doi:10.1037/h0050824

Sieb, R. A. (2016a). *Four-Dimensional Consciousness*. Retrieved from relativisticconsciousness.com

Sieb, R.A. (2016b). *Relativistic Consciousness*. Retrieved from relativisticconsciousness.com

Sieb, R. A. (2004). The emergence of consciousness. *Medical Hypotheses*, *63*(5), 900–904. doi:10.1016/j.mehy.2004.04.005 PMID:15488666

Sieb, R. A. (2013). The emergence of emotion. *Activitas Nervosa Superior*, *55*(4), 115–145. doi:10.1007/BF03379732

Sieb, R. A. (2015). Memory in four dimensions. In M. Sakakibara & I. Etsuro (Eds.), *Memory Consolidation* (pp. 259–313). New York: Nova Science Publishers, Inc.

Simon, S., & Bain, B. (Producers) & Ward, V. (Director). (1998). *What Dreams May Come* [DVD]. USA: PolyGram Films.

Simonite, T. (2015). A better way to design brain inspired chips. *MIT Technology Review*. retrieved from http://www.technologyreview.com/news/537211/a-better-way-to-build-brain-inspired-chips/

Simonite, T. (2015). A better way to design brain inspired chips. *MIT Technology Review*. Retrieved May 6, 2015 from http://www.technologyreview.com/news/537211/a-better-way-to-build-brain-inspired-chips/

Singh, K. (1983). *Hinduism. Religions of India*. New Delhi: Clarion Books.

Sinha, R., Lovallo, W. R., & Parsons, O. A. (1992). Cardiovascular differentiation of emotions. *Psychosomatic Medicine*, *54*(4), 422–435. doi:10.1097/00006842-199207000-00005 PMID:1502284

Sirina, N. F., & Bushueva, T. I. (2016). Sistemnye arkhetipy v upravlenii predpriyatiem zheleznodorozhnogo transporta [System archetypes in the management of railway transport organization]. Sovremennye nauchnye issledovaniya i innovatsii, 5, 94-99.

Sivananda, S. S. (1974). *Mind. Its mysteries and control.* Tehri-Garhwal, India: The Divine Life Society.

Smillie, I. (1995). *The Alms Bazaar: Altruism under fire, non-profit organizations and international development.* London: Intermediate Technology Publications. doi:10.3362/9781780446127

Smith, H. (1958). *The world's religions.* New York: Harper Collins.

Smith, J. B. (1994). *Collective intelligence in computer-based collaboration.* Hillsdale, NJ: Lawrence Erlbaum.

Smith, M. P. (2001). *Transnational urbanism: Locating globalization.* Malden, MA: Blackwell.

Smyser, W. R. (2003). *The humanitarian conscience: Caring for others in the age of terror.* Hampshire, UK: Palgrave Macmillan.

Sommer, T. (1998). Teure Frucht des wüsten Krieges. *Die Zeit (Hamburg)*, *22*, 94.

Sotirova-Kohli, M., Rosen, D. H., Smith, S. M., Henderson, P., & Taki-Reece, S. (2011). Empirical study of Kanji as archetypal images: Understanding the collective unconscious as part of the Japanese language. *The Journal of Analytical Psychology*, *56*(1), 109–132. doi:10.1111/j.1468-5922.2010.01893.x PMID:21241297

Steinvorth, S., Levine, B., & Corkin, S. (2005). Medial temporal lobe structures are needed to re-experience remote autobiographical memories: Evidence from H.M. and W.R. *Neuropsychologia*, *43*(4), 479–496. doi:10.1016/j.neuropsychologia.2005.01.001 PMID:15716139

Stilling, R. M., Dinan, T. G., & Cryan, J. F. (2016). The brain's Geppetto—Microbes as puppeteers of neural function and behaviour? *Journal of Neurovirology*, *22*(1), 14–21. doi:10.100713365-015-0355-x PMID:26047662

Strassman, R. (2001). DMT, the spirit molecule. Rochester, VT: Park Street Press.

Strauss, L. (1953). *Natural Right and History.* Chicago: University of Chicago Press.

Streedhar, N. (2017). Micro-organisms and their influence in the human body. *Financial Express*. Retrieved from https://www.financialexpress.com/lifestyle/health/micro-organisms-and-their-influence-in-the-human-body-find-out-how/547433/

Strogatz, S., & Stewart, I. (1993). Coupled oscillators and biological synchronization. *Scientific American*, *269*, 102–109. doi:10.1038cientificamerican1293-102 PMID:8266056

Stuart Hughes, H. (1952). *Consciousness and society: The reorientation of european social thought 1890 - 1930.* New York: Random House, Vintage Books.

Styhre, A. (2008). *Science-based innovation from modest witnessing to pipeline thinking.* Hampshire, UK: Palgrave Macmillan.

Sumner, B. (2009). *The effect of experience on the perception and representation of dialect variants. Journal of Memory and Language.*

Syutkina, E. V., & Solodukhina, O. S. (2016). Arkhetip ottsa v sisteme formirovaniya doveriya v obshchestve [The archetype of the father in the formation of the system of trust in society]. Vektory psikhologicheskoy bezopasnosti sovremennogo obshchestva: doverie i sub"ektivnoe blagopoluchie, 71-74.

Szekely, E. B. (1970). *The Essene teachings of Zarathustra*. Academy of Creative Living.

Talbot, M. (1991). *The holographic universe*. New York: Harper.

Taylor, T. (2014). Outline of the history of artificial life. *LinkedIn*. Retrieved from https://www.linkedin.com/grp/post/1039687-5900533809526378496

Taylor, A. (2003). *Animals & Ethics: An Overview of the Philosophical Debate*. Peterborough, Canada: Broadview Press.

Tegmark, M. (2014). *Our Mathematical Universe – My Quest for the Ultimate Nature of Reality*. Brockman, Inc.

Tester, K. (1995). *The inhuman condition*. London: Routledge.

The emerging science of human-data interaction. (2015a). *MIT Technology Review*. Retrieved January 11, 2015 from http://www.technologyreview.com/view/533901/the-emerging-science-of-human-data-interaction

The Secret Language of Plants. (n.d.). *National Geographic & Youtube*. The Secret Language Of Plants - National Geographic Documentary.

Thincube. (2015). Dr. Masaru Emoto's Water Experiment - Words are Alive! [YouTube video]. Retrieved from https://www.youtube.com/watch?v=au4qx_l8KEU

Tikhonov, A. V., & Duberman, Yu. E. (2015). Arkhetip [Archetype]. Sotsiologiya upravleniya: teoretiko-prikladnoy tolkovyy slovar', 28-29.

Tiller, W. A., McCraty, R., & Atkinson, M. (1996). Cardiac coherence: A new, noninvasive measure of autonomic nervous system order. *Alternative Therapies in Health and Medicine*, *2*(1), 52–65. PMID:8795873

Tomasino, D. (2007). The psychophysiological basis of creativity and intuition: Accessing "the Zone" of entrepreneurship. *Institute of Heartmath & e-Motion Institute*. Retrieved from https://www.bing.com/search?q=Tomasino,+Dana+(2007).+The+psychophysiological+basis+of+creativity+and+intuition%3A+Accessing+%E2%80%9Cthe+Zone%E2%80%9D+of+entrepreneurship.+Institute+of+Heartmath+%26+e-Motion+Institute.&src=IE-SearchBox&FORM=IESR4A&pc=EUPP_

Tomasino, D. (2007). The psychophysiological basis of creativity and intuition: Accessing the "zone" of entrepreneurship. *International Journal of Entrepreneurship and Small Business & Institute of Heartmath.*

Tonegawa, S., Liu, X., Ramirez, S., & Redondo, R. (2015). Memory Engram Cells Have Come of Age. *Neuron*, *87*(5), 918–931. doi:10.1016/j.neuron.2015.08.002 PMID:26335640

Tononi, G. (2004). An information integration theory of consciousness. *BMC Neuroscience*, *5*(1), 42. doi:10.1186/1471-2202-5-42 PMID:15522121

Trafton, A. (2014). New brain-scanning technique allows scientists to see when and where the brain processes visual information. *Medical Press*. Retrieved from http://medicalxpress.com/news/2014-01-brain-scanning-technique-scientists-brain-visual.html

Transcribed from the MCA exhibit. (n.d.). In *Youtube*. Retrieved from https://www.youtube.com/watch?v=316AzLYfAzw

Trinity in the World's Religions. (2013). *SPEAKINGTREE.IN*. Retrieved from https://www.speakingtree.in/blog/trinity-in-world-religions

Tulving, E. (1983). *Elements of Episodic Memory*. New York: Oxford University Press.

Turan, N. K. (2015). K voprosu ob izuchenii arkhetipov i mifov v reklame (na primere deyatel'nosti yuvelirnogo doma «freywille») [On the question of the study of archetypes and myths in advertising (for example, the activities of the jewelry house «freywille»)]. Molodoy uchenyy, 9, 1269-1274.

Turner, B. (1984). *The body and society*. Oxford, UK: Basil Blackwell.

Twist, J. (2004). Brainwaves control video game. *BBC*. Retrieved from http://news.bbc.co.uk/2/hi/technology/3485918.stm

Ullman, M. T. (2004). Contributions of memory circuits to language: The declarative/procedural model. *Cognition*, *92*(1-2), 231–270. doi:10.1016/j.cognition.2003.10.008 PMID:15037131

Uncapher, M. R., & Wagner, A. D. (2005). Posterior parietal cortex and episodic encoding: Insights from fMRI subsequent memory effects and dual-attention theory. *Neurobiology of Learning and Memory*, *91*(2), 139–154. doi:10.1016/j.nlm.2008.10.011 PMID:19028591

Universal Declaration of Human Rights. G.A. Res. 217A, art. 27, U.N. GAOR, 3d Sess., 1st plen. mtg., U.N. Doc. A/810 (Dec. 12, 1948).

Urry, J. (2000). *Sociology beyond societies: Mobilities for the twenty-first century*. London: Routledge.

Vagrecha, Y. S. (2016). Possession Behaviour. *Abnorm Behav Psychol, 2*, 119. Retrieved from https://www.omicsonline.org/open-access/possession-behaviour-2472-0496-1000119.php?aid=76421

van der Molen, M. W., Somsen, R. J. M., & Orlebeke, J. F. (1985). The rhythm of the heart beat in information processing. In Advances in Psychophysiology (Vol. 1). London: JAI Press.

Van Lammel, P. (2006). Near-death experience, consciousness, and the brain: A new concept about the continuity of our consciousness based on recent scientific research on near-death experience in survivors of cardiac arrest. *World Futures*, *62*(1-2), 134–151. doi:10.1080/02604020500412808

Vendantum, S. (2007). If it feels good to be good, it might be only natural. *The Washington Post*.

Vergerio, P. P. (2002). The character and studies befitting a free-born youth (C. W. Kallendorf, Trans. & Ed.). In *Humanist Educational Treatises* (p. 43). Cambridge, MA: Harvard University Press.

Vivekananda, S. (1973). *Raja yoga or conquering the internal nature*. Calcutta, India: Advaita Ashram.

Von Foerster, H. (1962). Bio-Logic. In E. E. Bernard & M. R. Kare (Eds.), *Biological Prototypes and Synthetic Systems* (Vol. 1, pp. 1-12). New York: Plenum Press. doi:10.1007/978-1-4684-1716-6_1

von Foerster, H. (1973). On Constructing a Reality. In *Understanding Understanding*. New York: Springer.

Von Foerster, H., Müller, A., Müller, K. H., Rooks, E., & Kasenbacher, M. (2014). *The Beginning of heaven and earth has no name: Seven days with second-order cybernetics*. Fordham University; doi:10.2307/j.ctt13wzw8d

Wagner, A. D., Shannon, B. J., Hahn, I., & Buckner, R. L. (2005). Parietal lobe contributions to episodic memory retrieval. *Trends in Cognitive Sciences*, *9*(9), 445–453. doi:10.1016/j.tics.2005.07.001 PMID:16054861

Walker, B. B., & Sandman, C. A. (1982). Visual evoked potentials change as heart rate and carotid pressure change. *Psychophysiology*, *19*(5), 520–527. doi:10.1111/j.1469-8986.1982.tb02579.x PMID:7134357

Wallach, J. R. (2005). Human rights as an ethics of power. In R. A. Wilson (Ed.), *Human rights in the 'war on terror'*. New York: Cambridge University Press. doi:10.1017/CBO9780511511288.006

Weaver, W. (1949). Recent Contributions to the Mathematical Theory of Communication. In *C. Shannon (Ed.), The Mathematical Theory of Communication* (pp. 3–28). Urbana, IL: University of Illinois Press.

Weissbrodt, D. (2001). *Selected International Human Rights Instruments* (3rd ed.). Cincinnati, OH: Anderson Publishing Co.

Weiten, W. (2010). *Psychology: Themes and Variations* (8th ed.). Boston: Wadsworth/Cengage Learning.

Wendelken, C. (2015). Meta-analysis: How does posterior parietal cortex contribute to reasoning? *Frontiers in Human Neuroscience*, *8*(1042). PMID:25653604

Weschsler, D. (1971). Concept of collective intelligence. *The American Psychologist*, *26*(10), 904–907. doi:10.1037/h0032223

Westen, D. (1999). The Scientific Status of Unconscious Process. *Journal of the American Psychoanalytic Association*, *47*, 1061–1106. doi:10.1177/000306519904700404 PMID:10650551

Whitehead, A. N. (1978) Process and Reality: An Essay in Cosmology (D.R. Griffin & D.W. Sherburne, Eds.). New York: The Free Press (Originally published 1929)

Whitehead, A. N. (1922). *The principle of Relativity with applications to Physical Science*. Cambridge University Press.

Whitehead, A. N. (1996). *Religion in the making: Lowell lectures 1926*. New York: Fordham University Press.

Wikibooks, The Free Textbook Project. (n.d.). Retrieved 11:18, June 24, 2016 from https://en.wikibooks.org/w/index.php?title=Special_Relativity/Spacetime&oldid=2766972

WilburK. (2018). Cultivating your superhuman body.

Wilkins, E. H. (1961). Life of Petrarch. University of Chicago Press.

Williams, B. (1985). *Ethics and the Limits of Philosophy*. Cambridge, UK: Cambridge University Press.

Wilson, C., Kalogrides, L., Cameron, J., & Landau, J. (Peoducers) & Cameron, James (Director). (2010). *Avatar* [DVD]. United States: Fox.

Wilson, R. A. (2005). *Human rights in the 'war on terror*. New York: Cambridge University Press. doi:10.1017/CBO9780511511288

Wilson, R. A. (Ed.). (1997). *Human Rights, Culture & Context: Anthropological Perspectives*. London: Pluto Press.

Wilson, R. A., & Mitchell, J. P. (Eds.). (2003). *Human Rights in Global Perspective: Anthropological Studies of Rights, Claims and Entitlements*. London: Routledge.

Wiltshire, A. (2016). History of AI beating humans. *PCGamer.* Retrieved from http://www.pcgamer.com/artificial-intelligence-beating-humans/

Witten, E. (2012). Fivebranes and Knots. *Quantum Topology*, *3*(1), 1–137. doi:10.4171/QT/26

Wolfe, J. M., Kluender, K. R., Levi, D. M., Bartochuk, L. M., Herz, R. S., Klatsky, R. L., & Lederman, S. J. (2008). *Gestalt grouping principles. Sensation and Perception* (2nd ed.; pp. 78–80). Sunderland, MA: Sinauer Associates.

Woodcock, J. C. (2015). The Peril In Thinking and Other Essays. Scotts Valley, CA: Createspace Publishing Company.

Wu, X., & Zhang, X. (2016). *Neural network correctly identify criminals and noncriminals with an accuracy of 89.5 percent. MIT Technology Review*. Retrieved from https://www.technologyreview.com/s/602955/neural-network-learns-to-identify-criminals-by-their-faces/

Yogananda, P. (1998). *Autobiography of a yogi* (13th ed.). Los Angeles, CA: Self-Realization Fellowship.

Yogi, M. M. (1990). *Maharishi Mahesh Yogi on the Bhagavad-Gita: A New Translation and Commentary*. Penguin Arkana.

Yoran, H. (2002). The humanist critique of metaphysics and the foundation of the political order. *Utopian Studies*, *13*, 1–19.

Yoran, H. (2007). Florentine civic humanism and the emergence of modern ideology. *History and Theory*, *46*(3), 326–344. doi:10.1111/j.1468-2303.2007.00413.x

Yoran, H. (2010). *Between utopia and dystopia: Erasmus, Thomas More, and the humanist republic of letters*. Plymouth, UK: Lexington Books.

Young, L., Bechara, A., Tranel, D., Damasio, H., Hauser, M., & Damasio, A. (2010b). Damage to the ventromedial prefrontal cortex impairs judgment of harmful intent. *Neuron*, *65*(6), 845–851. doi:10.1016/j.neuron.2010.03.003 PMID:20346759

Young, L., Camprodon, J. A., Hauser, M., Pascual-Leone, A., & Saxe, R. (2010a). Disruption of the right temporoparietal junction with transcranial magnetic stimulation reduces the role of beliefs in moral judgments. *Proceedings of the National Academy of Sciences of the United States of America*, *107*(15), 6753–6758. doi:10.1073/pnas.0914826107 PMID:20351278

Zahedi, R. (2015). *On Discrete Physics (Digital Philosophy/Digital Cosmology) and the Cellular Automaton: A Perfect Mathematical Deterministic Structure for Reality - as a Huge Computational Simulation*. rXiv:1501.01373v3

Zaltman, G., & Zaltman, L. (2008). *Marketing Metaphoria*. Harvard Business Press.

Zhang, J. X., Harper, R. M., & Frysinger, R. C. (1986). Respiratory modulation of neuronal discharge in the central nucleus of the amygdala during sleep and waking states. *Experimental Neurology*, *91*(1), 193–207. doi:10.1016/0014-4886(86)90037-3 PMID:3940875

Zlotnikova, T. S. (2015). Arkhetip v filosofsko-psikhologicheskoy traditsii [Archetype in the philosophical and psychological tradition]. Kody massovoy kul'tury: rossiyskiy diskurs, 34-48.

Zohar, D., & Marshall, I. (1993). The quantum society: Mind, physics, and a new social vision. New York: Bloomsbury Publishing.

Zohar, D. (1990). *The quantum self: Human nature and consciousness defined by the new physics*. New York: William Morrow and Company, Inc.

Zwiebach, B. (2009). *A First Course in String Theory*. Cambridge University Press. doi:10.1017/CBO9780511841620

About the Contributors

Stephen Shafer has a lifelong interest is survival of planet earth and on communications research relating to media influence on cultural transformation. He has been developing an analog between cognitive models and video game structure based on a multidimensional common denominator—narrative-metaphorical linguistic structure—that operates as an affective transducer between conscious and unconscious cognitive dimensions. Accordingly, a genre of Psychecology-based video games (PEG) may be used to access robust data relative to collective unconscious cognitive states. His current project is to generate a Global Humanitarian Advocacy Organism. Prof Schafer is Past Chairman of the Humanities & Social Sciences Department at Digipen Institute of Technology where he taught for fifteen years. In addition to teaching at several colleges and universities, he has twenty-five years of experience that includes senior management in both the public and private sectors, social and environmental activism, marketing, and journalism. The unusual breadth and depth of his practical experience supports his research on the media dynamics of cultural evolution.

* * *

Florin Gaiseanu, Ph.D. in Solid State Physics since 1984 from the Physics Faculty, University of Bucharest, is a Principal Senior Researcher (Professor) in Bucharest (Romania) and Barcelona (Spain). Born in a family of professors in a city (historical capital) near Bucharest, he demonstrated, besides the high level results in his general preparation, particular creative aptitudes in physics and literature (Olympic results), in music and arts (Music School). He was since 1971 an active scientist, principal researcher in the Research Institute of Semiconductor Components (ICCE), later became Institute of Microsystem Technology (IMT) in Bucharest, and in the National Center of Microelectronics in Barcelona and other companies of similar field. He was a visiting scientist in the Institute of Microelectronics, Athens, in the Physics Department of the University of Thessaloniki, Greece, in IV Physics Institute, University of Gottingen and in the Institute of Semiconductors Physics, Frankfurt/Oder, Germany. Within his main field of investigation, the Science and Technology of Information, specifically focused on the analytical simulation and characterization of technological processes in silicon and compound semiconductors, and more recently on the understanding/modeling of consciousness, cognitive centers and their activities from an informational perspective, he is the author of 14 patents and more than 140 published scientific communications and articles in national and especially in international Journals. He was a leading researcher in an European Project on the Capacitive Sensors for Biomedical Applications. Representative synthesis articles presenting his main research results on semiconductor technology were recently published in the Proceedings of the Romanian Academy, Series A (2013-4) and in the Annals of the Academy of Romanian Scientists, Series on Science and Technology of Information (2017). He was member of the Electrochemical Society, active as reviewer and author of published articles in the Journal of the Electrochemical Society (USA), member of the Conference Committee and Chairman of the International Semiconductor Conference (CAS) in Romania and in the Micromachining and Microfabrication International Conference in USA. The present activity is especially dedicated to understanding and modeling of consciousness and associated/manifested properties from an informational perspective. This is a new line of investigation in this field, where he has been opening/developing new directions, allowing to explain the normal (cognitive) and "extra"-normal (psy) properties of mind, like that associated to the near-death experiences (consciousness disembody, life retrovision process, tunnel crossing, etc.), premonition, clairvoyance, remote diagnostics, geographic exploring and others, all of them on the basis of the last discoveries in quantum physics and astrophysics. His original contributions were published in international journals like NeuroQuantology, Consciousness and Life Physics, Astrophysics and Cosmology Journal in Europe, in Gerontology and Geriatrics Studies, Philosophy Study Journal in USA, and in the Journal of Information Science and Systems Science. He is working also on a final preparation of two books on Physics of Consciousness and on Informational Circuits of Consciousness. He is member of the Editorial Board of the Journal of Information Science and Systems Science of the international Science Research Association and of the NeuroQuantology Journal.

Jan Holmgren is an Architect SAR/MSA, graduated at Chalmers University of Technology, Sweden. Retired 2000 after a long architectural career. I follow and take part in discussions on consciousness since before 1980. Also I am interested in research on birds, especially the Common Swift Apus apus. I have published on swift phylogeny, swift behaviour, and taken part in application of geolocator technique on the Common Swift.

Nami Lee is a psychiatrist and Jungian analyst, currently working at Seoul National University hospital. She is teaching psychotherapy for psychiatric residents and medical students, serving as the director of human rights center at same hospital. Her recent articles are: prenatal and postnatal exposure to di(2-ehtylhexyl) phthalate and neurodevelopmental outcomes ; A systematic review and meta-analysis. (Environemtnal Research Vol 167.Nov 2018 558-566) Urinary phthalate metabolites concentrations and symptoms of depression in an elderly population (Science of the Total Environment. 625(2018) 1191-1197 Poor work-life balance may lead to impaire cognitive function in bus drivers. (in process of publishing) She published 25 books and various articles on psychology, culture, and society. Her two books were translated into Chinese, Thai, and Vietnamese.

Denis Lvov Graduated from Siberian Federal University as a specialist in Public Relations in 2008. Defended at the same university a candidate thesis in Social Philosophy "The Archetypal Component of Organizational Culture" in 2012. Now is an associate professor at Department of Sociology of Siberian Federal University.

Rollin McCraty, Ph.D., is Director of Research of the HeartMath Research Center at the Institute of HeartMath. He is also a Professor at Florida Atlantic University. A psychophysiologist, Dr. McCraty's research interests include the physiology of emotion, with a focus on the mechanisms by which emotions influence cognitive processes, behavior, and health as well as the global interconnectivity between people and the earth's energetic systems. Findings from this research have been applied to the development of simple, user-friendly mental and emotional self-regulation tools and techniques that people of all ages and cultures can use in the moment to relieve stress and break through to greater levels of personal balance, stability, creativity, intuitive insight and fulfillment. Dr. McCraty has acted as Principal Investigator in numerous research studies examining the effects of emotions on heart–brain interactions and on autonomic, cardiovascular, hormonal, and immune system function. He is a member of the American Autonomic Society, the Pavlovian Society, National Association for Psychological Science, the Association for Applied Psychophysiology and Biofeedback, and Society for Scientific Exploration. His research team regularly participate in collaborative studies with U.S. and international scientific, medical and educational institutions; they have worked in joint partnership with research groups at Stanford University, Claremont Graduate University, Dalhousie University (Halifax, Nova Scotia), Prince Sultan Cardiac Center (Saudi Arabia), University of Lithuania, among many others. Dr. McCraty has been interviewed for numerous feature articles in publications such as Prevention, Natural Health, Men's Fitness and American Health magazines, and has appeared on television segments for CNN-Headline News, ABC-World News Tonight, ABC-Good Morning America, NBC-Today Show, PBS-Body & Soul and Discovery Channel- and has been featured in a number of documentary films such as I am, The Truth, The Joy of Sox movie, The Power of the Heart, Solar Revolution, and The Living Matrix among many others.

Syed Ismyl Mahmood Rizvi graduated from Patna University in 2009 with a degree in English literature. He earned his Master's Degree in English and Cultural Studies from the University of Mauritius on a scholarship from the Government of Mauritius. He now works as a university lecturer in India.

Brock Shafer is a lifelong student of Biblical Scripture. In that capacity, he has lectured and made presentations to numerous gatherings. The hallmark of his understanding of scripture is his applied experience of their teachings. This living experience and his close scrutiny of scripture has gradually qualified him as an expert in the field.

Manoj Sharma, MBBS, Ph.D., MCHES®, is a public health physician with a medical degree from the University of Delhi and a doctorate in Preventive Medicine from The Ohio State University. He is currently a tenured Full Professor in Behavioral & Environmental Health at Jackson State University in the School of Public Health. In his career, spanning over 30 years, he has trained/taught over 6,000 health professionals that include physicians, health education specialists, dietitians, community health nurses, health care administrators, epidemiologists, sanitarians, health and development workers, allied health professionals and others. He has designed and taught a variety of undergraduate and graduate courses in health behaviour, health education, and health promotion at nine national and international universities working full time or part-time. He has worked for local health departments, state health departments/ agencies, federal government agencies, invited keynote presenter at universities, non-profit agencies, professional organizations and international agencies including governments of other nations. He is a prolific researcher and as of December 2018 had published ten books, over 225 peer-reviewed research articles, and over 400 other publications and secured funding for over $4 million. His research interests are in developing and evaluating theory-based health behaviour change interventions, obesity prevention, stress-coping, community-based participatory research, and integrative mind-body interventions.

Changsoo Shin has been a professor at Soul National University for over twenty-two years and is a full Professor of Geophysics in the departments of Energy Systems Engineering and Energy Resources Engineering. His areas of expertise include high performance computing, signal processing, modeling, and numerical analysis. A National Science Foundation project will initiate a new international collaboration with Dr. Changsoo Shin at Seoul National University in South Korea to apply his method of obtaining background sound speed models in the solid earth from marine seismic data to the ocean itself. Extracting ocean sound speed models from conventional seismic data will enable us to calculate high-resolution 2-D temperature images of the ocean which can lead to a better understanding of ocean mixing–a process that is not well understood or quantified yet forms a key input to all global climate models because the oceans exchange heat and gases with the atmosphere.

John Woodcock was born in England and raised in Australia until 1979 when he left for further study in the USA. Dr. Woodcock gained a doctorate in Consciousness Studies in 1999 before returning to Australia in 2003. His thesis concerns the theme of "the end of the world", based on his own extreme experiences lasting many years. At first it seemed that he was undergoing a purely personal psychological/spiritual crisis but over time Dr Woodcock began to realize that he was also participating in the historical process of a transformation of the soul, as reflected in the enormous, even apocalyptic, changes occurring in our culture. During this difficult period of the author's life, he wrote two books to express his ordeal: Living in Uncertainty, Living with Spirit and Poems of Making. His next three books, Mouthpiece, The Imperative, and Hearing Voices, explore the experience of "the end of the world" more fully, while subsequent books, including Animal Soul and Manifesting Possible Futures, establish a firm theoretical ground for the claim that the soul is urging us towards the development of new inner capacities that can help us face the uncertainty of modern life and, as well, address the unknown future. Making New Worlds begins the work of articulating an art form emerging in response to the soul's movement of "incarnation in the real world". The author develops this theme more fully in The Coming Guest and the New Art Form. He has also written a story of four friends whose lives are interrupted by an intrusion of four possible futures, while its companion piece is a story of three friends caught up in our epochal transformation of Being. Both stories are found in his book Inception & The Vortex. Dr. Woodcock currently lives with his wife Anita Hansen in Sydney, where he teaches, writes, and consults with others concerning their own journey through the chaos of our present times, in his capacity as a depth psychotherapist. He and Anita also work with couples in a therapeutic setting. His books may be found at: amazon.com/author/johncwoodcock/.

Index

G

H

I

J

L

M

N

P

Q

R

S

V

W

www.ingramcontent.com/pod-product-compliance
Lightning Source LLC
LaVergne TN
LVHW062138300926
849161LV00010B/2567